Protein Biochemistry and Proteomics

T0229227

Protein Biochemistry and Proteomics

AMSTERDAM • BOSTON • HEIDELBERG • LONDON
NEW YORK • OXFORD • PARIS • SAN DIEGO
SAN FRANCISCO • SINGAPORE • SYDNEY • TOKYO

ELSEVIER
ACADEMIC
PRESS

Academic Press is an imprint of Elsevier
30 Corporate Drive, Suite 400, Burlington, MA 01803, USA
525 B Street, Suite 1900, San Diego, California 92101-4495, USA
84 Theobald's Road, London WC1X 8RR, UK

This book is printed on acid-free paper.

Copyright © 2006, Elsevier Inc. All rights reserved.

No part of this publication may be reproduced or transmitted in any form or by any means, electronic or
mechanical, including photocopy, recording, or any information storage and retrieval system, without permission
in writing from the publisher.

Permissions may be sought directly from Elsevier's Science & Technology Rights Department in Oxford, UK:
phone: (+44) 1865 843830, fax: (+44) 1865 853333, E-mail: permissions@elsevier.com. You may also complete
your request on-line via the Elsevier homepage (http://elsevier.com), by selecting "Support & Contact" then
"Copyright and Permission" and then "Obtaining Permissions."

Library of Congress Cataloging-in-Publication Data
Application Submitted

British Library Cataloguing-in-Publication Data
A catalogue record for this book is available from the British Library.

ISBN 13: 978-0-12-088545-9
ISBN 10: 0-12-088545-X

For information on all Academic Press publications visit our Web site at www.books.elsevier.com

Printed in the United States of America
05   06   07   08   09   10   9   8   7   6   5   4   3   2   1

Working together to grow
libraries in developing countries

www.elsevier.com | www.bookaid.org | www.sabre.org

ELSEVIER   BOOK AID   Sabre Foundation
            International

Contents

Preface

For I can tell thee, though composing it cost me some labour, I found none greater than the making of this Preface thou art now reading.

(Cervantes, 1605)[1]

You don't make any progress with your methodical repertoire? You want to give your work a new direction? You want to learn more about the methodical variety of a given field? Maybe the EXPERIMENTER can help you.

In writing the EXPERIMENTER, I wanted to write neither a textbook nor a pure method collection. Textbooks tell about everything that has already been investigated: they show the house of science. Method books (cookbooks) describe the individual stones and tools. However, the EXPERIMENTER aims to provide guidance in the construction, to convey what type of houses are currently in fashion, and to explore how much work the construction takes and how much is paid for it. It should indicate what methods are available, what they are or are not suited for, which research strategies can be pursued, how tasks are tackled, and how much work is likely to be involved. You don't put the EXPERIMENTER on the lab bench; you take it to bed with you to "philosophize" when the experiment has failed for the twelfth time or when you need new ideas. A strategy book, that's what the EXPERIMENTER should be. I have not achieved this completely, as in some sections it still resembles a method collection. Yet, the EXPERIMENTER does not overly delve into the details of the methods but recommends literature describing the methods in a comprehensive manner for practical work. This book provides only those techniques not found elsewhere.

The tactic of discussing the possibilities of methods while referring to other publications for their details has the advantage of brevity. It also encourages the reading and understanding of the literature and helps to develop a feeling for the right experiment at the right time. In addition, the inspirations of others inspire your own ideas. "Cookbooks" have their utility. However, it does not suffice to be able to perform, for example, gel filtration if you do not know when it makes sense to use it. And when you are inventing new recipes (i.e., methods), cookbooks are of little help anyway.

Admittedly, many authors document methods incompletely or incomprehensibly in their publications (from here referred to as papers). Some do this to keep the competition in their field at bay. Others do so because they are lazy or because of a lack of space due to page limitations for articles. I searched the paper jungle for the best, most current, most readable, and most reproducible methods, and I believe I have found useful instructions. I know many of these recipes or have tried them in my own kitchen. If you know new methods or better papers for a method, I would be grateful for a communication.

In any case, you should not approach papers as if they hold the holy revelation. Even the Lowry Protein Assay has not been examined and optimized in every direction, and a protocol that is optimal for protein 1 may not be so for protein 2. Treat methods circumspectly. Question them (why phosphate buffers, why incubate before with X-ase?). Upon first application it is actually advisable to follow a protocol exactly, but later a playful interaction with the recipe is more useful. Also, you should be wary of all assertions regarding advantages, speed, or sensitivity of the tests or methods. On the other hand, warnings about disadvantages should be taken seriously. Researchers write papers not to assist their colleagues but to receive their recognition and to extend their publication list. The tone of the EXPERIMENTER is pointedly unacademic.

---

1. These and all other quotes found in the book come from Miguel de Cervantes Saavedra's *The Life and Deeds of the Keen Nobleman Don Quixote of la Mancha*.

Finally, books on scientific research talk a lot about fascination, thirst of knowledge, and enthusiasm. You find little on endless pipetting, failed experiments, and lack of funding. Many years' work often leads only to the discovery that there is nothing to discover in the direction taken. Only luck, diligence, ingenuity, good mentoring, and intelligence—in this order—bring success at the lab bench. Thus, your adventures will resemble those of the knight errant: every day and Sunday you take your tottery Rosinante to a castle where the king is difficult, the efforts great, the results bewildering, the food bad, and setbacks the rule. However, this time of trial strengthens your soul. From the abyss of desperation you rise to the level of indifference, where gray rough streets lead to seemingly endless horizons. Don't let yourself be discouraged! The others also slave away unsuccessfully. It is normal at first that no result is in sight. Hang in there!

Hubert Rehm

Preface to the Fourth Edition

Again a new edition of the EXPERIMENTER! The reason? It struck a chord with the readers: the EXPERIMENTER does not pedantically list method instructions, it only tells where these are to be found. It concentrates on strategical pieces of advice: For what can you use this or that method? What can you gain from it? Which problems can it help you solve? The EXPERI-MENTER is no cookbook, but guides you "through battle"; it is no Emeril, but a Patton. Researchers seem to need this: the book sells faster than protein biochemistry is developing.

That does not mean, however, that protein biochemistry was standing still last year. It is astonishing what the method crafters have found out in their diligence. Who would have thought that the Lämmli system could still be improved on? Who would have guessed that somebody could develop a new method of protein determination with obvious advantages compared to the existing half-dozen protocols? And would you have believed that it would take researchers until the year 2000 to figure out how to draw blood from lab mice in a convenient and painless way—after centuries of contact with these rodents? Things were also happening in the biochemistry of oligosaccharides as well as in c-terminal microsequencing and—not to overlook the big guys—I added a chapter on the purification of His-tagged proteins.

But why do you just want to read about others? Don't you have your own tricks and success strategies? Send them to *hr@laborjournal.de* and you'll find your name in the fifth edition. Of course, you'd really enjoy reading about yourself!

Hubert Rehm
Spring 2002

Acknowledgments

For their ideas, tips, corrections, and support I'd like to give my warmest thanks to Harald Backus, Siegfried Bar, Inge Bliestle, Cord-Michael Becker, Jurgen Dedio, Heiko Herwald, Willi Jahnen, Dieter Langosch, Stefan Mertens, Werner Muller-Esterl, Jaques Paysan, Andre Schrattenholz, Sabine Schrimpf, Christian Schroder, Andreas Trindl, Rando Wiech, Angelika Zengerle, and Jasminka Zimmermann.

Also, the EXPERIMENTER would have quietly passed away in the course of writing if it had not been for the courage and tough patience of the program planner of the Spektrum Akademischer Verlag, Ulrich Moltmann. And the care with which the editors Inga Eicken, Jutta Hofmann, and Bettina Saglio edited the text will (hopefully) create the (unwarranted) illusion that I understand something of grammar and punctuation.

"There is no book so bad but it has something good in it," said the bachelor.

Abbreviations

BSA	bovine serum albumin
CF	chromatofocusing
CFA	complete Freund's adjuvant
DMSO	dimethylsulfoxide
DSK	differential scanning calorimeter
DTT	dithiothreitol
EDTA	ethylenediaminetetra acetic acid
ESI	electrospray ionization
FA	Freund's adjuvant
FPLC	fast-performance liquid chromatography
SEC	size exclusion chromatography
HA	hydroxyapatite
HPLC	high-performance liquid chromatography
IEC	ion exchange chromatography
IEF	isoelectric focusing
IP	immunoprecipitations
ITC	isothermal titration calorimeter
3C network	three-component network
kD	kiloDalton
LUV	large unilamellar vesicle
MALDI	matrix-assisted laser-desorption ionization
MALDI-TOF	matrix-assisted laser-desorption ionization time of flight
MLV	multilamellar vesicle
MW	molecular weight
PAL	photoaffinity ligand
PBS	phosphate-buffered saline solution
PEG	polyethylene glycol
PEI	polyethylenimine
PIC	phenylisocyanate
PICT	phenylisothiocyanate
PMSF	phenylmethylsulfonyl fluoride
PVDF	polyvinyliden difluoride
RT	room temperature
SDS	sodium dodecylsulfate
SUV	small unilamellar vesicle
TFA	trifluoroacetic acid
TFEITC	trifluoroethylisothiocyanate
WGA	wheat germ agglutinin

Chapter 1 Daily Bread

Hold thy peace and have patience; the day will come when thou shalt see with thine own eyes what an honorable thing it is to wander in the pursuit of this calling.

1.1 Making Buffers

When you start in a lab as a Ph.D. student, you will spend the first weeks filling out forms, reading apartment ads, and making buffers. This last task is especially dear to the heart of the protein biochemist, because nobody else works with so many different buffers and solutions.

However, the composition of the buffers in most protein biochemical recipes is not the product of careful contemplation but of chance (this or that bottle happened to stand at arm's length). In fact, the following rules are usually all you need to keep in mind when making buffers.

- The buffer's pKa should be near the selected pH (Figure 1.1).
- The buffer capacity should be sufficiently high.
- The buffer should not react with any other molecules in the solution and it should not precipitate.

Frequently used buffers are acetate, phosphate, Tris, triethanolamine, HEPES, PIPES, and MOPS. The pKa of acetate and phosphate buffers is independent of the temperature. This advantage comes with the disadvantage of the narrow range of acetate buffers (pH 4.5 to 5.5) and of the tendency of phosphate to fall out of solution with divalent cations. The pKa of Tris and triethanolamine buffers is highly temperature dependent.

HEPES, PIPES, and MOPS belong to the "Good buffers." Good et al. synthesized these substances to get buffers with the following characteristics: nontoxic, good water solubility, no penetration of phospholipid membranes, negligible complex formation with cations, chemically stable, no effect on biochemical reactions, low temperature dependency, and no UV absorption. These goals were not always reached. For example, HEPES increases the growth of some cell lines (Ferguson et al. 1980) and influences biochemical reactions. The temperature dependency of the pKa of buffers such as CHES or MOPS (Figure 1.1) is considerable, and Good buffers are also expensive. You can find an overview of the most important buffer systems according to characteristics, production, usage, temperature dependency, and working pH range in Stoll and Blanchard (1990) and the information contained in Figure 1.1.

Sources
1. Ferguson, J., et al. (1980). "Hydrogen Ion Buffers for Biological Research," *Anal. Biochem.* 104: 300–310.
2. Stoll, V., and Blanchard, J. (1990). "Buffers: Principles and Practice," *Methods Enzymol.* 182: 24–38.

1.2 Protein Determination

The vagueness of protein determination methods causes grief for the beginning biochemist. The identical protein solution gives one result with assay A and another with assay B. Similarly, the identical assay yields different values with identical concentrations of different proteins (e.g., 1 mg/ml of BSA, ovalbumin, cytochrome).

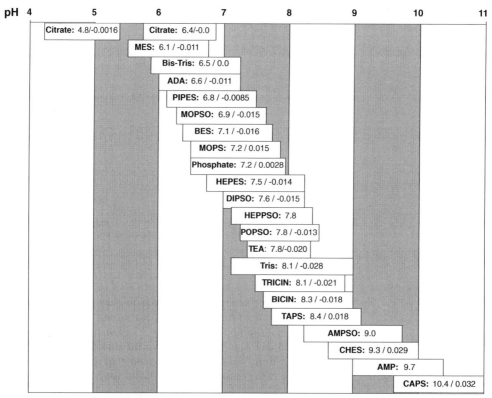

Figure 1.1. Working range, pKa, and temperature dependence of buffers. The numbers behind the buffer names are pKa and its temperature dependence (change of pKa per °C). **MES**: 2-(N-morpholino)-ethanesulfonic acid; **Bis-Tris**: [bis(2-hydroxyethyl)imino]Tris(hydroxymethyl)methane; **ADA**: N-(2 acetamidoimino)diacetic acid; **PIPES**: piperazine-N, N-bis(2-ethane sulfonic acid); **MOPSO**: 3-(N-morpholino)-2-hydroxy propanesulfonic acid; **BES**: N,N-bis(2-hydroxyethyl)-2-aminoethane sulfonic acid; **MOPS**: 3-(N-morpholino) propanesulfonic acid; **HEPES**: N-2-hydroxyethylpiperazine-N-2-ethanesulfonic acid; **DIPSO**: 3-[N-bis(hydroxyethyl)amino]-2-hydroxypropanesulfonic acid; **HEPPSO**: N-(2-hydroxyethyl) piperazine N-(2-hydroxypropanesulfonic acid); **POPSO**: piperazine-N,N-bis(2-hydroxypropanesulfonic acid); **TEA**: triethanolamine; **TRIS**: Tris(hydroxymethyl)aminomethane; **TRICIN**: N-[Tris(hydroxymethyl) methyl]glycine; **BICIN**: N, N-bis(2-hydroxyethyl)glycine; **TAPS**: 3-{[Tris(hydroxymethyl)methyl]amino}pro-panesulfonic acid; **AMPSO**: 3-[(1,1-dimethyl-2-hydroxy-ethyl)amino]-2-hydroxypropanesulfonic acid; **CHES**: cyclohexylaminoethanesulfonic acid; **AMP**: 3-aminopropanesulfonic acid; **CAPS**: 3-(cyclohexyla-mino)propanesulfonic acid.

Furthermore, every protein determination is sensitive to detergents or certain ions. Hence, when presenting a concentration value it is good practice to also mention the assay that was used as well as the benchmark protein. The methods of choice are the Bradford assay and the BCA (bicinchoninine acid) assay. Anyone working with membrane proteins and detergents should use the BCA assay. Otherwise, the choice between the BCA and the Bradford assays seems to be a question of taste.

1.2.1 BCA Assay

Proteins form a complex with Cu^{2+} ions in alkaline solution (Biuret reaction). The Cu^{2+} ions of the complex are presumably reduced to Cu^+ ions, which form a violet color complex with BCA.

Advantages: The test is quick (10 minutes at 65° C), sensitive (detection threshold 0.5 µg of protein), and resistant to detergents such as TRITON-X-100. The reaction takes place in an alkaline environment in which almost all proteins remain in solution. The reagents are available via retail (Pierce) and yield wonderful colors.

Problems: It is disrupted by high concentrations of complex-forming reagents (e.g., EDTA) and by ammonium sulfate; N-acteyl-glucosamine, glycine, reducing materials such as glucose, DTT, or Sorbitol; and a host of pharmaceuticals such as chlorpromazine, penicillin, and vitamin C (Marshall and Williams 1991).

Sources
1. Marshall, T., and Williams, K. (1991). "Drug Interference in the Bradford and 2,2′bicinchoninic Acid Protein Assays," *Anal. Biochem.* 198: 352–354.
2. Smith, P. K., et al. (1985). "Measurement of Protein Using Bicinchoninic Acid," *Anal. Biochem.* 150: 75–85.
3. Wiechelmann, K., et al. (1988). "Investigation of the Bicinchoninic Acid Protein Assay: Identification of the Groups Responsible for Color Formation," *Anal. Biochem.* 175: 231–233.

1.2.2 Bradford Assay

When Coomassie brilliant blue G-250 binds to proteins, the absorption maximum of the color changes (465 nm without protein; 595 nm with protein). The increase of the absorption to 595 nm is a measure of the solution's protein concentration.

Advantages: The color is completely developed after 2 minutes, the coloring varies little between different proteins, and there is only minimal pipetting work. The reagents are available via retail (e.g., Bio-Rad and Pierce).

Problems: The reaction takes place in an acidic environment in which many proteins fall out of solution. Strong lyes and commonly used detergents such as TRITON-X-100, SDS (sodium dodecylsulfate), or CHAPS interfere.

Sources
1. Bradford, M. (1976). "A Rapid and Sensitive Method for the Quantitation of Microgram Quantities of Protein Utilizing the Principles of Protein-dye Binding," *Anal. Biochem.* 72: 248–254.
2. Read, S., and Northcote, D. (1981). "Minimization of Variation in the Response to Different Proteins of the Coomassie Blue G Dye-binding Assay for Protein," *Anal. Biochem.* 116: 53–64.

1.2.3 Lowry Assay

Cu^+ ions from the Biuret reaction (see Section 1.2.1) form an unstable blue complex with the Folin-Ciocalteau reagent. This complex serves as a measure of the protein concentration.

Advantages: A reliable procedure.

Disadvantages: The Lowry assay requires a lot of pipetting work and many buffer components can interfere, including mercaptoethanole, HEPES, TRITON-X-100, and other detergents (they precipitate). Hence, it is advisable to precipitate the protein from the samples (see Section 1.5) and to perform the Lowry assay on the protein pellet. You have to make the reagents yourself, except for the Folin-Ciocalteau reagent available from (Merck). The Lowry protein values of different labs are difficult to compare, because every lab carries out the assay a little differently.

Sources
1. Larson, E., et al. (1986). "Artificial Reductant Enhancement of the Lowry Method for Protein Determination," *Anal. Biochem.* 155: 243–248.
2. Legler, G., et al. (1985). "On the Chemical Basis of the Lowry Protein Determination," *Anal. Biochem.* 150: 278–287.
3. Markwell, M., et al. (1978). "A Modification of the Lowry Procedure to Simplify Protein Determination in Membrane and Lipoprotein Samples," *Anal. Biochem.* 87: 206–210.

1.2.4 Starcher Assay

There are men who have the courage to overthrow kings. The Texan Barry Starcher is such a man. His goal is—at least in my estimation—to dethrone the citation king Lowry with a new protein determination assay. Starcher introduced (Starcher 2001) a protein determination method that determines the concentration of soluble as well as insoluble proteins (i.e., including the protein of tissue samples). The method is more sensitive than the Lowry or Bradford assays and has the additional advantage that identical amounts of different proteins yield an identical reading (even gelatine supposedly results in almost the same extinction coefficient as BSA). Starcher achieves this by applying two old reactions to the old problem of the variation of the specific coloring: the acidic hydrolysis of the proteins to the amino acids and the reaction of the amino acids with ninhydrine.

In its details, this works as follows. The sample (tissue or protein solution) is hydrolyzed to the amino acids over 24 h in microfuge tubes with 0.5 ml 6-NHCl at 100° C. Starcher dries the hydrosylate on the speed-vac and dissolves it again in water. An aliquot is then pipetted onto a microtiter plate and ninhydrine reagent is added. Starcher lets the plate float for 10 minutes in a boiling water bath, whereupon the assay is done and can be read in the microplate reader.

The ninhydrine solution can usually be kept for only a few hours and must be dissolved and stored under nitrogen. Thus, the reagent would have to be prepared anew every day. No worries: Starcher has found a solvent in which the reagent remains stable for several weeks. This certainly not insignificant relief for the researcher was Starcher's original contribution to his assay. After all, the other two reactions had been known for a long time, and it was also known that the most exact determination of the protein can be achieved via the amino acids. Starcher simply put one and one together. We learn this: with some thought, even modest experimental efforts can lead to a worthwhile paper.

Disadvantages: It takes 25 or 26 h until you can read your result. In addition, you have to manipulate the sample repeatedly: hydrolyze, speed-vac, pipette, centrifuge, pipette, heat up. This is not only a test of your patience; exactness also suffers. Hence, I fear that the Starcher assay will not be able to reach a similar number of citations as the Lowry. However, it makes the impression of being reliable, and if you need a protein determination with low protein-to-protein variation I would recommend it.

Source
1. Starcher, B. (2001). "A Ninhydrin-based Assay to Quantitate the Total Protein Content of Tissue Samples," *Anal. Biochem.* 292: 125–129.

1.2.5 Protein Concentration

The concentration of a pure protein in solution can be determined via its extinction at 280 nm, as long as the experimenter knows the extinction coefficient (rule of thumb: 1 mg/ml of BSA has about 1 OD) and the solution contains no other UV-absorbing substances. Gill and Hippel (1989) calculate the extinction coefficient of a protein from its sequence.

Sources
1. Gill, S., and Hippel, P. (1989). "Calculation of Protein Extinction Coefficients from Amino Acid Sequences," *Anal. Biochem.* 182: 319–326.
2. Stoschek, C. (1990). "Quantitation of Protein," *Methods Enzymol.* 182: 50–68.

1.3 Gels

Gel electrophoresis is used for the analysis of protein mixtures and allows for quick MW determinations. Hence, SDS gels belong to the protein biochemist like forms to the accountant. It is desirable to establish a gel system that takes only 30 to 45 minutes per gel run. Beautiful bands are desirable, but not necessary. In my experience, prolonged futzing with gel lengths or gradients brings little knowledge gain.

1.3.1 SDS Gels

Most proteins bind the detergent SDS to negatively loaded SDS protein complexes with a constant charge-to-mass ratio (1.4 g SDS/g protein in 1% SDS solutions). SDS denatures the proteins—especially after previous reduction with mercaptoethanole or DTT—and prevents protein-to-protein interactions (quarternary structures). For the purposes of many measuring methods, the SDS-protein complexes of different proteins thus differ only in their size and have comparable hydrodynamic qualities.

During SDS electrophoresis, the SDS-protein complex moves in the electric field toward the positive pole. The molecular sieve effect of a porous polyacrylamide matrix separates the SDS protein complexes according to their Stokes radius and thus according to their MW.

The various SDS gel electrophoresis systems differ among other things in the buffers they use. The discontinuous Lämmli system with Tris-glycine buffers is the most widely used. A stacking gel (Tris-glycine buffer pH 6.8; 3 to 4% acrylamide) is poured over a separation or running gel (Tris-glycine buffer pH 8.8; 5 to 20% acrylamide). The longer the running gel the better the separation. The thinner the gel the nicer the bands and the less can/may be loaded. With 1.5-mm-thick gels and 0.5-cm-wide pockets, the upper limit of the load is 1 mg of protein/pocket.

Fifteen percent separation gels are suited for proteins of an MW of 10 to 60 kd, 10% gels for proteins of an MW of 30 to 120 kd, and 8% gels for proteins of an MW of 50 to 200 kd (Figure 1.2). 18% gels with 7 to 8 M of urea can separate mixtures of small proteins and peptides (MW 1.5 to 10 kd) (Hashimoto et al. 1983). However, urea crystallizes from concentrated solutions and at temperatures less than RT it carbamylizes proteins and interferes with the binding of SDS. The alternative is a Tricin gel system after Schägger and Jagow (1987). It separates peptides between 1 and 100 kd and does not require urea (Figure 1.2).

Gradient gels (e.g., 8 to 15%) have a broader separation range and bands that are slightly more defined. Gradient mixers are suited for pouring linear gradients. However, it is easier to first draw the light solution into a glass pipette using a Peleus ball and then the heavy solution. Allowing a few air bubbles to pass between the two layers transforms them into a gradient that is poured between the glass plates of the electrophoresis apparatus. Perfectionists add another 5% cane sugar and dye to the heavy phase (the higher-percent acrylamide) and visually follow the course and extent of the gradient formation. Smith and Bell's (1986) machine pours good exponential gradients. Storage is possible. Gels wrapped in wet tissues can be kept in a sealed plastic bag for up to two weeks.

Problems:
- Acrylamide is toxic! Unwashed gels are also toxic, because they still contain unpolymerized acrylamide.
- Bad polymerization? Make fresh ammoniumpersulfate, ideally from a new bottle. Still no luck? Then a new buffer, a new acrylamide stock solution. The polymerization process is also very temperature sensitive. Polymerize in the oven at 40° C.
- The sample is usually heated in test buffer for 5 minutes at 95° C so that the proteins in the sample dissolve completely, potentially present proteases are inactivated, and tertiary structures are prevented. However, longer boiling cleaves unstable proteins, and membrane proteins often aggregate with SDS and mercaptoethanole. If you want to avoid the latter, heat only to 40° C.
- Incomplete reduction of the proteins? Make your DTT in 0.5-ml aliquots. Store the aliquots at −20° C. Thaw the DTT just before use.
- In SDS extracts of cells or nuclei, DNA and/or RNA form a slime that is difficult to load onto the gel and interferes with the run. Preprocessing of these samples with DNAse I and RNAse A works wonders (both enzymes still work in 0.3% SDS; see detailed protocols for different samples in the User Guide for the 2-D electrophoresis by Millipore).
- In an attempt to do everything correctly, the beginner often uses HCl to set the pH of the Lämmli Tris-glycine running buffer. This is wrong. Exact weighing of Tris and glycine results in the correct pH and smaller divergences don't matter. However, a Cl⁻ run buffer mixes up the ion system and leads to blurry bands.

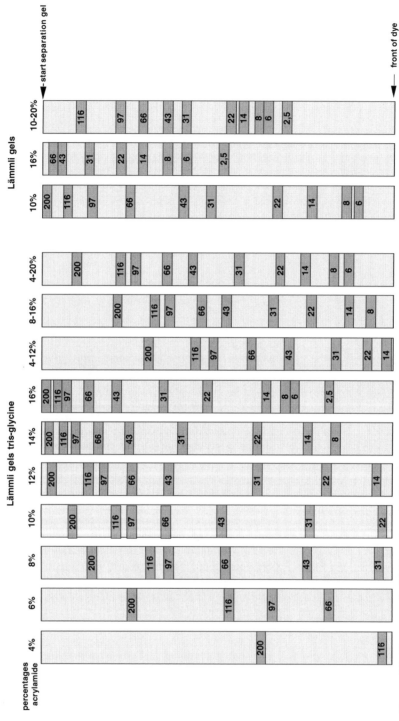

Figure 1.2. Run speed of MW markers in SDS gels.

- Special proteins behave in special ways. The hydrophile sugars of glycosylated proteins do not bind SDS. Hence, the charge-to-mass ratio of the SDS complexes of glycosylated proteins differs from those of non-glycosylated proteins. Glycosylated proteins run atypically and show a wide band in the gel (microheterogeneity). Ca^{2+}-binding proteins such as calmoduline run faster in SDS gels in the presence of 1 mM Ca^{2+} than in the presence of 1 mM EDTA, and the phosphorylation of proteins likewise changes their run behavior (evidence of phosphorylation).
- Band distortion? This is often due to high ion concentrations in the sample applied on the gel. Precipitate the sample following Wessel and Flügge (see Section 1.5.1), dry the pellet, and dissolve it in a test buffer.
- Potassium dodecyl sulfate is difficult to dissolve and SDS precipitates below 10° C.

Sources

SDS Gel Electrophoresis System (Standard)
1. Lämmli, U. K. (1970). "Cleavage of Structural Proteins During Assembly of the Head of Bacteriophage *T4*." *Nature* 227, pp. 680–685. This original protocol by Lämmli is difficult to follow. However, because almost every good lab has a good working protocol for the Lämmli gel, I don't think it's necessary to provide a detailed protocol.

Other SDS Gel Electrophoresis Systems
1. Dewald, D., et al. (1986). "A Nonurea Electrophoretic Gel System for Resolution of Polypeptides of MW 2000 to MW 200,000," *Anal. Biochem.* 154: 502–508.
2. Hashimoto, F., et al. (1983). "An Improved Method for Separation of Low-molecular Weight Polypeptides by Electrophoresis in Sodium Dodecyl Sulfate-polyacrylamide Gels," *Anal. Biochem.* 129: 192–199.
3. Schägger, H., and Jagow, G. (1987). "Tricine-sodium-dodecylsulfate Polyacrylamide Gel Electrophoresis for the Separation of Proteins in the Range from 1–100 kDa," *Anal. Biochem.* 166: 368–379.

Gradient Maker
1. Smith, T., and Bell, J. (1986). "An Exponential Gradient Maker for Use with Minigel Polyacrylamide Electrophoresis Systems," *Anal. Biochem.* 125: 74–77.

1.3.2 For the Impatient: SDS Electrophoresis Without Stacker

As you are able to infer from the previous section, the Lämmli system is a mature method that nevertheless has its weaknesses. Thus, time and again there are researchers who attempt to improve it, presumably with Lämmli's citation count in mind. Ahn et al. (2001) claim to be able to do without the stacker by simply switching the buffer system. The resolution supposedly remains identical or is even a little better than with Lämmli because the separation gel can be made longer. Above all, however, less (making of) solution is required.

For the Ahn gel, you need acrylamide stock solution, running buffer, test buffer, and separation gel buffer. Acrylamide stock solution, running buffer, and test buffer are identical to the Lämmli system. The secret lies in the separation gel buffer. In Ahn et al. it consists of 76 mM Tris-HCl and 0.1 M serine, 0.1 M glycine, and 0.1 M asparagine, with a pH of 7.4. Thus, the separation gel buffer does not contain any SDS.

Separation gel without SDS for the SDS electrophoresis? No problem! The SDS from test buffer and running buffer runs faster than the proteins (i.e., they remain in an environment that contains SDS). The advantage of the SDS-free separation gel: with SDS-free test buffer and running buffer, the Ahn system becomes a native gel electrophoresis.

This and the labor savings due to the missing stacker do not yet exhaust the advantages of the Ahn gel. Because the pH of the gel is only at 7.4, the gel can be stored at 4° C for about half a year without being damaged. In contrast, the acrylamide slowly hydrolyzes at the 8.8 pH of the Lämmli system.

Ahn et al. assure that in spite of the missing stacker the electrophoresis in the Ahn gel is insensitive to sample volume, to NaCl concentrations up to 0.5 M, and to 2% CHAPS or TRITON. The proteins also allow blotting.

This sounds useful. This sounds good. However, I have not yet checked the method. At the moment I only use SDS for brushing my teeth—and some things made me suspicious. For example, Ahn et al. do not show a direct comparison with Lämmli gels in their paper, and a precise description of the gel system is also missing.

It would not be difficult, however, to check Ahn's assertions. You would simply have to make the new separation gel buffer, and because you make so many buffers anyway one more is hardly an issue. You would than run your Lämmli gel and an Ahn gel in parallel, with identical samples and identical dimensions. Then staining, then destaining—and then you send me your result under *hr@laborjournal.de*.

Source

1. Ahn, T., et al. (2001). "Polyacrylamide Gel Electrophoresis Without a Stacking Gel: Use of Amino Acids as Electrolytes," *Anal. Biochem.* 291: 300–303.

1.3.3 Native Gels

Native gel systems do not contain SDS. Thus, the charge of the proteins in the gel is dependent on their isoelectric point and the pH of the buffer used. With native gels, the protein solution is loaded directly on the gel with cane sugar and an indicator, but without pretreatment. In the electric field, some of the proteins then move to the positive pole and some to the negative pole. The speed with which a protein moves depends on its size, its charge, the porosity of the gel matrix, and the pH of separation gel and running buffer.

Maurer (1971) describes a half-dozen native systems. Native gels have fallen out of fashion, because neither MW nor isoelectric point can be determined and they are not suited for membrane proteins (see, however, the following). They are occasionally used to check the purity of soluble proteins.

Advantages: Many proteins survive the electrophoresis and can be eluded again from the gel in active form. Some enzymes can be detected directly in the gel via the enzyme reaction.

Problems: Running a native gel takes hours. Hydrophobic (integral) membrane proteins smudge in conventional native gels, even if nonionic detergents such as TRITON-X-100 have been added to buffer and gel.

Schägger gels: Schägger and Jagow (1991) describe native gel electrophoreses for membrane proteins. The proteins are brought into solution with nonionic detergents (e.g., TRITON-X-100) with addition of aminocaproic acid. They are then stained with negatively charged hydrophobic Serva Blue G. The stain transforms the dissolved membrane proteins into negatively charged stain/detergent/protein complexes. The complexes do not aggregate because of the electrostatic repulsion, and their tertiary structure and function (e.g., enzymatic activity) are preserved. The complexes are cleft on a gel containing aminocaproic acid, whereas the cathode buffer supplies Serva Blue G (blue gel). The proteins of cow heart mitochondriae serve as orientation markers. The blue gel is suited for membrane proteins in the MW range of 10^5 to 10^6 for mitochondrial membrane proteins and integral cell membrane proteins such as the nicotinic acetylcholine receptor. For smaller membrane proteins, Schägger recommends substituting the stain with taurodeoxycholate.

The blue gel is both a new purification method and potentially a means of determining the quaternary structure of proteins (see Chapter 8), especially in combination with an SDS gel—which, as a second dimension, cleaves the natively separated oligomers into their subunits.

Problems: Serva Blue G presumably prefers to attach to trans-membrane regions. Large membrane proteins thus show a lesser charge density than small membrane proteins. Soluble proteins bind still less stain. The charge-to-mass ratio of different protein-stain complexes is thus not constant, and the native electrophoresis does not separate the protein complexes by MW (personal communication by A. Schrattenholz, Mainz). Soluble marker proteins such as thyroglobulin, ferritin, and the like smudge in the gel or partially disintegrate into subunits.

Serva Blue G binds tightly to nitrocellulose and PVDF membranes. The blot of a blue gel is thus blue, and this interferes with the immunostain or ligand coloring of the blot (see Section 1.6.3). Finally, the run of a blue gel lasts 3 to 6 h, and the bands are blurry.

Lämmli-taurodeoxycholate gels: If the experimenter substitutes the 0.1% SDS in buffers and gel with 5 mM of taurodeoxycholate in the Lämmli system (see Section 1.3.1), many membrane proteins focus in 5 to 7.5% separation gels to sharp bands. 5% separation gels separate protein complexes with an MW up to 106. Subunit structure and native conformation of many membrane proteins is preserved (personal communication by A. Schrattenholz, Mainz).

The electrophoresis in the detergent cetyltrimethylammoniumbromide lies somewhere between native and denaturing gel electrophoresis. Akin et al. (1985) claim that this allows an MW determination of the proteins while preserving the biological activity.

Sources
1. Akin, D., et al. (1985). "The Determination of Molecular Weights of Biologically Active Proteins by Cetyltri-methylammonium bromideMinus-Polyacrylamide Gel Electrophoreses," *Anal. Biochem.* 145: 170–176.
2. Maurer, H. (1971). *Disc Electrophoresis and Related Techniques of Polyacrylamide Gel Electrophoresis.* Berlin/New York: Walter de Gruyter.
3. Schägger, H., and Jagow, G. (1991). "Blue Native Electrophoresis for Isolation of Membrane Protein Complexes in Enzymatically Active Form," *Anal. Biochem.* 199: 223–231.
4. Thomas, J., and Hodes, M. (1981). "A New Discontinuous Buffer System for the Electrophoresis of Cationic Proteins at Near-neutral pH," *Anal. Biochem.* 118: 194–196.

1.4 Staining Gels

1.4.1 Fixing

Before or during the staining, the proteins are fixed in the gel (i.e., denatured and precipitated). Fixing is typically done with mixtures of ethanol, acetic acid, water, and a dye such as Coomassie Blue. Trichloroacetic acid is sometimes used instead of acetic acid, because some people believe that trichloroacetic acid fixes the proteins better.

Basic proteins with low MW often cannot be fixed either with acetic acid or trichloroacetic acid/ethanol/water mixtures, and their bands diffund over time. Fixing the proteins with formaldehyde, as in Steck et al. (1980), helps. If you want to stain the gel with silver afterward, you first have to completely wash out the formaldehyde.

Source
1. Steck, G., et al. (1980). "Detection of Basic Proteins and Low-molecular-weight Peptides in Polyacrylamide Gels by Formaldehyde Fixation," *Anal. Biochem.* 107: 21–24.

1.4.2 Staining

For protein amounts over 1 μg/band, the Coomassie stain is recommended, then the silver stain. If you want to blot the protein afterward, you can stain the gel with Eosin, although it is better to stain the protein only on the blot. For larger amounts of protein, which need to treated with care (e.g., for a sequencing), acetate staining is the means of choice.

The Coomassie stain is easy and requires little work, but even with thin gels (0.75 mm) the destaining takes 2 to 3 h. The sensitivity is moderate (lower threshold of 200 to 400 ng per band; see Table 1.1).

With silver stain, the Ag^+ ion forms complexes with Glu, Asp, and Cys residues of the proteins. Alkaline formaldehyde reduces the Ag^+ of the complexes to Ag. The details of this reaction are unknown. Prestaining the gel with Coomassie should strengthen the silver stain (Moreno et al. 1985). The advantage of silver stain lies in its high sensitivity. The reagents are available via retail.

The number of silver stain protocols amounts to several dozen. With the standard protocol developed by Merril et al. (1981), approximately 5 ng additional protein can be detected per band. Heukeshoven and Demick (1988) improved the stain by adding a reduction step with thiosulfate and thus pushed the detection threshold to 50 to 100 pg protein per band.

Table 1.1. Protein staining in gels.

Method	Sensitivity (Thresholds in ng/0.5 cm Bands)	Time	Variability of the Stain*	Blotting Possible After Staining?
Silver	5–30	1–2 h	High	No
Eosin Y	10	30 min.	Moderate	Yes[†]
SYPRO Orange	30–50	1 h	Low	Yes[†]
Stains all	100–200	3–4 days	Low	
Coomassie	200–400	2–4 h	Low	Yes[†]
Nitroblue tetrazolium	200–400	20 min.		
Na-acetate	1,000–3,000	20 min.		Yes[†]
PCI	2,000–4,000	10–40 min.		Yes[†]

* Color intensity with identical amounts of different proteins.
[†] Before blotting, the stained gels must be treated; for example, by washing out salts or incubating in 0.1% SDS solution (Perides et al. 1986; see Section 1.6).

The Heukeshoven and Dernick (1988) method was optimized by Pharmacia for the Phast system, which runs at varying temperatures. At low sensitivity, the experimenter can also perform all steps at RT (staining time of approximately 75 minutes). In some protocols, the proteins develop specific colors (e.g., sialoglycoproteins turn yellow, BSA blue). The basis for these color reactions remains unknown, and their experimental utility was low until recently (Dzandu et al. 1984; Nielsen and Brown 1984). If your silver gel turns out too dark, you can destain the gel again as follows.

1. Wash the gel for 5 minutes with water. Incubate the gel in destaining solution until the desired degree of decolorization has been reached (note that the destaining process is fast), and then stop it with 5% acetic acid.
2. Solution A: dissolve 37 g NaCl and 37 g $CuSO_4 \times 5H_2O$ in approximately 800 ml H_2O, add concentrated NH_4OH solution until precipitation occurs and the solution turns deep blue, and fill with H_2O to 1 l.
3. Solution B: dissolve 684 g $NaS_2O_3 \times 5H_2O$ in approximately 900 ml H_2O and fill to 1 l.
4. Destaining solution: mix 10 ml of solution A and 10 ml of solution B and fill with H_2O to 200 ml. If the color becomes a little greenish, add some drops of concentrated NH_4OH until the color is deep blue again.

Problems: Silver staining is complicated, takes a long time (1 to 2 h), is difficult to reproduce exactly, and it is not quantifiable because different proteins stain with different intensity (Poehling and Neuhoff 1981). In addition, the stained protein cannot be used for anything else (no blot, no elution, ^3H is quenched). In addition, the silver stain by no means targets proteins specifically, and it stains nucleic acids, lipopolysaccharides, lipids, and glycolipids.

If the Lämmli test buffer contains mercaptoethanole or DTT, two artifact bands often develop in BSA height during the silver staining of the gel. These bands disappear if the sample is treated with iodacetamide after the reduction (Hashimoto et al. 1983).

Sources
1. Dzandu, J., et al. (1984). "Detection of Erythrocyte Membrane Proteins, Sialoglycoproteins and Lipids in the Same Polyacrylamide Gel Using a Double Staining Technique," *Proc. Natl. Acad. Sci. USA* 81: 1733–1737.
2. Hashimoto, F., et al. (1983). "An Improved Method for Separation of Low-molecular-weight Polypeptides by Electrophoresis in Sodium Dodecyl Sulfate-polyacrylamide Gel," *Anal. Biochem.* 129: 192–199.
3. Heukeshoven, J., and Dernick, R. (1988). "Improved Silver Stain Procedure for Fast Staining in Phastsystem Development Units: Staining of Sodium Dodecylsulfate Gels," *Electrophoresis* 9: 28–32.
4. Merril, C., et al. (1981). "Ultrasensitive Stain for Proteins in Polyacrylamide Gels Shows Regional Variation in Cerebrospinal Fluid Proteins," *Science* 211: 1437–1438.
5. Moreno, M., et al. (1985). "Silverstaining of Proteins in Polyacrylamide Gels: Increased Sensitiviy Through a Combined Coomassie Blue-Silver Stain Procedure," *Anal. Biochem.* 151: 466–470.
6. Nielsen, B., and Brown, L. (1984). "The Basis for Colored Silver-protein Complex Formation in Stained Polyacrylamide Gels," *Anal. Biochem.* 141: 311–315.

7. Poehling, H., and Neuhoff, V. (1981). "Visualization of Proteins with a Silver Stain: A Critical Analysis," *Electrophoresis* 2: 141–147.

Stains-all staining: Stains-all is a cationic carbocyanin dye. It stains sialoglycoproteins blue, Ca^{2+}-binding proteins deep blue to violet, proteins red, and lipids yellow-orange. Stains-all is useful for very acidic and/or highly phosphorylized proteins (e.g., from dentin or bone). These stain badly or not at all with Coomassie or silver, but they shine in deep blue with Stains-all (but not for long). In daylight, Stains-all fades within minutes—while you are studying the gel. Myers et al. (1996) claim to have overcome this disadvantage with a double stain. They stain the gel first with Stains-all, and then with silver. This apparently makes highly phosphorylized proteins (and others) permanently visible.

Advantages: Stains-all is colorful and may be useful for special purposes (e.g., during the purification of proteins that bind Ca^{2+} or that are highly phosphorylized).

Problems: SDS interferes even in small quantities. The reagent is photosensitive, and thus the stain keeps for minutes only (but see Myers et al. 1996).

Sources
1. Campbell, K., et al. (1983). "Staining of the Ca^{2+} Binding Proteins, Calsequestrin, Calmodulin, Troponin C and S-100 with the Cationic Carbocyanine Dye Stains-all," *J. Biol. Chem.* 258: 11267–11273.
2. King, L., and Morrison, M. (1976). "The Visualization of Human Erythrocyte Membrane Proteins and Glycoproteins in SDS Polyacrylamide Gels Employing a Single Staining Procedure," *Anal. Biochem.* 71: 223–230.
3. Myers, J., et al. (1996). "A Method for Enhancing the Sensitivity and Stability of Stains-all for Phosphoproteins Separated in Sodium Dodecyl Sulfate-polyacrylamide Gels," *Anal. Biochem.* 240: 300–302.

Nitroblue tetrazolium: This is a negative stain (i.e., nitroblue tetrazolium stains only the gel and not the protein bands). The stain is more sensitive than Coomassie and takes just 20 minutes.

Source
1. Leblanc, G., and Cochrane, B. (1987). "A Rapid Method for Staining Proteins in Acrylamide Gels," *Anal. Biochem.* 161: 172–175.

Eosin staining: After Lin et al. (1991), the fluorescein derivative Eosin Y stains proteins down to 10 ng protein/band in SDS gels. Eosin Y also detects sialoglycoproteins, which are not stained by Coomassie. The antigenicity of the proteins stained with Eosin Y is preserved and the proteins can be blotted after staining.

Source
1. Lin, F., et al. (1991). "Eosin Y Staining of Proteins in Polyacrylamide Gels," *Anal. Biochem.* 196: 279–283.

The fluorescent SYPRO stains are praised as a breakthrough in the protein staining of gels. These bind to SDS protein complexes and are hence suited for the staining of SDS gels. Usually, SYPRO Orange is used.
- SYPRO Orange primarily stains proteins. It does not stain DNA, lipopolysaccharides, or lipids, and glycolipids are only stained a little.
- SYPRO Orange stains with a sensitivity similar to that of silver. Over long measuring times, rare proteins can be especially emphasized.
- After a SYPRO Orange stain, the proteins can still be blotted. To the delight of the friends of proteom research, they can still be digested and sequenced.
- The staining is not complicated: dilute SYPRO 1:5000 in 7.5% acetic acid, let the gel sway in the soup for about 1 h, and wash it briefly.

So much for the good news. Now the bad hews.
- Native gels are difficult to stain with SYPRO dyes, unless the gel is incubated in running buffer with 0.05% SDS before the staining—but then you might just as well have run an SDS gel.
- SPYRO stains are fluorescent stains (i.e., you have to induce fluorescence and take a photo). This means: carry the gel tub to the transilluminator, put on your protection mask, attach the filter (now where could it be?) to the camera, inspect the film box (empty, of course), take a new film, induce with 300 nm, take the photo, and develop the photo—only to find out afterward that you had forgotten to properly set the aperture.

- If you stain with 7.5% acetic acid as described previously, the proteins are difficult to transfer onto the blot. It works, but with about 75% reduced efficiency in comparison to unstained gel. If you make the buffer without acetic acid, the transfer works better but the staining is worse.

A hint: SYPRO Orange stains proteins less than SDS protein complexes. Everything that removes the SDS from the protein also weakens the stain. Which means what? It means: do not stain the gel for too long, eliminate any additional fixation step with 7.5% acetic acid (especially not overnight), and do not wash the gel with non-ionic detergents such as TRITON or Tween.

And another hint: let the front SDS run out of the gel. Otherwise, too much SDS diffunds from the gel into the staining solution. The SDS micelles bind SYPRO Orange, which degrades its actual concentration. SDS SYPRO Orange micelles also seem to raise the background.

Source

1. Steinberg, T., et al. (1996). "Applications of SYPRO Orange and SYPRO Red Protein Gel Stains," *Anal. Biochem.* 239: 238–245.

Acetate staining: 4 M Na acetate for 20 to 60 minutes precipitates the free (not bound to proteins) SDS in Lämmli gels. The gel becomes murky, but the protein bands remain clear. Light from the side against a black background makes the bands visible. Acetate staining is gentle on the protein, and is reliable and easy. The protein remains unchanged and can be processed further (e.g., for partial amino acid sequences). However, only larger amounts of protein (upward of 5 µg per band) become visible. The stain disappears when the Na acetate is washed out.

Source

1. Higgins, R., and Dahmus, M. (1979). "Rapid Visualization of Protein Bands in Preparative SDS-polyacrylamide Gels," *Anal. Biochem.* 93: 257–260.

PCl staining: Prussak et al. (1989) precipitate the SDS in Lämmli gels with 250 mM PCl. This makes the protein bands visible as white P-SDS-protein complexes. The authors claim that staining with PCl is gentle, fast, and robust. However, it seems to me that Na-liacetate staining is more easily reproducible than PCl staining. In addition, the proteins are easier to elude from the gel.

Source

1. Prussak, C. E., et al. (1989). "Peptide Production from Proteins Separated by Sodium Dodecylsulfate Polyacrylamide Gel Electrophoresis," *Anal. Biochem.* 178: 233–238.

Enzyme staining: Sometimes (e.g., with native gel electrophoreses), the activity of an enzyme survives the electrophoresis. Then the experimenter can try to selectively stain the enzyme bands in the gel using the enzyme activity. Proteases or enzymes that release phosphate or CO_2 are well suited.

Sources

1. Lynn, K., and Clevette-Radford, N. (1981). "Staining for Protease Activity on Polyacrylamide Gels," *Anal. Biochem.* 117: 280–281.
2. Nimmo, H., and Nimmo, G. (1982). "A General Method for the Localization of Enzymes that Produce Phosphate, Pyrophosphate, or CO2 After Polyacrylamide Gel Electrophoreses," *Anal. Biochem.* 121: 17–22.

1.4.3 Drying

Cleanly dried and filed gels adorn the protocol book and are solid proof of the activity of the experimenter. You can then always go back and check whether this or that result was really as unequivocal as you believed, and you can shoot a second photo or cut out bands and process them further.

Gels are dried on Whatman paper or (better) on cellophane. The latter is transparent and allows for scanning of the dried gel. High-percent gels (15 to 20%) and TRICIN gels after Schägger tear easily during drying. The following are countermeasures in this situation.

- Do not use a water jet pump, but a mechanical vacuum pump.
- Allow no air bubbles between the gel and the rubber lid of the dryer.
- Turn off the vacuum only when the gel is completely dry (rule of thumb: 0.75-mm gels need 1 h to dry, 1.5-mm gels need 1.5 h).
- Preincubate high-percent gels for 1 h with 5 to 10% glycerin.

I have tried different dryers and the one from Bio-Rad works best. You can forego the gel drying if you scan the band patterns with a frame grabber and store them on your PC. This also gives you extensive picture processing possibilities. However, this has the disadvantage that it entices you to "beautify."

1.5 Precipitation and Concentration

Thou must take notice, brother Sancho, that this adventure and those like it are not adventures of islands, but of cross-roads, in which nothing is got except a broken head or an ear the less.

Protein is precipitated to get rid of ions or agents that interfere with the protein determination or gel electrophoresis, and/or to concentrate the protein. The method of choice for samples less than 500 µl is the chloroform/methanol precipitation. However, the native conformation of the proteins gets lost.

1.5.1 Denaturing Precipitation

Chloroform/methanol precipitation: Wessel and Flügge (1984) dilute watery protein solutions (volume of 10 to 150 µl) in Eppendorf tubes with methanol and precipitate the proteins with chloroform. Addition of water separates the water/methanol/chloroform solution into two phases. The precipitated proteins collect in the interphase. Test volumes of 0.2 to 2 ml can also be processed with Corex glass tubes.

Advantages: In spite of its complexity, the method works reliably also for low protein amounts (20 ng) and in the presence of detergents or high salt concentrations. The rest of the chloroform/methanol/water mixture can be removed easily and quickly on the speed-vac (see Section 1.5.3). This (largely invisible) pellet contains the dry protein, largely free of residues from precipitation agent or buffer components.

Problems: When you take off the upper (watery) phase, the protein easily goes down the drain. Chloroform is a liver poison (vent hood!).

Alternatives to Wessel and Flügge (1984) are precipitation with 10% trichloroacetic acid in the presence of yeast RNA as a carrier following Polachek and Cabib (1981) or acetone precipitation, as follows. The watery sample is mixed with four parts acetone and cooled for 1 h to −20° C. Then the precipitated proteins are extracted by centrifugation. Finally, proteins in solutions of higher concentration (> 0.1 mg/ml) can also be precipitated with perchloric acid, trichloroacetic acid, or by applying heat. Rests of trichloroacetic acid in the pellet are removed by repeated rinsing with ice-cold 80% acetone.

Sources
1. Polachek, I., and Cabib, E. (1981). "A Simple Procedure for Protein Determination by the Lowry Method in Dilute Solutions and in the Presence of Interfering Substances," *Anal. Biochem.* 117: 311–314.
2. Wessel, D., and Flügge, U. (1984). "A Method for the Quantitative Recovery of Protein in Dilute Solution in the Presence of Detergents and Lipids," *Anal. Biochem.* 138: 141–143.

1.5.2 Native Precipitation

If the biological activity of the protein needs to be preserved, it should be precipitated at low temperatures with ammonium sulfate or polyethylene glycol (PEG) or organic solvents (acetone, ethanol, or methanol).

Problems: Ammonium sulfate and PEG do not precipitate the protein or do so incompletely from diluted solutions, and the precipitation with ammonium sulfate results in large amounts of often undesirable ions. The organic solvents denature some proteins even at low temperatures.

Source
1. Englard, S., and Seifter, S. (1990). "Precipitation Techniques," *Methods Enzymol.* 182: 285–300.

1.5.3 Concentration

Freeze-dry: Freeze-drying removes the water from frozen protein solutions by sublimation. You remember: in vacuum, ice does not change to water. Rather, the ice turns directly into steam (sublimation). Freeze-drying is typically done at RT (the solution remains frozen because sublimation continually channels energy away). Freeze-drying is suited for large amounts of solutions whose proteins are stable enough to withstand freezing, thawing, and low ion concentrations.

The solution is frozen by holding it (in a round glass flask) in a tub containing liquid N2 and using a circular swinging motion to create an ice surface that is as large as possible and has an even thickness. At the freeze-dryer the flask is immediately attached to a strong vacuum (oil pump). In a round flask with a diameter of 10 cm, the sample loses 20 to 30 ml of water per hour.

You run into trouble when the osmolarity of the solution is too high. Because only the water sublimates, salt, cane sugars, glycerin, and so on accumulate over time and lower the melting point of the solution. If the concentration of nonsublimating materials is high enough, the ice thaws and the broth starts boiling. The concentration of nonsublimating materials should be at less than 10 mM, especially with larger amounts of liquid.

Source
1. Pohl, T. (1990). "Concentration of Proteins and Removal of Solutes," *Methods Enzymol.* 182: 68–83.

Speed-vac: Under high vacuum, watery solutions already start boiling at room temperature. Simultaneous centrifuging of the solution drives out the gas bubbles and prevents foaming. This makes it possible to use the speed-vac to extract smaller (> 500 µl) amounts of solvent (water, ethanol, and so on) quickly and gently. Because only the solvent evaporates, nonvolatile salt and detergents accumulate in the solution.

Important: At the start, turn on the centrifuge first, then the vacuum. At the end, turn off the vacuum before the centrifuge. Larger volumes can be restricted with vacuum dialysis: the solution is filled into a dialysis tube, which is inserted into a vacuum bottle.

Source
1. Pohl, T. (1990). "Concentration of Proteins and Removal of Solutes," *Methods Enzymol.* 182: 68–83.

Centrifugation: With the concentration devices from the Amicon company, the solution to be concentrated is centrifuged in a fixed-angle rotor over a filter (max. 5,000 g). The concentrate collects above the filter and can be transferred into an Eppendorf tube in a second centrifugation step. The different systems (Microcon, Centricon, Centriprep) concentrate 0.5 to 15 ml of source solution into 5 to 500 µl. Depending on the filter, the process takes between 10 minutes and 3 h. The company offers filters with MG limits from 3 to 100 kd.

Advantages: Fast, easy, and gentle.

Problems: Even diluted protein solutions often plug the filters completely. If the sample is also viscous (e.g., buffered with 10% glycerin), even centrifuging for several hours is of no use. The sample does not even think about passing the filter. The filter rips at higher centrifugation speeds. To remove blocking aggregates, it sometimes helps to centrifuge the sample to be concentrated for 1 h at 100,000 g before application. The Microcon system allows you to run it through a coarse filter first. Finally, the filters adsorb protein, which becomes apparent in diluted protein solutions.

Dialysis: Dialysis against 5 to 15% high-molecular polyvinylpyrrolidon (e.g., T360) gently concentrates watery protein solutions within a few hours. However, the contaminants of the concentrated polyvinylpyrrolidon solution diffund into the protein solution.

It is expensive but effective to lay the dialysis tube into dry Sephadex G-100 to G-300. A (somewhat older) trick for the dialysis of many samples with volumes of 100 to 500 µl is to cut a big hole in the lid of an Eppendorf tube (e.g., with a hot needle). Add the sample, lay a piece of soaked dialysis membrane between lid and tube, and close the lid. Put the tube into the stand and lay the stand upside down on the dialysis liquid. Remove air bubbles beneath the lid of the tube using a Pasteur pipette. Let it float.

1.6 Blotting

During blotting, the proteins of an SDS gel or a native gel are transferred onto a membrane electrophoretically. The blot is the most versatile and popular tool of the protein biochemist, because on the blot the protein is naked and helplessly exposed. You can stain it, sequence it, let it react with antibodies, expose it to enzymes, determine its derivatization (phosphate groups? sugar residues?), and check for binding of ligands and ions. In comparison to the gel, the membrane is easily manipulable, and reactions and washing processes run faster, unhindered by diffusion problems.

Blot membranes consist of nitrocellulose (BA 85 from Schleicher and Schüll), polyvinyliden difluoride (PVDF) (Immobilon from Millipore), positively charged nylon ($^{+}$nylon) (Zetaprobe from Bio-Rad), or glass fiber coated with polybrene (GF/C from Whatman). The membranes bind the proteins through hydrophobic (nitrocellulose) or hydrophobic and ionic interaction ($^{+}$nylon, polybrene-coated glass fiber). Even peptides with only 20 amino acids still stick to nitrocellulose.

The popular nitrocellulose membranes have a high protein binding capacity and are suited for protein staining, immunostaining, lectin staining, or $^{45}Ca^{2+}$ staining. Its chemical instability forbids its application to amino acid analyses and sequencing, and when dry the membranes are friable and highly flammable. To activate the protein binding sites of the nitrocellulose membrane, 20% methanol is added to the blot buffer.

Compared to the negatively charged nitrocellulose, $^{+}$nylon membranes bind three to four times more protein per cm^2 (up to 500 µg/cm^2) and have better mechanical qualities. With $^{+}$nylon membranes, it is not necessary to add 20% methanol to the blot buffer. This makes the protein transfer (from the gel onto the membrane) quicker and more efficient. The saturation of the free protein binding sites of the $^{+}$nylon membrane is cause for grief (see Section 1.6.2).

The hydrophobic PVDF membranes are mechanically and chemically stable. They are suited for protein staining, immunostaining, lectin staining, and $^{45}Ca^{2+}$ staining, as well as for amino acid sequencing and analyses. Similar conditions apply to blot and development as to nitrocellulose, but in contrast to nitrocellulose the PVDF membrane does not completely bind even tiny amounts of protein. Depending on the protein, 10 to 50% passes the membrane unbound. Watery solutions do not moisten dry PVDF membranes. The membranes first have to soak in methanol. Moist PVDF membranes are light-permeable and transparent in dioxan isobutanol. Stained bands can also be measured in the scanner.

Membranes from glass fiber coated with polybrene (GF/C of Whatman) are suitable for amino acid sequencing because the membranes are inert against the chemicals used for the sequencing. Although the protein binding capacity of the coated glass fiber membranes (10 to 30 µg/cm^2) is comparable to that of nitrocellulose, most sequencers prefer PVDF membranes.

Sources

Nitrocellulose
1. Towbin, H., et al. (1979). "Electrophoretic Transfer of Proteins from Polyacrylamide Gels to Nitrocellulose Sheets: Procedure and Some Applications," *Proc. Natl. Acad. Sci. USA* 76: 4350–4354.

Nylon
1. Gershoni, J., and Palade, G. (1982). "Electrophoretic Transfer of Proteins from Sodium Dodecylsulfate-polyacrylamide Gels to a Positively Charged Membrane Filter," *Anal. Biochem.* 124: 396–405.

PVDF
1. Gültekin, H., and Heermann (1988). "The Use of Polyvinylidendifluoride Membranes as a General Blotting Matrix," *Anal. Biochem.* 172: 320–329.

Glass-fiber/polybrene
1. Vandekerckhove, J., et al. (1985). "Proteinblotting on Polybrene-coated Glass-fiber Sheets," *Eur. J. Biochem.* 152: 9–19.

The semi-dry cell from Bio-Rad with platinum/stainless steel electrodes (instead of the charcoal electrodes used previously with semi-dry cells) is a good blot chamber. The chamber blots 0.75-mm-thick gels within 15 to 20 minutes. The membrane moistened with blotting buffer lies on three layers of filter paper (Whatman) soaked in blotting buffer (for nitrocellulose and PVDF, e.g., 15 to 25 mM Tris-glycine pH 8.3 with 20% methanol). The gel is laid onto the membrane in such a way that no air bubbles can settle between membrane and gel. Generally, the gel is neither fixed nor stained before the blot, but it is also possible to blot fixed and Coomassie-stained gels (Perides et al. 1986). Three layers of filter paper soaked in blotting buffer are put on the gel. The careful blotter wipes the paper gently from left to right to remove any air bubbles and then applies electricity. Dehydrated protein blots can be kept for months and are usable after rehydration (nitrocellulose, e.g., with blot buffer) for protein or immunostains.

Blotting problems: The greater the MW of the protein the less efficient the transfer and the longer the blotting process. This effect becomes noticeable with proteins with an MW of more than 150 kd. In difficult situations, the method from Gibson (1981) may help.

When nitrocellulose or PVDF blots are extensively washed with detergentcontaining buffers, the bound proteins separate over time. Small proteins (MW < 6,000) can even disappear from $^+$nylon blots during blocking, washing, and so on (Karey and Sirbasku 1989). Fixing the proteins on the blot prevents them from becoming loose. Fixing is done with heat, glutaraldehyde, trichloroacetic acid, or, with mixtures of acetic acid and ethanol. Some protein-staining solutions (e.g., Ponceau red solution; see Section 1.6.1) contain a fixing agent.

Sources

1. Gibson, W. (1981). "Protease-facilitated Transfer of High-molecular-weight Proteins During Electrotransfer to Nitrocellulose," *Anal. Biochem.* 118: 1–3.
2. Karey, K., and Sirbasku, D. (1989). "Glutaraldehyde Fixation Increases Retention of Low-molecular-weight Proteins (Growth Factors) Transferred to Nylon Membranes for Western Blot Analysis," *Anal. Biochem.* 178: 255–259.
3. Perides, G., et al. (1986). "Protein Transfer from Fixed, Stained and Dried Polyacrylamide Gels and Immunoblot with Protein A Gold," *Anal. Biochem.* 152: 94–99.
4. Swerdlow, P., et al. (1986). "Enhancement of Immunoblot Sensitivity by Heating of Hydrated Filters," *Anal. Biochem.* 156: 147–153.

Proteins bind noncovalently to blot membranes. You can take the blotted protein off the membrane (e.g., for proteolytic digestion or for analysis in the MALDITOF). For nitrocellulose, Lui et al. (1996) have systematically examined the interaction of protein and blot membrane. According to them, Zwittergent 3–16 (1% in 100 mM NH_4HCO_3) removes between 60 and 90% of the blotted protein from the nitrocellulose. The detergent also works with PVDF membranes, albeit not as well.

There are other detergents (e.g., Tween 20, Tween 80, hydrogenated TRITON-X-100) that remove proteins from the membrane. If you want as much protein as possible to remain bound to the membrane (e.g., for immunostaining or ligand staining), you should not wash the blot too often with such detergent solutions.

Source

1. Lui, M., et al. (1996). "Methodical Analysis of Protein-nitrocellulose Interactions to Design a Refined Digestion Protocol," *Anal. Biochem.* 241: 156–166.

1.6.1 Protein Staining on Blots

For blots with protein amounts less than 50 ng/band, the Ponceau red staining is the method of choice. Aurodye and copper iodide can be used for protein amounts less than 50 ng/band. Li et al. (1989) compare different protein staining methods. Ponceau red stains protein on nitrocellulose blots (see Table 1.2). The staining is reversible and allows subsequent immunostaining. The blot is incubated for 1 to 2 minutes at RT in Ponceau red (2% in 3% trichloroacetic acid), and the surplus stain is then washed away with water.

Ponceau red staining of the gel is more sensitive than a Coomassie staining (lower threshold of about 50 ng/band), and the trichloroacetic acid in the staining solution fixes the proteins on the blot at the same time. The stain disappears during subsequent saturation of the membrane's protein binding sites (blocking). If you want to keep track of the position of certain proteins (e.g., the marker) until after blocking and further test reactions, you need to mark the bands with pencil on the nitrocellulose before blocking.

India ink stains with about the same sensitivity as Ponceau red, but the procedure takes several hours and the staining intensity is quite different for different proteins (nitrocellulose, Hancock and Tsang 1983; PVDF, Gültekin and Heermann 1988).

Colloidal gold particles stain the protein bands on nitrocellulose with a sensitivity that is comparable to silver staining of gels. Gold stain is not reversible and is incompatible with later immunostaining. It is also a somewhat awkward and lengthy (2 to 18 h) procedure and has quite a low linearity. The linear range of gold stain lies between 2 and 200 ng/mm^2 and has a high background: the staining intensity of 200 ng is only about 10 to 20% stronger than that of 2 ng.

Moeremans et al. (1986) describe a method of staining proteins on nitrocellulose and [+]nylon membranes. The reagents are available via retail under the name FerriDye (Janssen Life

Table 1.2. Protein staining on blots.

Membrane	Stain	Sensitivity (Thresholds in ng/0.5-cm Bands)	Required Time	Reversible	Compatible with Immunostaining?	Compatible with MALDI?
Nitrocellulose	Aurodye	1–5	2–18 h	No	No	No
	SYPRO Ruby	2–8	45 min.	Yes	Yes	Yes
	Copper iodide	5–20	5 min.	Yes		Yes
	Bathophenanthroline disulfonate/ europium	15–30		Yes	Yes	Yes
	Amido black	15–60		No	No	Yes
	Congo red	30–60		No		
	Ponceau red	50–150	5 min.	Yes	Yes	Yes
	FerriDye	50–150	2 h	No		
	India ink	80–200	ca. 18 h	No	No	No
PVDF	Aurodye	1–5	2–18 h	No	No	No
	SYPRO Ruby	2–8	45 min.	Yes	Yes	Yes
	Coomassie	10–30	20 min.	No	No	Yes
	Bathophenanthroline disulfonate/ europium	15–30		Yes	Yes	Yes
	Congo red	50–60		No	No	Yes
	India ink	30–60		No		
[+]nylon	Copper iodide	80–200	ca. 18 h	No	No	No
	FerriDye	5–20	5 min.	Yes		
Glass fiber polybrene- coated	Fluorescamine	50–150	2 h	No		

Sciences). The reaction with some toxic reagents lasts approximately 2 h, must be carried out under the vent hood, and is not reversible.

Copper iodide stains proteins on nitrocellulose and $^+$nylon with a sensitivity comparable to silver stain in the gel. Root and Reisler (1989) claim that this stain is inexpensive, quick (5 minutes), can easily be removed again, and maintains the immunoreactivity of the blotted proteins.

As with the gel stain, fluorescent colors have also caused a small revolution in blot staining. SYPRO Ruby stains proteins on PVDF or nitrocellulose with almost the same sensitivity as gold—certainly with more sensitivity than Coomassie and India ink. Staining takes just about three quarters of an hour: bathe nitrocellulose membranes for 2×10 minutes in 7% acetic acid and 10% methanol, and then stain for 15 minutes with SYPRO Ruby and finally wash with water for 6×1 minutes.

SYPRO Ruby is best visible in light with a wavelength of 302 or 470 nm. The emission maximum lies at 6l8 nm (stimulation wavelength 470 nm). The lower sensitivity threshold is 2 to 8 ng of protein per band, or 0.25 to 1 ng BSA/mm^2. The advantages of SYPRO Ruby lie in its high sensitivity, wide measuring range (linear from 2–8 to 1,000–2,000 ng per band), good evenness, and compatibility with immunostaining, microsequencing, and MALDI. SYPRO Ruby is fairly specific to proteins and does not stain DNA or RNA well.

Does the wonder substance also have disadvantages? It has. It is expensive and in some cases is difficult to get rid of. The fluorescence signal only disappears in the course of immunostaining. Presumably, it is distributed onto the proteins of the block solutions. In addition, you need a UV box or (better) a laser gel scanner to see something.

Sources
1. Berggren, K., et al. (1999). "A Luminescent Ruthenium Complex for Ultrasensitive Detection of Proteins Immobilized on Membrane Supports," *Anal. Biochem.* 276: 129–143.
2. Gültekin, H., and Heermann, K. H. (1988). "The Use of Polyvinylidendifluoride Membranes as a General Blotting Matrix," *Anal. Biochem.* 172: 320–329.
3. Hancock, K., and Tsang, V. (1983). "India Ink Staining of Proteins on Nitrocellulose Paper," *Anal. Biochem.* 133: 157–162.
4. Li, K., et al. (1989). "Quantification of Proteins in the Subnanogram and Nanogram Range: Comparison of the Aurodye, Ferridye, and India Ink Staining Methods," *Anal. Biochem.* 182: 44–47.
5. Moeremans, M., et al. (1985). "Sensitive Colloidal Metal (Gold or Silver) Staining of Proteins Blots on Nitrocellulose Membranes," *Anal. Biochem.* 145: 315–321.
6. Moeremans, M., et al. (1986). "Ferri-dye: Collodial Iron Binding Followed by Perls Reaction for the Staining of Proteins Transferred from Sodium Dodecyl Sulfate Gels to Nitrocellulose and Positively Charged Nylon Membranes," *Anal. Biochem.* 153: 18–22.
7. Root, D., and Reisler, E. (1989). "Copper Iodide Staining of Protein Blots on Nitrocellulose Membranes," *Anal. Biochem.* 181: 250–253.

1.6.2 Blocking

Before letting a blot react with antibodies, lectins, or protein ligands, the remaining protein binding sites of the blot membrane are saturated with a blocker (BSA, skim milk powder, fetal calf serum, Tween 20, gelatin, and so on). One can argue about what the best blocking agent may be, and it probably also depends on the respective antibodies. I had good experiences with BSA, but BSA gets expensive at the amounts needed for blotting. In addition, its dusty consistency and the gigantic amounts mean that adjacent protein cleaners always find a BSA band in their silver gels. Milk powder is inexpensive and blocks well, but its solutions quickly become infested with bacteria.

Hauri and Bucher (1986) recommend the combination gelatin/Nonidet P40. Some people also block with Tween 20. However, the blots that were blocked only with Tween 20 show a high background stain and unspecific bands with the immunostaining. Nevertheless, Tween 20 complements other blockers (such as BSA) well. But careful: Tween partially separates proteins from the blot. Too much Tween and they float away.

Blockers that are good for nitrocellulose are also suited for PVDF. To block $^+$nylon membranes, however, the experimenter has to resort to rough conditions such as 1% hemoglobin or 10% BSA in PBS overnight at 45° C (Gershoni and Palade 1982).

Sources
1. Batteiger, B., et al. (1982). "The Use of Tween 20 as a Blocking Agent in the Immunological Detection of Proteins Transferred to Nitrocellulose Membranes," *J. Immunol. Methods* 55: 297–307.
2. Gershoni, J., and Palade, G. (1982). "Electrophoretic Transfer of Proteins from Sodium Dodecylsulfate-polyacrylamide Gels to a Positively Charged Membrane Filter," *Anal. Biochem.* 124: 396–405.
3. Gültekin, H., and Heermann, K. H. (1988). "The Use of Polyvinylidendifluoride Membranes as a General Blotting Matrix," *Anal. Biochem.* 172: 320–329.
4. Hauri, H., and Bucher, K. (1986). "Immunoblotting with Monoclonal Antibodies: Importance of the Blocking Solution," *Anal. Biochem.* 159: 386–389.

1.6.3 Immunostaining

Blotted antigens can be stained with antibodies. The blocked blot is first incubated with anti-antigen antibody (first antibody). Then the experimenter washes away the unbound anti-antigen antibody and incubates the blot with a marked antibody (second antibody, e.g., ^{125}I-anti-rabbit IgG antibody). This binds to the anti-antigen antibody. After further washing, the blot is developed using the mark of the second antibody as guidance and the position of the antigen becomes visible.

The specificity of the immunostaining depends on the specificity of the first and second antibody (see Chapter 6) and also on the dilution with which the antibodies are used. The incubation times of the first and second antibody with the blot depend on the amount of antigen and the affinity of the antibodies. For most purposes, an incubation time of 1 to 2 h (RT) may be sufficient for the first one and 1 h (RT) for the second.

The second antibodies are marked with ^{125}I, alkaline phosphatase, or peroxidase, and quality antibodies are available in retail. Peroxidase has edged out the other markers over the last years. Peroxidase-marked antibodies catalyze the oxidation of Luminol and thus trigger chemiluminescence (the emitted light is measured with film). This ECL reaction is about 10 times more sensitive than blot development with alkaline phosphatase (supposedly it measures antigens in the pg range). In addition, the ECL reaction takes seconds or minutes, its solutions are nontoxic and are available in retail or are easy to make, and the reaction is quantifiable (via scanning of the film, or counting the light flashes in the β-counter).

Finally, the experimenter can stain the blot repeatedly with different antibodies. After developing the film, the first and second antibodies are washed out either with glycine HCl pH 1.8 or with 3 M sodium thiocyanate, 0.5% mercaptoethanol, 0.05% Tween 20 pH 9.5. Then block once more and stain the blot again with another antibody combination (Heimer 1989). Although this is also possible with ^{125}iodine-marked second antibodies, these are nowhere near as sensitive as ECL (and are radioactive and thus a health risk). In contrast, when the blot is developed with alkaline phosphatase or peroxidase/diaminobenzidine the precipitated reaction products prevent a second staining of the blot. In addition, the peroxidase substrate diaminobenzidine is said to be a carcinogen and must be decontaminated in Na-hypochlorite solutions (a less sensitive but nontoxic alternative would be the peroxidase substrate chloronaphthol; Ogata et al. 1983).

Special problems: On nitrocellulose, the ECL reaction peters out after 15 minutes. On PVD/membranes it is sustained longer, but with a higher background. The high sensitivity of the ECL reaction can become a problem: to keep the background low, you need to block and wash well and highly dilute antibody solutions (at least 1:3000).

On blots blocked with milk powder (5%), the ECL reaction apparently runs only weakly. Blots blocked with gelatin, on the other hand, exhibit a good ECL reaction, but also a higher background. The best blocking agents for ECL blots are said to be serums. Important: acid inhibits peroxidases.

Short ECL protocol: Block blot (1 to 2 h), incubate first antibody (1 h), wash well, incubate second (peroxidase-marked) antibody, wash well, detect reagent for 1 minute, wrap blot in cellophane, and incubate with film for 30 seconds to 30 minutes.

Detection reagent: A. 250 mM luminol (Fluka 09253) in DMSO; B. 90 mM p-Coumar acid (Fluka 28200) in DMSO; C. 1 M of Tris-Cl pH 8.5; D. 30% H_2O_2 solution. Mix: 200 µl A,

89 µl B, and 2 ml C, and fill with water to 20 ml. This solution is stable at RT in a brown bottle. Detection reagent is created with 6.1 µl D.

General problems: If the experimenter incubates the blot long enough with the film or the substrate solution, pretty much all protein bands appear on the film or the blot. This is because almost all bands adsorb unspecific small quantities of first and second antibody, albeit in different magnitudes. The effect cannot be suppressed completely, regardless of how much you block and wash. Sometimes spots and bands that have nothing to do with the antigen appear at the beginning of the blot development. According to Girault et al. (1989), this is due to the fact that human skin keratins make it into the gel or onto the blot and serums contain antibody against keratins. Experimenters with dandruff problems may think about whether a purification of the serums with keratin affinity columns would improve the quality of their immunoblots. If several bands appear during immunostaining, there are three possibilities.

- The antigen sits on several proteins.
- All bands are stained unspecifically.
- Only one band is stained specifically; the others are artifacts.

The latter two possibilities are the most likely, especially when the development takes relatively long. One could start profound guessing games about which band could be the right one, and this could certainly be entertaining, but the clever experimenter carries out controls.

- A parallel strip is developed with preimmuneserum at the same time and under identical conditions.
- A blot strip is stained (to which a sample was applied that does not contain the antigen, but otherwise resembles the questionable sample). Both strips are treated the same (blocked, washed, developed, and so on). If bands likewise appear on the control strip, the experimenter has drawn a blank, even if the bands are different.
- The band must not appear if the first antibody was incubated with the blot strip in the presence of the antigen (e.g., antipeptide antibody with peptide).
- Does the band also appear with affinity-purified antibody?

Sources

125Iodine-marked Second Antibody
1. Dunn, S. (1986). "Effects of the Modification of Transfer Buffer Composition and the Renaturation of Proteins in Gels on the Recognition of Proteins on Western Blots by Monoclonal Antibodies," *Anal. Biochem.* 157: 144–153.
2. Girault, J. A., et al. (1989). "Improving the Quality of Immunoblots by Chromatography of Polyclonal Antisera on Keratin Affinity Columns," *Anal. Biochem.* 182: 193–196.
3. Hauri, H., and Bucher, K. (1986). "Immunoblotting with Monoclonal Antibodies: Importance of the Blocking Solution," *Anal. Biochem.* 159: 386–389.
4. Heimer, R. (1989). "Proteoglycan Profiles Obtained by Electrophoresis and Triple Immunoblotting," *Anal. Biochem.* 180: 211–215.
5. Swerdlow, P., et al. (1986). "Enhancement of Immunoblot Sensitivity by Heating of Hydrated Filters," *Anal. Biochem.* 156: 147–153.

Peroxidase-marked Second Antibody
1. Gültekin, H., and Heerman, K. H. (1988). "The Use of PVDF Membranes as a General Blotting Matrix," *Anal. Biochem.* 172: 320–329.
2. Ogata, K., et al. (1983). "Detection of Toxoplasma Membrane Antigens Transferred from SDS-polyacrylamide Gel to Nitrocellulose with Monoclonal Antibody and Avidin-biotin, Peroxidase, Anti-peroxidase, and Immunoperoxidase Methods," *J. Immunol. Methods* 65: 75–82.

Alkaline Phosphatase-marked Second Antibody
1. Blake, M., et al. (1984). "A Rapid, Sensitive Method for Detection of Alkaline Phosphatase-conjugated Anti-antibody on Western Blots," *Anal. Biochem.* 136: 175–179.

1.6.4 Ca^{2+} Binding

Some proteins bind Ca^{2+} with high affinity. This Ca^{2+} binding is astoundingly stable and survives even SDS-gel electrophoresis and blot, especially that of the EF-hand proteins (calmodulin, troponin C, Parvalbumin, and so on). The Ca^{2+} binding of proteins on nitrocellulose or

PVDF blots can thus be detected with $^{45}Ca^{2+}$ (Maruyama et al. 1984; Garrigos et al. 1991). The dot blot by Hincke (1988) measures the $^{45}Ca^{2+}$ binding of purified native proteins. Blots and dot blots can also be stained with other radioactive ions. Schiavo et al. (1992) describe an example for $^{65}Zn^{2+}$.

Problems: Sc^{3+}, the decomposition product of $^{45}Ca^{2+}$, interferes (Rehm et al. 1986; Hincke 1988). Also, the $^{45}Ca^{2+}$ blot works only with proteins that bind Ca^{2+} with high affinity.

With acidic proteins, the $^{45}Ca^{2+}$ blot and the dot blot often give false-positive signals (Rehm et al. 1986). Controls with acidic proteins that do not bind Ca^{2+}, such as albumin and ovalbumin, are thus a necessity. In addition, the putative Ca^{2+} binding protein should run faster in SDS gels, which contain Ca^{2+}, than in gels with EDTA (Garrigos et al. 1991).

Sources
1. Garrigos, M., et al. (1991). "Detection of Ca^{2+}-binding Proteins by Electrophoretic Migration in the Presence of Ca^{2+} Combined with $^{45}Ca^{2+}$ Overlay of Protein Blots," *Anal. Biochem.* 194: 82–88.
2. Hincke, M. (1988). "Routine Detection of Calcium-binding Proteins Following Their Adsorption to Nitrocellulose Membrane Filters," *Anal. Biochem.* 170: 256–263.
3. Maruyama, K., et al. (1984). "Detection of Calcium Binding Proteins by ^{45}Ca Autoradiography on Nitrocellulose Membrane After Sodium Dodecyl Sulfate Gel Electrophoresis," *J. Biochem.* 95: 511–519.
4. Rehm, H., et al. (1986). "Molecular Characterization of Synaptophysin, a Major Calcium Binding Protein of the Synaptic Vesicle Membrane," *EMBO J.* 5: 535–541.
5. Schiavo, G., et al. (1992). "Botulinum Neurotoxins Are Zinc Proteins," *J. Biol. Chem.* 267: 23479–23483.

1.6.5 Ligand Staining

SDS is easy to wash away from blots. In principle, this allows, us to renature SDS-denatured blotted proteins and to detect them via ligand binding or enzymatic activity. Protein/protein interactions should also be detectable, because (for example) a radioactively marked protein recognizes its binding partner on the blot of a cell lysate or a membrane preparation (Carr and Scott 1992).

The detection of a blotted protein—for example, via a radioactively marked ligand (ligand staining)—succeeds with small stable proteins or with ligands whose binding is independent of the conformation of the protein (which recognize, e.g., an amino acid sequence). If the binding is conformation dependent, a denaturing/renaturing cycle should help. The blotted proteins are completely denatured in 8 M urea or 6 M guanidine and DTT and then renatured again (Celenza and Carlson 1986; Ferrel and Martin 1989). However, there will also be proteins whose disulfide bridges are best left intact by the experimenter. The precise conditions are different from protein to protein, and the blot membrane also plays a role.

Min Li et al. (1992) achieved impressive success with the denaturing/renaturing technique. They were able to narrow down the sequences involved in the oligomer formation from K^+ channels. As a rule, however, the native conformation of bigger proteins and membrane proteins can rarely be restored on a blot. Thus, there are only few examples of successful renaturings of blotted receptors and their affinity to the ligands is lower than that of native protein by orders of magnitude. However, little work is involved and it is worth a try.

Ligand staining is more promising with proteins that are not denatured with SDS (i.e., blotted by native gels) for membrane proteins, one would have to use, for example, Schägger gels (see Section 1.3.2).

Sources
1. Carr, D., and Scott, J. (1992). "Blotting and Band-shifting: Techniques for Studying Protein-protein Interaction," *TIBS* 17: 246–249.
2. Celenza, J., and Carlson, M. (1986). "A Yeast Gene That Is Essential for Release from Glucose Repression Encodes a Proteinkinase," *Science* 233: 1175–1178.
3. Daniel, T., et al. (1983). "Visualization of Lipoprotein Receptors by Ligand Blotting," *J. Biol. Chem.* 258: 4606–4611.
4. Estrada, E., et al. (1991). "Identification of the Vasopressin Receptor by Chemical Crosslinking and Ligand Affinity Blotting," *Biochemistry* 30: 8611–8616.
5. Ferrel, J., and Martin, S. (1989). "Thrombin Stimulates the Activities of Multiple Previously Unidentified Protein Kinases in Platelets," *J. Biol. Chem.* 264: 20723–20729.
6. Min Li, et al. (1992). "Specification of Subunit Assembly by the Hydrophilic Amino-terminal Domain of the Shaker Potassium Channel," *Science* 257: 1225–1230.

Table 1.3. Autoradiography/fluorography of gels (following Laskey 1980).

Isotope	Method	dpm/cm² for Visible Fluorography Signal Within 24 h	Magnification via Direct Autoradiography
^{125}I	Screen	100	16
^{32}P	Screen	50	10.5
^{14}C	Scintillator	400	15
^{35}S	Scintillator	400	15
^{3}H	Scintillator	8,000	>1,000

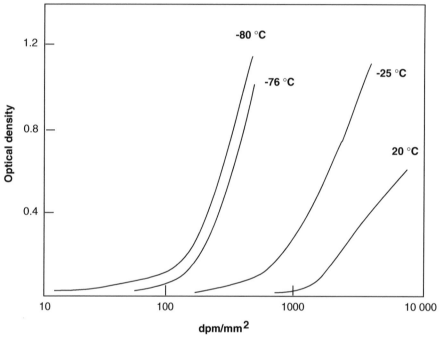

Figure 1.3. Temperature and film sensitivity. Diluted ^{3}H leucin was applied on silica gel thin layer plates that were then coated with EN^{3}HANCE spray. Then the plates were incubated at different temperatures for 24 h with Kodak X-Omat film (following "A Guide to Fluorography," DuPont).

1.7 Autoradiography of Gels and Blots

There are two problems with autoradiography of gels (Laskey 1980).
1. Isotopes such as ^{3}H or ^{14}C emit low-energetic β-radiation. This radiation is already absorbed in the gel and does not reach the film.
2. Isotopes such as ^{32}P or ^{125}I emit high-energetic β-radiation or γ-radiation. This penetrates the film without blackening it.

Fluorography (see Table 1.3) solves both problems. In case 1, the experimenter soaks the gel with a scintillator liquid (Enhance, Enlightning, Entensify). This converts the β-radiation into light. The light is not absorbed in the gel and thus reaches the film. However, the gels should not be stained with Coomassie Blue because the stain partially absorbs the light. Fluorography is also applicable to nitrocellulose blots. You spray the blot with Enhance Spray by NEN.

For a ^{3}H/^{14}C fluorogram, the gel is fixed (without Coomassie!) and washed three times with 40% methanol and 10% acetic acid (15 minutes each), and then incubated for 15 to 30 minutes in a scintillator (EN^{3}HANCE or ENLIGHTENING of NEN) and afterward dehydrated. ENLIGHTENING and EN^{3}HANCE are aggressive liquids.

Table 1.4. Relief for film problems.

Problem	Reasons	Remarks
Gel sticks to the film	Gel is not dry	Warm film cassette to room temperature before opening.
No picture	Not enough radioactivity Exposition time too brief Exposition temperature too high Bad contact between gel/blot/thin layer plate and film Gel is still moist Does the gel contain EN^{3}HANCE and was the drying temperature above 70° C?	 See Fig. 1.3. Moisture in gel/blot reduces the sensitivity. Drying temperature should lie between 65 and 70° C.
Spots on the film	Contaminated film cassette The gel was not properly equilibrated with the scintillator (could not swim freely, not enough scintillator)	 Gel must not stick to the floor of the tub, allow to swim freely in 5 times more scintillator than gel.
High background	Film too old Light leak Wrong preexposure of the film External radiation (e.g., your colleagues keep ^{125}I or ^{32}P samples in the same fridge)	
Film too bright	Developer used up Developer too diluted Wrong film	
Film too dark	Contaminated developer Developer too concentrated	
Stripes	Fixing solution exhausted	
Half-moon stripes	Film was bent after the exposition and before the development	

The development of a fluorogram can take weeks and months. If you do not want to wait that long, cut the run strip of the dehydrated gel into small sections. With ^{3}H/^{14}C gels that already contain scintillator (EN^{3}HANCE or ENLIGHTENING), the sections can be counted directly. Gels with ^{32}P or ^{125}I are dried (without previous soaking in scintillator), cut into sections, and counted directly (^{125}I) or after adding normal scintillator (^{32}P). The same basically applies to blots or thin layer plates. The experimenter measures the absolute amount of the available radioactivity to get a rough idea of the radioactivity distribution.

In case 2, the high-energetic radiation is converted into light by a fluorescent screen (intensifying screen). Two screens in the arrangement screen 1, gel, film, screen 2 further magnify the signal, but at the expense of resolution. Screens work with blots and thin layer plates just as well as with gels.

In both cases, preexposing the film raises the sensitivity. At the same time, preexposure creates a linear relationship between the amount of radioactivity and the blackening of the film (Laskey and Mills 1975). Preexposing is useful with tiny amounts of radioactivity (i.e., with gels that must be exposed over a longer period of time). The film is flashed once for 1/2,000 to 1/1,000 second. Laskey (1980) describes the precise conditions. Storage at −70° C also strengthens the signal of the fluorogram (see Figure 1.3 and Table 1.4).

Preexposing the film or very low temperatures only strengthen fluorography (light) signals. The signals of the direct autoradiography are not influenced. Fluorography has a slightly lower resolution than direct autoradiography.

Sources
1. Laskey, R. (1980). "The Use of Intensifying Screens or Organic Scintillators for Visualizing Radioactive Molecules Resolved by Gel Electrophoresis," *Methods Enzymol.* 65: 363–371.
2. Laskey, R., and Mills, A. (1975). "Quantitative Film Detection of ^{3}H and ^{14}C in Polyacrylamide Gels by Fluorography," *Eur. J. Biochem.* 56: 335.

Chapter 2 Ligand Binding

Ligand binding and binding assay are dull, according to common prejudice, which is correct. Nevertheless, these simple assays often provide the key to interesting results. This is because the binding between two molecules is the first step in every biochemical reaction and thus every function.

It takes two to bind: ligand and binding site. A ligand is a molecule that reversibly binds to a binding site. A binding site is the "locus" of a protein—the binding protein. What are the steps in a typical binding project? It goes something like this. A ligand (e.g., a toxin) shows an effect in low concentrations in a physiological assay. The experimenter thinks: This effect must be communicated via cell molecules, and thus the ligand must be interacting with a cell molecule (and it can do so only if it binds to the cell molecule). The binding site on this cell molecule is the first link in the ligand's chain of effects and is thus the starting point for the experimenter. He radioactively marks the ligand, develops a binding assay, and characterizes the binding site and the binding protein. Then the experimenter purifies the binding protein, characterizes it more thoroughly, clones it, and along the way (with luck) learns something about its function and thus about the effect mechanism of the ligand. The binding assay thus opens a new research direction for the experimenter in which he can be active for years. He can even distinguish himself with binding experiments if the binding protein performs an important function in the cell, the molecular basis of which was unknown until then.

The search for the binding protein requires the radioactively marked ligand, a binding assay, and tissue that contains the binding protein in a high enough concentration. The amount of work and time involved depends on the ligand's characteristics. First, the experimenter has to isolate the ligand, because it must be as pure as possible for binding assays. The second hurdle is the radioactive marking of the ligand. In the process, the biological characteristics of the ligand easily get lost, even with large peptides or proteins with their multiple possibilities for marking. Finally, the experimenter must develop one more binding assay. Without luck and with 10 h of daily work, the inexperienced doctoral candidate needs maybe half a year for marking the ligand (including purification and functionality tests), developing the binding assay, and producing results.

Many proteins do not have any ligands. No problem. You make one yourself. In epitope libraries you can find for every protein a host of peptides that bind to it with almost any affinity and at different sites (Scott and Smith 1990). The precondition is that the protein is pure and available in sufficient quantity.

A warning: As simple as the binding of two partners seems to be, relationships often fail because of little things that can become astoundingly complicated. Maybe the ligand loses its biological activity during radioactive marking, or it does not bind specifically (i.e., to several different binding sites), or it requires elaborate chemical syntheses, or it turns out that the binding site you found with such trouble lies on a known protein. If the latter is the case, you merely found a new ligand for an old protein—a result that yields a publication and then leaves the experimenter in the rain.

Enough of this pessimism. Binding projects are more predictable than others, and although the methods belong to the classics and thus provide little prestige they are nevertheless indispensable.

Source

1. Scott, J., and Smith, G. (1990). "Searching for Peptide Ligands with an Epitope Library," *Science* 249: 386–390.

Faith, senor, it's my opinion the poor man should be content with what he can get, and not go looking for dainties in the bottom of the sea.

2.1 Radioactive Ligand Marking

Before you set foot into the isotope lab, you should characterize the ligand pharmacologically (i.e., you should measure a biological reaction). This can be the effect of a neurotoxin at the nerve-muscle endplate or even a tedious LD_{50}. You learn in which concentrations, where, and how the ligand has its effect. This is important not only for the choice of the isotope but for the binding assay to be developed later. Above all, however, a quantitative pharmacological assay allows you to test the ligand marked with (for example) ^{127}iodine (not radioactive) for biological activity. Without a pharmacological assay, you grope in the dark. The most common isotopes for ligand marking are ^{125}iodine and ^{3}H. ^{125}iodine is the isotope of choice. Proteins, peptides, and compounds with phenyl groups or primary amino groups are marked with ^{125}iodine.

2.1.1 Iodination of Peptides and Proteins

For iodination, proteins and peptides must have tyrosine, histidine, or primary amino groups (Figure 2.1). In addition, the biological activity must survive the iodination procedure. Tyrosine residues are often iodined with chloramine T. Chloramine T in watery solution slowly

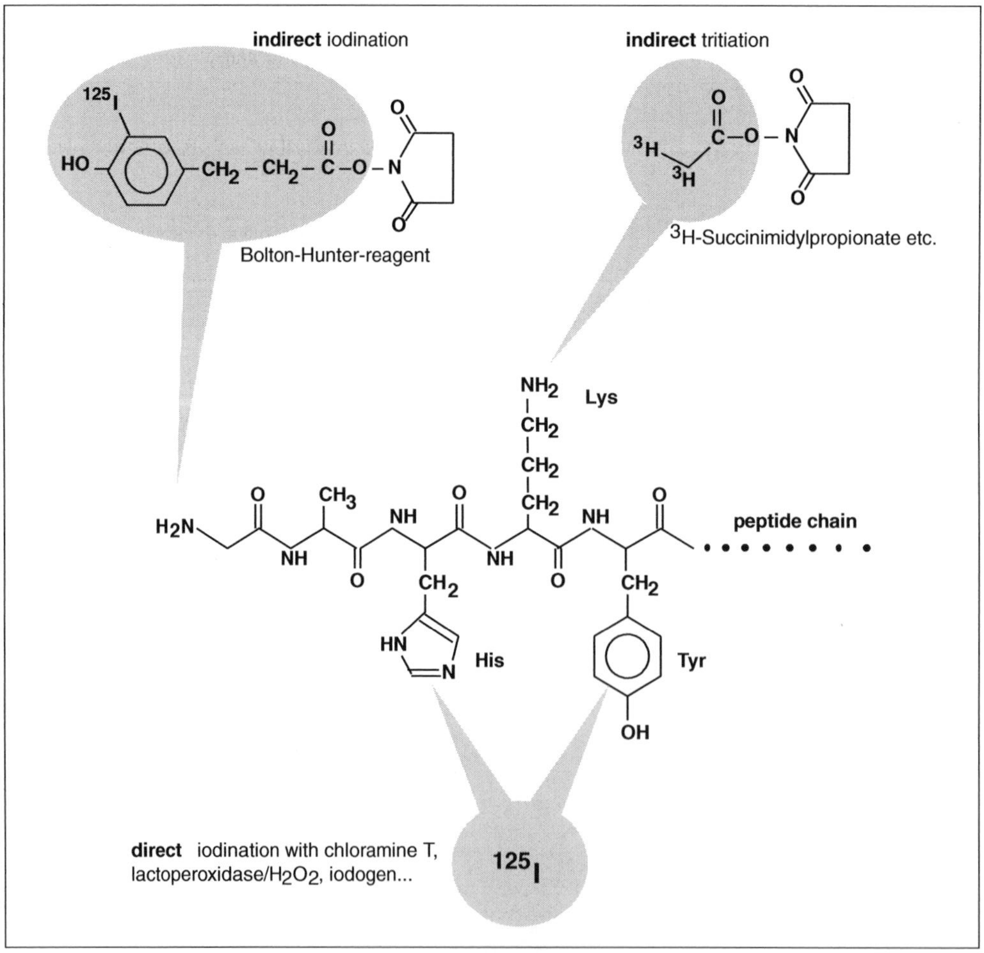

Figure 2.1. Radioactive marking of peptides and proteins.

decomposes to hypochloric acid, and it oxidizes $^{125}I^-$ to $^{125}I^+$. $^{125}I^+$ reacts with the anionic form of tyrosine to ^{125}I-tyrosine. The reaction has a pH optimum of 7.5 and is inhibited by thiocyanate in micromolar concentrations. Because companies deliver $Na^{125}I$ in NaOH, the experimenter neutralizes the reaction mixture with a strong buffer, usually sodium phosphate. At the end of the reaction, the oxidizer chloramine T is destroyed by a reductant (largely bisulfite). The different tyrosine residues of a protein react to a varying degree with $^{125}I^+$, because of different accessibility, different neighbors, and so on.

Rehm and Lazdunski (1992) describe an iodination protocol developed for proteins and peptides that also delivers good results with molecules of small MW and a phenyl group. The $^{125}I^-$ is oxidized with low chloramine T concentrations (molar ratio of the molecule to be iodined/chloramine T/^{125}I about 1/1–2/0.5–1). The iodination reaction is stopped either by diluting the reaction mixture (Rehm and Lazdunski 1992) or (if the protein can handle it) by addition of $Na_2S_2O_5$ (30 mg/ml). An ion exchanger separates peptides or proteins from the remaining $^{125}I^-$. If you want to avoid the ion exchanger, use a gel filtration column (e.g., Sephadex G-25 or Biogel P-60). TRITON-X-100 and BSA prevent the adsorption of the iodined peptide to reaction vessels and pipettes. After the iodination reaction, you have a mixture of simply iodined, multiply iodined, and non-iodined molecules, where the mono-iodined compound predominates.

Source
1. Rehm, H., and Lazdunski, M. (1992). "Purification, Affinity Labeling, and Reconstitution of Voltage-sensitive Potassium Channels," *Methods Enzymol.* 207: 556–564.

During the iodination with the oxidizer chloramine T, all reactants are present in solution (one-phase system). Pierce offers oxidizers that were applied to a solid phase (two-phase system: iodobeads, iodogen). Iodobeads are N-chlorobenzene sulfonamides attached to polystyrene beads. Iodogen is a hydrophobic chloramine T derivative applied to the wall of the reaction vessel. After the reaction with iodobeads and iodogen, the solid phase with the oxidizer can easily be separated from the reaction mixture. Hence, the addition of reducing agent (bisulfite) is unnecessary, which spares the sensitive disulfide bridges of some proteins. In addition, N-chlorobenzene sulfonamide is a milder oxidizer than chloramine T.

Even milder is the iodination of the tyrosine residues with lactoperoxidase and H_2O_2. The experimenter repeatedly adds small quantities of H_2O_2 to the reaction mixture with $^{125}I^-$, lactoperoxidase, and the protein/peptide. The lactoperoxidase splits the H_2O_2 into water and O_2 and oxidizes the $^{125}I^-$ at the same time. Instead of repeatedly adding H_2O_2, the experimenter can achieve a steady production of H_2O_2 in the reaction mixture with a combination of glucoseoxidase and glucose. Bio-Rad offers the glucoseoxidase-lactoperoxidase reaction chain as a two-phase iodination system.

Advantages: The reaction mixture remains free of enzymes. The iodined glucoseoxidase/ lactoperoxidase molecules (self-iodination) remain in the reaction vessel. *Important:* Azid inhibits the lactoperoxidase. If the peptide/protein to be marked has no tyrosine residue, the experimenter tries to iodize a histidine residue. Above pH 8 to 8.5, iodine preferentially substitutes the imidazole ring of histidine.

The experimenter can convert primary amino groups of the protein/peptide with the Bolton-Hunter reagent or other ^{125}I marked N-succinimidyl compounds (indirect iodination, Figure 2.1) into an acid amid (i.e., the positive charge of the primary amino group gets lost). ^{125}I Bolton-Hunter reagent is available from NEN or Amersham. Imidoesters can also be used instead of N-hydroxysuccinimidyl compounds (Bright and Spooner 1983). These react specifically with lysin residues and receive the positive charge of the derivatized lysin. Finally, there is still the possibility of converting the primary amino group with 4-hydroxybenzaldehyde into an imine and to reduce that to a secondary amine (careful with disulfide bridges) (Su and Jeng 1983). Charge and nucleophili of the primary amino group are preserved. Afterward, the phenyl group is iodined (e.g., with chloramine T).

After iodination, the free ^{125}iodine needs to be separated from iodined (and uniodined) protein or peptide. Traditionally, small ion exchangers or gel filtration columns (for proteins) are used for this purpose, but HPLC (for smaller stable proteins and peptides) is gaining popularity. HPLC is worth using if you iodine often, because after a run with 1 mCi ^{125}iodine the machine cannot be used for cold work for at least one year. The HPLC is often able to distin-guish between monoiodized and di-iodized peptides/proteins. For iodination reactions with good incorporation, you can skip separating the free [125]iodine, especially because free [125]iodine develops again during storage of the reaction product (back reaction).

If you have to iodize a peptide that can withstand pH 3 and 100° C, you can iodize it first with [127]iodine. After separating out the free [127]iodine, you substitute [125]iodine for the [127]iodine you incorporated into the peptide (Breslav et al. 1996). The advantage: The products of the [127]I iodination (see Section 2.1.4) can comfortably be separated on the (cold) reversed-phase HPLC and then examined for their biological activity. For the active [127]iodine derivative, you then switch the [127]iodine with [125]iodine. You can also skip the iodination reaction and instead synthesize the peptide with suitable [127]iodine amino acid at the desired site (peptide synthesizer).

The method is limited to acid-stable peptides. In addition, only 3,5-di-iodotyrosin willingly releases its [127]iodine for [125]iodine. However, even with peptides containing iodotyrosin Breslav et al. only achieved a specific activity of 10 Ci/mM. However, the method may not be com-pletely mature yet, both with respect to milder exchange conditions and to higher specific activities. Breslav et al. (1996) separate free [125]iodine with a C18 Sep-Pak cartridge: [125]iodine runs through, peptide binds, and is eluded after washing with methanol.

Sources

Tyrosine Iodination with Iodobeads
1. Salacinski, P., et al. (1981). "Iodination of Proteins, Glycoproteins, and Peptides Using a Solid Phase Oxidizing Agent, 1,3,4,6-tetrachloro-3α,6αdiphenyl Glycoluril (Iodogen)," *Anal. Biochem.* 117: 136–146.
2. Tuszynski, G., et al. (1983). "Labeling of Platelet Surface Proteins with ^{125}I by the Iodogen Method," *Anal. Biochem.* 130: 166–170.

Tyrosine Iodination with H_2O_2 and Lactoperoxidase
1. Rehm, H., and Betz, H. (1982). "Binding of β-bungarotoxin to Synaptic Membrane Fractions of Chick Brain," *J. Biol. Chem.* 257: 10015–10022.
2. Sutter, A., et al. (1979). "Nerve Growth Factor Receptors: Characterization of Two Distinct Classes of Binding Sites on Chick Embryo Sensory Ganglia Cells," *J. Biol. Chem.* 254: 5972–5982.

Histidine Iodination
1. Chisholm, D., et al. (1969). "The Gastrointestinal Stimulus to Insulin Release I Secretin," *J. Clin. Invest.* 48: 1453–1460.
2. Taylor, J., et al. (1984). "The Characterization of High-affinity Binding Sites in Rat Brain for the Mast Cell Degranulating Peptide from Bee Venom Using the Purified Monoiodinated Peptide," *J. Biol. Chem.* 259: 13957–13967.

Indirect Iodination
1. Bolton, A., and Hunter, W. (1973). "The Labeling of Protein to High Specific Radioactivities by Conjugation to a ^{125}I-containing Acylating Agent," *Biochem. J.* 133: 529–539.
2. Breslav, M., et al. (1996). "Preparation of Radiolabeled Peptides Via an Iodine Exchange Reaction," *Anal. Biochem.* 239: 213–217.
3. Bright, G., and Spooner, B. (1983). "Preparation and Reaction of an Iodinated Imidoester Reagent with Actin and α-actinin," *Anal. Biochem.* 131: 301–311.
4. Su, S., and Jeng, I. (1983). "Conversion of a Primary Amine to a Labeled Secondary Amine by the Addition of Phenolic Groups and Radioiodination," *Anal. Biochem.* 128: 405–411.

2.1.2 The Day After

No incorporation of [125]iodine? If you did not forget anything in the iodination mix and did not add anything wrong, the protein either does not contain any tyrosine residue or the oxida-tion conditions were too weak. If other or stronger oxidizers (e.g., more chloramine T) do not help, you could try to convert with Bolton-Hunter reagent or ^{3}H-marked N-succinimidyl com-pounds or you could attempt to iodize a histidine of the protein.

The iodized protein does not exhibit specific binding? Often, the incorporated [125]iodine is to blame, but the reagents used for the iodination can also inactivate the protein. For example, chloramine T oxidizes not only I⁻ but the α-amino group of peptides and amino acids to nitrile. Sensitive phenol derivatives are also destroyed. By the way, these reactions were the basis for the application of chloramine T as a disinfectant during World War I (Dakin et al. 1996).

Milder oxidation conditions (e.g., H_2O_2 or iodobeads instead of chloramine T) or other reaction conditions (pH, ion strength, ligands) often turn the situation around.

The number of the iodine atoms incorporated per protein heavily influences its binding ability. The greater the ratio of molecules of iodine atoms to molecules of protein in the reaction mixture the more iodine is incorporated into the protein, and the less immune reactivity and biological activity of the iodized protein. The stoichiometric relation of the reactants in the iodination mixture is an important factor: too much iodine and the protein is dead. It is advisable to perform the reaction first with [127]iodine (nonradioactive) and to check the iodized species, especially the simply iodized compound, for biological effects.

If the experimenter wishes for monoiodized proteins, however, just a single iodine can keep the protein from binding. In this case, the experimenter changes the reaction conditions to try to iodize another tyrosine residue or a histidine residue, in the hope that iodine in the new position does not interfere with the binding. If this hope turns out to be unfulfilled, or you do not want to get involved in endless screening of reaction conditions, Bolton-Hunter reagent or ^{3}H marking remain an option.

Source

1. Dakin, H. D., et al. (1916). "The Antiseptic Action of Substances of the Chloramine Group." *Proc. Royal Soc. London Ser. B* 89: 232–242.

The protein has disappeared? In this worst-case scenario, the protein has probably been denatured by the iodination procedure and as a result was aggregated, precipitated, or adsorbed (e.g., to the column that was supposed to separate the iodized protein from free [125]iodine). A milder iodination procedure brings rescue.

2.1.3 Iodination of Molecules with Low MW

Iodized molecules with low MW serve as ligands in binding assays, for the indirect iodination of proteins, or as photoaffinity ligands. Typically these are phenol compounds.

In the iodination of molecules with low MW, $Na^{125}I$ usually serves as the source of ^{125}I and chloramine T as the oxidizer (molar ration of the three reactants about 1:1:1). A slightly basic pH of the reaction mixture favors the monoiodination. With compounds that are sensitive to water (e.g., N-succinimidyl derivatives), the iodination reaction is carried out in organic solvents (e.g., acetone, acetonitrile). The iodination efficiency depends on the solvent being used. HPLC or thin-layer chromatography separates free [125]iodine, iodized molecule, and noniodized molecule.

The derivatization of primary amino groups is similar to that of proteins/peptides, but molecules with low MW can often also be derivatized in organic solvents. The characteristics (e.g., the solubility) of molecules with small MW change more during the derivatization with (for example) Bolton-Hunter reagent than those of a large protein. As a precaution against radiolysis (see Section 2.1.5), the iodized molecule should be kept in diluted solution, together with radical catchers such as ethanol, ascorbate, and mercaptoethanol.

Sources

1. Ji, J., et al. (1985). "Radioiodination of a Photoactivable Heterobifunctional Reagent," *Anal. Biochem.* 151: 358–349.
2. Kometani, T., et al. (1985a). "Iodination of Phenols Using Chloramine-T and Sodium Iodide," *Tetrahedron Letters* 26: 2043–2046.
3. Kometani, T., et al. (1985b). "An Improved Procedure for the Iodination of Phenols Using Sodium Iodide and Tert-butyl Hypochlorite," *J. Org. Chem.* 50: 5384–5387.
4. Masson, J., and Labia, R. (1983). "Synthesis of a ^{125}I-radiolabelled Penicillin for Penicillin Binding Proteins," *Anal. Biochem.* 128: 164–168.

2.1.4 Isolation of Iodized Species

An iodination reaction often creates a mixture of different molecules. For example, the iodination of proteins that contain several tyrosine residues yields monoiodized, di-iodized,

The amount A of a radioactive compound at time t is calculated as follows:

$$A = A_0 \cdot \exp\left[-\frac{\ln 2 \cdot t}{T_{1/2}}\right]$$

A_0 is the amount of radioactive compound at time $t = 0$, $T_{1/2}$ is the half-life of the isotope. The activity (decays per time unit) of a radioactive compound at time t is thus:

$$\frac{dA}{dt} = -\frac{A_0 \cdot \ln 2}{T_{1/2}} \cdot \exp\left[-\frac{\ln 2 \cdot t}{T_{1/2}}\right]$$

Let A_0 be 1 mM (e.g., ^{125}I). Then, dA/dt is at time $t = 0$:

$$\left[\frac{dA}{dt}\right]_{t=0} = -\frac{A_0 \cdot \ln 2}{T_{1/2}} = \frac{6.02 \cdot 10^{20} \cdot \ln 2}{59.6 \cdot 24 \cdot 60 \text{ min}} = 4.83 \cdot 10^{15} \text{ decays/min.}$$

With 1 Ci = 2.22×10^{12} decays/min, you get: $\left[\dfrac{dA}{dt}\right]_{t=0} = 2176 \text{ Ci}$

That is, 1 mM [125]iodine has an activity of 2,176 Ci. Its specific radioactivity is thus 2,176 Ci/mM.

Figure 2.2. Time dependence of the activity of a radioactive compound.

and uniodized proteins, where the monoiodized proteins can still be iodized on different tyrosines. After all, phenyl groups and imidazole rings can be mono- and bisubstituted with iodine.

Monoiodized molecules can be separated from uniodized and multi-iodized molecules via HPLC (Bidard et al. 1989) or isoelectric focusing (Rehm and Betz 1982). The isolation of the simply iodized molecule (from a mixture of simply iodized and uniodized molecules) not only provides for a higher specific activity and a better signal/space ratio but simplifies the interpretation of the binding results. The amount A of a radioactive compound at time t is calculated as follows:

If the monoiodized molecule species cannot be isolated, you have a mixture of iodized and uniodized molecules (i.e., you have to determine the specific radioactivity). This is calculated as the amount of incorporated radioactivity divided by the amount of molecules (iodized + uniodized) in mol. The specific activity is given in Ci/mM. On the basis of the specific activity, you can calculate how many atoms of the radioactive isotope are contained in a molecule (Figure 2.2). Thus, if you have a protein derivatized with [125]iodine and with a specific radioactivity of 2,200 Ci/mM, each protein molecule contains one [125]iodine atom on average.

Sources
1. Bidard, J., et al. (1989). "Analogies and Differences in the Mode of Action and Properties of Binding Sites (Localization and Mutual Interactions) of Two K+ Channel Toxins, MCD Peptide and Dendrotoxin I," *Brain Res.* 495: 45–57.
2. Rehm, H., and Betz, H. (1982). "Binding of β-bungarotoxin to Synaptic Membrane Fractions of Chick Brain," *J. Biol. Chem.* 257: 10015–10022.

2.1.5 Advantages and Disadvantages of Iodination

[125]Iodine has a high specific activity (Table 2.1) and is quickly and easily detected (e.g., in autoradiograms). For counting, a γ-counter is sufficient, which does not need an expensive

Table 2.1. The most important radioactive isotopes.

Isotope

Half-life

Type of Radiation

Measurement

Specific Activity Ci/mM

Energy Max. (MeV)

Reach Max. (cm)

Critical Organ

Daughter Nucleid
  Years
  Days
  Air
  Plexiglas
  Lead
  Eye
  Bones
  Thyroid

LSC: liquid scintillation counter.

and toxic scintillator. The time required for an iodination, including the separation of the iodized molecule of the free iodine, is about 2 to 3 h. The preparations, such as making buffers and setting up the iodine lab, take about a day. In addition, iodination is relatively inexpensive (2 mCi Na^{125}iodine cost about $150).

The usual hand measuring instruments react even to traces of ^{125}iodine with frightening crackling. Hence, working with ^{125}iodine rarely leads to the extensive contamination of tools and workplace as can often be observed with ^{3}H. After all, ^{3}H can only be detected with time-consuming wipe tests, and in my experience researchers do this at most once a year.

During everyday work with tiny amounts (< 100 μCi) of ^{125}iodine, protection is provided by 1 mm high-grade steel plates. These block the weak γ-radiation almost as well as lead plates, and they are easier to handle and are nontoxic.

Handling ^{125}iodine has its concerns. During iodination, the experimenter has to work with large amounts of radioactivity (1 to 5 mCi) and must take special safety measures, such as using vent hoods with iodine filters. The biological activity of many proteins does not withstand the oxidizing conditions of the iodination protocols, and the large ^{125}iodine atom changes the qualities of the molecules.

^{125}Iodine has a half-life of 60 days, but the specific activity of an iodized compound decreases faster, because in storage the back reaction to free ^{125}iodine takes place. The radiation generated by decaying ^{125}iodine destroys other molecules of the iodized compound and with water generates free radicals. This radiolysis can destroy iodized molecules of low MW within days. Dilution of sample and radical catcher slows down the radiolysis. A reduction of the temperature barely influences this process, but it keeps fungi and bacteria in check. Finally, many proteins aggregate during iodination and subsequent storage. The nerve-wracking iodination procedure thus has to be repeated more often than researchers would like.

The bane of indirect iodination methods is the low specific activity of the product. For a good yield of the conjugation reaction, you need to add large amounts of the molecule to be derivatized. However, if the specific activity is too low, you have to separate iodized from uniodized species, which can be tricky (recommendation: make the ^{127}iodine product with ^{127}I Bolton Hunter and a trace of ^{125}I Bolton Hunter and use it to calibrate the separation method beforehand). Bolton Hunter reagent, for example, is expensive and unstable in watery solution, and the incorporation of the large lipophile group can have even more disastrous effects on the activity of a compound than the conversion of a phenyl residue into a ^{125}I-phenyl residue.

2.1.6 Tritiation

Tritiation is an option when the molecule to be marked has free amino groups. Tritiations via halogen-tritium exchange, reduction of double bindings with ^{3}H gas, and so on should remain limited to emergencies and to specialty labs. ^{3}H marking has the advantages of low radiation, which allows for work without fear; of longer durability in comparison with ^{125}iodine (the half-life of ^{3}H amounts to more than 12 years; see Table 2.1); and the ability to perform a tritiation with small side chains without an aromatic ring. NEN and Amersham offer about one dozen different ^{3}H-marked N-succinimidyl compounds that react with free amino groups. Nevertheless, molecules tritated with N-succinimidyl compounds have a low specific radioactivity (Table 2.1) and differ in isoelectric point and solubility from the source compound.

During the incubation of the molecule that is to be marked with the ^{3}H-marked N-succinimidyl compound, you need to keep in mind the solubility of two molecules and the water sensitivity of the N-succinimidyl compound. Finally, the buffer or solvent being used may not contain any molecules with free amino groups.

To carry out the conversion with the N-succinimidyl compound as completely as possible and to avoid wasting the expensive reagent, you add large amounts of the molecule to be marked. After the reaction, the experimenter has to separate the non-marked from the marked molecule, or be contented with a low specific activity. Herein lies the problem. Carrying out complex separations with large amounts of ^{3}H without contaminating the work environment is like moving an anthill without losing an ant. A simple and clear separation step must do it. An ion exchanger or isoelectric focusing is recommended for proteins. For peptides and molecules with low MW, reversed-phase HPLC or thin-layer chromatography is preferred.

Sources

1. Dolly, J., et al. (1981). "Tritiation of α-bungarotoxin with N-succinimidyl(2,3-^{3}H)proprionate," *Biochem. J.* 193: 919–923.
2. Othman, I., et al. (1982). "Preparation of Neurotoxic ^{3}H-β-bungarotoxin: Demonstration of Saturable Binding to Brain Synapses and Its Inhibition by Toxin I," *Eur. J. Biochem.* 128: 267–276.

> Peace, friend, greater secrets I mean to teach thee and greater favours to bestow upon thee.

2.2 Binding

2.2.1 Isolation of Membranes

Many important proteins (e.g., ion channels, neurotransmitter receptors, transporters, and ion pumps) are integral membrane proteins. You need membranes to examine the binding sites of these proteins. If appropriate to the question being researched, you can obtain the basic material (e.g., muscle, liver, brain) from the slaughterhouse. Pig brain is much less expensive than rat brain.

Is it better to isolate membranes directly from the freshly extracted tissue. An alternative is to freeze the tissue at −80° C, allowing you to make membranes from the frozen material when needed. However, freezing the tissue damages the cell organelles. Thus, you need fresh tissue for experiments that depend on the physiological function and content of the organelles. For binding assays, however, in my experience (with neural ion channels and receptor proteins) it is irrelevant whether the membranes come from fresh or deep-frozen tissue. The membrane seems to stabilize the integral membrane proteins against freeze/thaw cycles. Damaged lysosomes do release proteases, and a clean separation of the organelles is not possible with tissue that has been frozen. However, fresh tissue has similar problems with lysosomes being damaged (e.g., during homogenization and when the membranes suffer an osmotic shock).

The first step in obtaining membranes is the homogenization of the tissue. It is suspended in a homogenization buffer and then chopped up. The homogenization buffer is iso-osmotic to the tissue fluid, has a low ion strength, and contains cane sugar (250 to 320 mM), buffer (often 5 to 10 mM tris-Cl or Na-HEPES pH 7.4), and a mixture of different protease inhibitors (Bacitracin, PMSF, EDTA, and so on). But how do you chop it up?

- With its rotary knife, the polytron chops up large amounts of material, tough tissues (muscle), or deep-frozen tissues. To avoid warming up the homogenate, you chop the tissue in bursts (four to five times for 15 seconds at medium rotation speed, with 3-minute breaks between). Important: Foam denatures proteins.
- Some people crush the tissue in liquid nitrogen into powder form and homogenize the tissue dust with a potter or polytron. Pulverizing with a porcelain mortar is laborious and dangerous, and takes several hours (protective glasses!). It is better to use a closed iron pipe in which the deep-frozen tissue is crushed with iron stick and hammer. The additional trouble of pulverization may be worth it if you are after the membranes of special organelles (e.g., synaptic vesicles). Pulverization is not necessary for membranes for the usual binding assays.
- The potter homogenizes tissues that are easy to chop, such as brain or liver. A potter is a tube-shaped glass vessel with a fitted Teflon plunger that rotates at adjustable speeds. The potter homogenizes more gently than the polytron, but it creates more work. You potter when you would like to extract metabolically active organelles (mitochondria, synaptosomes, and so on) or their highly purified membranes.
- You can also grind tissues by hand between the roughened surfaces of a conical glass pestle and a matching glass potter. This technique is largely used for small tissue samples. I remember it as laborious and conducive to carpal tunnel syndrome, and I broke the pestle's glass handle more than once.
- Large numbers of small tissue samples are homogenized with (disposable) plastic pestles in microcentrifuge tubes. That is, one would like to do so, but often it does not work. Sometimes the surfaces of microcentrifuge tubes and pestle are too smooth, and sometimes they do not fit together well. In addition, "sample evasion" often occurs with small samples and (relatively) large buffer volumes. The sample does not even think about letting itself be crushed by the pestle, but elegantly evades the threat. If the pestle goes down, the sample swims upward, and vice versa. Kusumoto et al. (2001) solve this problem by cutting a slit in the elastic polypropylene pestle. Apparently, the slit offers the following advantages. While pushing up and down into the conical end of the microcentrifuge tube the slit opens and closes. This thoroughly cuts the sample to pieces, assure the authors. Also, the sample cannot escape as easily anymore because it is held by the split pestle as if by a pair of pliers. In fact, the authors show that their split pestle allows them to isolate twice as much RNA from mouse liver as with a normal pestle, which grinds the tissue only between pestle and vessel walls. With small tissue samples (<50 mg) the customary pestle fails completely, whereas the split pestle still yields RNA. The authors do not say whether this works similarly well with other tissues. In any case, one cannot deny that the method has a certain appeal.

Homogenization is followed by differential centrifugation. The differential centrifugation separates the homogenate into several pellets (P1, P2, P3) and supernatant (S1, S2, S3). The P1 pellet contains cell fragments and nuclei. The P2 pellet contains mitochondriae, terminals, lysosomes, and (with brain tissue) synaptosomes. These pellets are sufficient for most binding assays. If you want it finer, purify further with sucrose, Ficoll, or Percoll gradients. In my experience, these gradients yield modest purification factors (twofold to fourfold), and they are labor intensive and lead to heavy losses. All homogenization and purification steps are carried out at 4° C.

Isolation of P2 membranes:

- The tissue is coarsely cut up (e.g., with scissors) in the homogenization buffer (hb; e.g., 5 mM tris-Cl pH 7.4, 300 mM sucrose, 0.1 mM EDTA, 10 µM PMSF). Volume ratio of tissue to hb 1:10.
- Homogenization of the tissue with polytron or potter.
- First centrifugation of the homogenate for 8 to 16 minutes at 800 to 1,000 g to generate pellet P1 and supernatant S1. It is recommended that you wash P1 once with hb to increase the yield. For this, P1 is resuspended in three- to fivefold volumes of hb and then centrifuged

again for 8 to 16 minutes at 800 to 1,000 g. The supernatant is combined with S1, and the washed P1 is rejected.

The combined S1 are centrifuged for 30 minute at 10,000 to 20,000 g to generate pellet P2 and supernatant S2. If you would like to remove the soluble proteins in the membrane vesicles, subject P2 to an osmotic shock. For this, P2 is resuspended in a tenfold volume of hb without sucrose and incubated for 30 minutes on ice, and the membranes are then centrifuged for 30 minutes at 20,000 g. The pellet is resuspended in buffer (e.g., 5 mM tris-Cl, pH 7.4), aliquoted, and frozen at −80° C.

Sources

Plasma membranes from cell lines and cell cultures: It is usually sufficient to scrape off the cells from the petri dish, to homogenize in PBS in the potter, and to make a P2 pellet from the homogenate. Finer methods appear in the following papers.

1. Harms, K., et al. (1981). "Purification and Characterization of Human Lysosomes from EB-virus Transformed Lymphoblasts," *Exp. Cell Res.* 131: 251–266.
2. Mersel, M., et al. (1987). "Isolation of Plasma Membranes from Neurons Grown in Primary Culture," *Anal. Biochem.* 166: 246–252.
3. Miskimins, W. K., and Shimuzu, N. (1982). "Dual Pathways for Epidermal Growth Factor Processing After Receptor-mediated Endocytosis," *J. Cell Physiol.* 112: 327–338. This gentle method breaks cells open by triturating in an EDTA buffer. Cell fragments and nuclei are pelleted. If you triturate the cells or the pellet often enough, with occasional changes of the EDTA buffer, nice nuclei stay behind in the pellet (distinguishing mark: the pellet becomes transparent).
4. Record, M., et al. (1985). "A Rapid Isolation Procedure of Plasma Membranes from Human Neutrophils Using Self-generating Percoll Gradients: Importance of pH in Avoiding Contaminations by Intracellular Membranes," *BBA* 819: 1–9.

Membranes from Tissue
1. Kusumoto, M., et al. (2001). "Homogenization of Tissue Samples Using a Split Pestle," *Anal. Biochem.* 294: 185–186.

Mitochondriae
1. Bustamante, E., et al. (1977). "A High Yield Preparative Method for Isolation of Rat Liver Mitochondria," *Anal. Biochem.* 80: 401–408.

Synaptosomes
1. Dodd, P. R., et al. (1981). "A Rapid Method for Preparing Synaptosomes: Comparison with Alternative Procedures," *Brain Research* 226: 107–118.
2. Nagy, A., and Delgado-Escueta, A. V., (1984). "Rapid Preparation of Synaptosomes from Mammalian Brain Using Nontoxic Isoosmotic Gradient Material (Percoll)," *J. Neurochem.* 43: 1114–1123.

Synaptosomal Membranes
1. Rehm, H., and Betz, H. (1982). "Binding of β-bungarotoxin to Synaptic Membrane Fractions of Chick Brain," *J. Biol Chem.* 57: 10015–10022.
2. Taylor, J., et al. (1984). "The Characterization of High-affinity Binding Sites in Rat Brain for the Mast Cell Degranulating Peptide from Bee Venom Using the Purified Monoiodinated Peptide," *J. Biol. Chem.* 259: 13957–13967.

Synaptic Vesicles
1. Hell, J., et al. (1988). "Uptake of GAB A by Rat Brain Synaptic Vesicles Isolated by a New Procedure," *EMBO J.* 7: 3023–3029.
2. Huttner, W., et al. (1983). "Synapsin I (Protein I), a Nerve Terminal-specific Phosphoprotein. III Its Association with Synaptic Vesicles Studied in a Highly Purified Synaptic Vesicle Preparation," *J. Cell Biol.* 96: 1374–1388.

Membranes from Chromaffine Granules and Other Secretion Vesicles
1. Cameron, R., et al. (1986). "A Common Spectrum of Polypeptides Occurs in Secretion Granule Membranes of Different Exocrine Glands," *J. Cell Biol.* 103: 1299–1313.
2. Reiffen, F., and Gratzl, M. (1986). "Ca^{2+} Binding to Chromaffin Vesicle Matrix Proteins: Effect of pH, Mg^{2+}, and Ionic Strength," *Biochemistry* 25: 4402–4406.

2.2.2 Binding Assay

In the binding assay, the binding protein is incubated with the radioactive ligand at a certain temperature in the binding buffer until the binding protein/ligand complex has formed (equilibrium). Then the experimenter separates the unbound (free) ligand from the binding protein/ligand complex and determines its quantity (Figure 2.3).

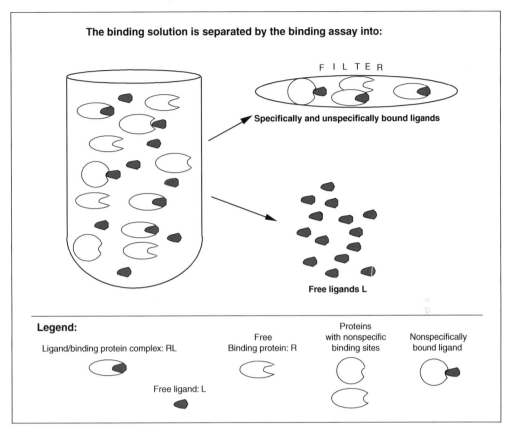

The binding solution is separated by the binding assay into:

F I L T E R

Specifically and unspecifically bound ligands

Free ligands L

Legend:

Ligand/binding protein complex: RL

Free ligand: L

Free Binding protein: R

Proteins with nonspecific binding sites

Nonspecifically bound ligand

Figure 2.3. Binding assay.

The dissociation constant K_D is a quantitative measure of the affinity (stickiness) between ligand and binding site (see Figure 2.4). The smaller the K_D the higher the affinity of the binding. A binding is called highly affine if the K_D is smaller than 10 nM.

A central concept is the specificity of the binding. One thinks of a specific binding as a binding with high affinity. For the other, it is the binding that mediates the physiological effect. A binding is called specific if it can be displaced by cold ligands, and for the fourth one it is the binding with the right pharmacology. However, a ligand binds specifically if it binds to just one species of binding site. This may occur with high or low affinity. The affinity has nothing to do with specificity, although high specificity often goes hand-in-hand with high affinity (counterexample: glutamate binds to the NMDA receptor with higher affinity than the more specific N-methyl-D-aspartate). Specificity also has nothing to do with physiological effect, because a ligand can bind specifically without showing an effect. If the binding of the radioactive ligands can be displaced by cold ligands, this merely means that the ligand binds to a restricted number of somewhat affine binding sites. However, these can be several different binding sites. The more specifically a ligand binds to a binding site the lower its affinity is to other binding sites and the smaller the number of these species.

In real life, every ligand binds to a nearly infinite number of binding sites with very low affinity. This binding is called nonspecific binding. The nonspecific binding is determined by performing the binding assay in the presence of a large amount of cold ligand. The difference between overall binding (binding without cold ligands) and nonspecific binding (binding in the presence of a 100–1,000-fold excess of cold ligand) is the saturable binding (or displaceable binding). It is saturable because this difference reaches a plateau with increasing con-

Definitions:

- **R:** Concentration of the free binding protein R.

- **L:** Concentration of the free (unbound) ligands L. This quantity is calculated from the (known) overall concentration of the ligands in the binding solution minus the (measured) overall concentration of the (saturatably and nonspecifically) bound ligands.

- **(RL):** Concentration of the binding protein-1/ligand complex RL.

- **K_D:** Dissociation constant. With binding mechanism I it is a measure of the ligand concentration at which the concentration of the free binding protein is equal to that of the binding protein/ligand complex. The dissociation constant is only dependent on the temperature and has the dimension of a concentration.

- **B:** Concentration of saturatably bound ligand. B is calculated from the saturatable binding (overall binding minus unspecific binding) and the specific activity of the ligands. For simple mechanisms, B is (like I) identical to (RL).

- **B_{max}:** The maximally attainable concentration of saturatably bound ligands (overall concentration of the binding sites).

- **R_0:** Overall concentration of the binding protein. R_O is equal to B_{max} if there is only one binding protein with only one binding site.

- **L_0:** Sum of the concentrations of free and saturatably bound ligands (equal to the overall concentration of ligands in the binding solution, disregarding the unspecifically bound ligands).

Binding mechanism I:

$$R + L \rightleftharpoons RL$$

$$K_D = \frac{R \cdot L}{(RL)} \quad \text{Mass-action law}$$

$$R_0 = R + (RL)$$
$$L_0 = L + (RL) \quad \text{Retention equations}$$
$$(RL) = B$$
$$R_0 = B_{max}$$

Derivation of the Scatchard equation (B/L as a function of B):

$$R = R_0 - B \longrightarrow K_D = \frac{[R_0 - B] \cdot L}{B} \longrightarrow K_D \cdot B = [R_0 - B] \cdot L \longrightarrow K_D \cdot \frac{B}{L} = R_0 - B$$

$$\longrightarrow \quad \frac{B}{L} = \frac{R_0}{K_D} - \frac{B}{K_D}$$

Axis sections of the Scatchard plot: $\quad \left(\frac{B}{L}\right)_{B=0} = \frac{R_0}{K_D} \quad \left(B\right)_{B/L=0} = R_0$

Figure 2.4. Concept definitions and binding mechanism I.

centrations of radioactive ligand, and that means that this difference corresponds to a limited number of binding sites. The nonspecific binding, on the other hand, is not saturatable and increases linearly with the ligand concentration. In a good assay the nonspecific binding amounts to less than 20% of the overall binding.

The saturatable binding is not necessarily specific, nor is the binding you are looking for nor one with a biological effect. To prove specificity, you need to show that the saturatable binding corresponds to just one species of binding site. The pharmacology of the binding often helps, and shows whether the binding is the right one.

2.2.3 Binding Assays with Membranes

The purpose of the binding assay is to determine the amount of bound ligand. For this, the assay separates the bound from the free ligand. For membranes this is easy, because the physical difference is large between membrane vesicles and the radioactive ligand (from simple molecule to protein).

In the centrifugation assay the membranes settle with the ligand, which is bound to them. The free ligand remains in the supernatant.

Often one starts the assay in Eppendorf containers and centrifuges out in the Eppendorf centrifuge. The radioactivity in the pellet is a measure of the amount of bound ligand. Washing of the pellet reduces the nonspecific binding. Washing is possible when the bound ligand only slowly dissociates from the membranes. For washing, the pellet is resuspended in cold binding buffer and the suspension is centrifuged again.

The experimenter resuspends soft pellets by vortexing. If the pellets stick to the wall of the Eppendorf vessel, he resuspends by repeatedly pulling pellet and binding buffer into a Gilson pipette or scrapes the tip of the vessel a few times along a steel net (e.g., autoclave basket). For a binding solution without membranes, the careful researcher makes sure that her ligand does not bind to the Eppendorf vessels themselves.

If washing is not possible, the binding solution is layered over a 5% sucrose cushion in binding buffer (a laborious procedure) and the membranes centrifuged through the cushion. The supernatant is siphoned off, and the pellet is counted. During siphoning, some of the free ligands move in the meniscus down to the pellet and increase the nonspecific binding. You avoid this source of error if you freeze the Eppendorf vessel with the centrifuged solution in fl. N_2, cut off the tip, and then count. Instead of a 5% sucrose cushion, Mackin et al. (1983) use a silicone oil.

The centrifugation assay is complicated, labor intensive, takes a long time, and is often not quantitative because the low gravity fields of an Eppendorf centrifuge do not completely sediment the membranes and/or the ligand partially dissociates during the protracted washing process. If you do not wash, you obtain a bad signal-to-noise ratio.

Advantages of the centrifugation assay are reliability and completeness. It detects the binding from low-affinity ligands, even if you skip the washing and use a high-speed centrifuge (Airfuge or Sorvall centrifuge). Nevertheless, researchers use the centrifugation assay only in dire need (e.g., with ligands with low affinity or to check the values of another assay).

Filter assays stand out due to their speed and simplicity. They allow for easy and thorough washing and yield a good signal-to-noise ratio. I use the following technique. The binding assay is started in a volume of 200 to 400 µl in 5 ml hemolysis tubes. When binding equilibrium is reached (or for kinetics at the predetermined time) the binding solution is diluted with 4 ml cold washing buffer and filtered on a filtration device (Sartorius, Millipore, Hoefer) using the vacuum of a water jet or a membrane pump.

Immediately after the supernatant is completely siphoned off, the filters are washed once or twice with cold washing buffer. The washing buffer can be identical to the binding buffer, but if possible you should leave out the expensive components of the binding buffer. With ^{3}H-marked ligands, the filters have to equilibrate with the scintillator before the counter is applied. Equilibrating takes hours. Shaking of the samples accelerates it. The filtration devices from Millipore and Hoefer are expensive ($1,000 to $1,500). An alternative is to build similar devices.

Adsorption on glass fiber filters is simple, inexpensive, and popular. The large internal surface of the filters adsorbs membrane vesicles via weak electrostatic and hydrophobic interactions. Bigger vesicles (mitochondriae, synaptosomes) are also filtered mechanically (pore size of glass fiber filters: 1 to 3 µm; Ø mitochondriae: 1 µm). Molecules in solution largely pass through. Glass fiber filters rarely block and have a large binding capacity for membranes. If it is deemed necessary, the filters can also be coated (e.g., with BSA or PEI). This influences the retention behavior of the membranes and the nonspecific binding of the ligands to the filters. Glass fiber filters are available in different degrees of thickness and porosity. Filters marked GF/C and GF/B (Whatman) are well suited for binding assays.

Filtering larger numbers of binding assays is quite boring (sample transfer, washing, take off filter, put filter on, insert sample), which stimulates philosophizing and leads to the most abstruse trains of thought. Therefore, people who need to filter more than 200 assays per day should use the filtration device from Brandel. This device eliminates the insertion of filters, makes washing easier and more reproduceable, and processes 48 samples at once. Ion exchange filters that distinguish between the charges of membrane vesicles and ligands or dimension filters (e.g., filter made from cellulose acetate) have fallen out of fashion.

Problems:
- Uncoated glass fiber filters do not completely adsorb the membranes of a binding solution, and their capacity is limited.
- During washing, specifically bound ligand gets lost, especially with faster off rate of the ligand (see Section 2.3.2).
- The large internal surface of the glass fiber filters can lead to high nonspecific binding of the ligands to the filter. This is especially the case with peptide and protein ligands. Coating of the filters often helps, but not always. Therefore, test once without membranes.

Source
1. Mackin, W. M., et al. (1983). "A Simple and Rapid Assay for Measuring Radiolabeled Ligand Binding to Purified Plasma Membranes," *Anal. Biochem.* 131: 430–437.

2.2.4 Development of Membrane Binding Assays

The protein biochemist is interested in binding sites that largely distinguish themselves by low concentration and low stability. Hence, the development of a membrane binding assay requires:
- Radioactive ligands with high specific activity and affinity for the binding site.
- Membranes from a tissue or a cell line that contain the binding site in high concentration.
- A suitable binding buffer. Initially, the choice of ions, ion strength, and pH of the binding buffer is dependent on the physiological environment.

The experimenter incubates membranes in the binding solution and radioactive ligands in binding buffer until binding equilibrium is reached. I recommend a volume of 200 to 400 µl for the binding solution. The membrane concentration can vary between 10 and 200 µg/solution, and the ligand concentration between 0.1 and 100 nM (the lower the expected affinity the higher the concentration). Most people choose 4° C for the incubation temperature, and the incubation duration depends on how quickly the binding partners associate. At first, the latter is unknown, but 1 h is typical.

First, for simplicity's sake, you use a binding assay with glass fiber filters and measure the binding of the ligands with and without membranes. If the binding of the ligands to the filter is small enough (< 1% of the total counts), you continue with the filter assay and vary the binding conditions (i.e., the ions in the binding buffer, the pH, and so on). If no saturatable binding to the membranes is measurable (no difference in the binding with and without excess of non-marked ligand), you can still try the centrifugation assay. If this does not produce any binding, the affinity of the ligands is too low there is no binding protein in the membranes, or the membrane preparation is saturated with endogenous ligands. Section 2.2.6 provides further help. It is advisable to measure everything three times from the beginning on (i.e., record three values for each unspecific binding and overall binding). This provides relatively certain conclusions and checks the quality of the measuring method at the same time.

If there is saturatable binding (the unspecific binding is substantially smaller than the overall binding), the experimenter has proven that the saturatable binding is specific (i.e., the difference between overall binding and unspecific binding) because being saturatable and specific is not the same! In addition, he must show that it is this binding that mediates the biological effect. For this, he determines the pharmacology of the binding (i.e., he examines whether a host of molecules with known specificity and effect have an influence on the binding of the ligands). It is persuasive when these molecules and the ligand have a different chemical structure and still show inhibition and, vice versa, when molecules that are chemically similar to the ligand and have neither a pharmacological effect nor binding specificity also do not influence its binding.

The organ distribution of the binding sites is also a clue regarding the sense or nonsense of the binding: A ligand that works exclusively in the brain should have binding sites in the brain and not in the liver. Binding proteins often appear only in certain developmental stages or—for cell lines—under certain conditions, and it is a strong argument if the binding has a parallel distribution. If the molecule with the binding site is a protein, the binding must disappear if the membranes are treated with proteases (e.g., Pronase from *Streptomyces griseus*). In addition, the binding must increase linearly with the amount of membranes used in the assay.

Finally, the kinetics of the binding should be roughly the same as the kinetics of the biological effect. If there are enough indicators pointing to the desired binding, the binding is characterized. First, the experimenter determines the speed with which the ligand binds to the binding site ($T_{1/2}$ of the "on rate," see Section 2.3.2). This is because many interesting parameters (the affinity constant K_o, the concentration of the binding protein B_{max}, K_i, and Hill coefficients of the inhibitors) are measures of equilibrium. For their determination, the binding solutions must be incubated until the reaction has reached an equilibrium (approximate $4 T_{1/2}$). You also need to know $T_{1/2}$ to determine the kinetic constant k_{off}. It is also useful to know how dependent the binding is on ion strength, univalent and bivalent cations, and so on.

2.2.5 Binding Assays with Soluble Proteins

2.2.5.1 Low-tech Assays

With soluble proteins, the separation of free ligands from the binding protein/ligand complex is more difficult than with membranes. This is due to the fact that physical (e.g., size) differences between the soluble binding protein and the ligand are smaller than between membrane vesicles and ligands. Thus, the ligand can be a protein with an MW and charge similar to that of the binding protein. Binding assays with soluble proteins use:

- Size differences between ligand and binding protein (gel filtrations columns or PEG precipitation)
- Adsorption differences between ligand and binding protein (coated glass fiber filters, nitrocellulose/PVDF membranes, polylysin-coated microtiter plates)
- Charge differences between ligand and binding protein (ion exchange filter)

Gel filtration column: The basis of the column assay is the size difference between free ligand and binding protein/ligand complex (e.g., ligand MW < 1,000; complex MW > 60,000). The gel material is so chosen that the ligand penetrates into the gel pores but not into the binding protein/ligand complex (Figure 2.5).

A 2-ml syringe is sealed with a suitable paper filter (Whatman), placed at the bottom of the barrel, and filled with 2-ml gel filtration material soaked in binding buffer (Sephadex, Biogel). Hanging freely in a 10-ml test tube, the column with the gel is centrifuged for 2 minutes at 1,000 g. The binding solution (total volume 200 to 250 l) is carefully loaded on the still-moist gel and then centrifuged again (column hanging in the test tube) for 2 minutes at 1,000 g. Afterward, the bound ligand is in the test tube and the free ligand is in the column.

The column assay is indispensable during the introduction of a binding assay, in spite of the complicated setup and handling of the columns and the expensive gel filtration material. If a tissue extract shows no binding, the possible causes are countless and difficult to discern. Among these are the possibilities that the binding assay does not work, the binding protein is

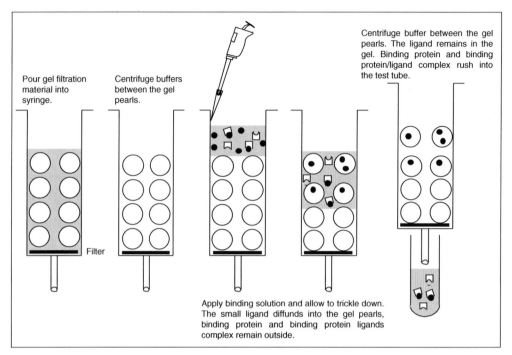

Figure 2.5. The column assay.

denatured, the affinity of the ligands is too low, there is no binding protein in the tissue extract, and so on. With some certainty, the column assay excludes the binding assay as a source of error. The column assay fails only if:

- The binding protein adsorbs to the gel material (unlikely with raw protein mixtures of high concentration).
- The ligand does not penetrate into the gel pores. This is the case when the ligand incorporates into high-molecular buffer components (e.g., small hydrophobic ligands in TRITON micelles). This error source can be eliminated by a simple experiment: if you perform the column assay without binding protein, the applied radioactivity must remain in the column.

Sources
1. Parcej, D., and Dolly, J. (1989). "Oendrotoxin Acceptor from Bovine Synaptic Plasma Membranes," *Biochem. J.* 257: 899–903.
2. Rehm, H., and Betz, H. (1984). "Solubilization and Characterization of the β-bungarotoxin-binding Protein of Chick Brain Membranes," *J. Biol. Chem.* 259: 6865–6869.

Coated glass fiber filters: Glass fiber filters (e.g., GF/C from Whatman) are coated with 0.25 to 1% polyethylenimine (PEI) (i.e., soaked for at least 5 minutes in binding buffer containing 0.25 to 1% PEI). The filters thereby bind soluble proteins (especially negatively charged ones). The filtration through coated glass fiber filters allows us to quickly and efficiently separate the free ligands from the binding protein/ligand complex—under the assumption that the binding protein/ligand complex adsorbs to the filter and the free ligand does not. This is the case for unexpectedly many ligand/binding-protein pairs, because bigger binding proteins adsorb almost always, whereas pharmaceuticals, peptides, amino acids, and smaller and positively charged proteins pass through the filters. The method gives a good signal and is reliable and inexpensive.

Problems appear when the radioactive ligand unspecifically adsorbs in larger quantities to the coated glass fiber filters. This behavior is exhibited by (for example) peptides with negative total charge. The assay with PEI-coated glass fiber filters was first and best described by Bruns et al. (1983).

Source

1. Bruns, R., et al. (1983). "A Rapid Filtration Assay for Soluble Receptors Using Polyethylenimine-treated Filters," *Anal. Biochem.* 132: 74–81.

If the ligand is a protein, is sized similarly to the binding protein, and has similar qualities, both column assay and glass fiber filter fail. Do not despair. Dot the binding protein or the tissue extract on a blot membrane (e.g., nitrocellulose) (Petrenko et al. 1990). With some luck, the binding protein survives the adsorption to the nitrocellulose. Of course, the dot buffer may not contain any denaturing additions and the blotted proteins may not be fixed. Adding 10 to 20% methanol to the dot buffer can be beneficial, because it raises the binding capacity of the nitrocellulose. Microtiter plates, small stamping machines, and filtration devices alleviate the lot of the dotter (for the dot blot technique, see Section 6.5). Once the binding protein is applied, the dot blot is blocked and incubated with the radioactive ligand.

A related technique adsorbs the binding protein to polylysine-coated microtiter plates instead of dotting it on nitrocellulose. After blocking the unsaturated polylysine, the wells are incubated with radioactive ligand (Scheer and Meldolesi 1985).

Sources

1. Petrenko, A., et al. (1990). "Isolation and Properties of the X-latrotoxin Receptor," *EMBO J.* 9: 2023–2027.
2. Scheer, H., and Meldolesi, J. (1985). "Purification of the Putative α-latrotoxin Receptor from Bovine Synaptosomal Membranes in an Active Binding Form," *EMBO J.* 4: 323–327. This protocol is incomplete (Scheer incubates the wells for 2 h at 4° C) with radioactive α-latrotoxin. He also fails to tell how the radioactivity gets out of the wells, but with 150 μl 0.1 M NaOH this should be no problem.

PEG precipitation: PEG (polyethylene glycol) precipitates high-molecular substances (e.g., proteins). Low-molecular substances remain in solution. The threshold between high-molecular and low-molecular depends on the conditions of the precipitation (PEG type, PEG concentration, temperature).

During the PEG assay, the proteins of the binding solution are precipitated with 6 to 8% PEG. In diluted protein solutions, a carrier protein such as γ-globulin guarantees complete precipitation. Filtration through glass fiber filters separates the precipitate and the binding protein/ligand complex from the free ligand, and the radioactivity of the filters is a measure of the bound ligands.

You are in trouble when the PEG precipitates the ligand (e.g., with radioactive proteins as ligands), when the ligand adsorbs to the precipitate, or the binding protein/ligand complex dissociates during the PEG precipitation. In addition, the assay requires two additional pipetting steps (addition of γ-globulin and PEG) in comparison with the filter assay.

Sources

1. Atha, D., and Ingham, K. (1981). "Mechanism of Precipitation of Proteins by Polyethylene Glycols," *J. Biol. Chem.* 256: 12108–12117.
2. Demoliou-Mason, C., and Barnard, E. (1984). "Solubilization in High Yield of Opioid Receptors Retaining High-affinity Delta, Mu, and Kappa Binding Sites," *FEBS Lett.* 170: 378–382.
3. Pfeiffer, F., and Betz, H. (1981). "Solubilization of the Glycine Receptor from Rat Spinal Cord," *Brain Res.* 226: 273–279.

Ion exchanger: This method uses the charge difference between free ligand and binding protein/ligand complex. One of the two binds to the ion exchanger. Ion exchanger filters were used with toxin ligands. The filters held back the binding protein toxin complex, whereas the free toxin passed the filter (Schneider et al. 1985). The adsorption to PEI-coated glass fiber filters superceded the ion exchanger filter, because the latter is complicated to use and binding buffers and washing buffers must have a low ion strength. The latter limits the possible applications of the ion exchanger filter, because some proteins require a high ion strength for the preservation of their binding conformation.

In contrast, the assay of Hingorani and Agnew (1991), which presses the binding solution through a short ion exchanger column, has the advantage of being very quick. The separation

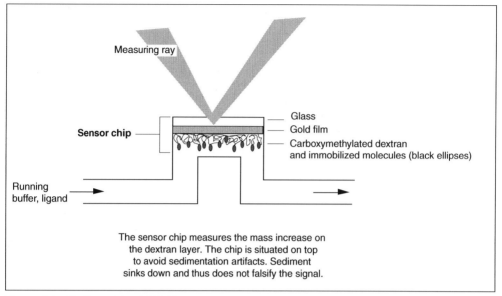

Measuring ray

Sensor chip

Glass
Gold film
Carboxymethylated dextran
and immobilized molecules (black ellipses)

Running
buffer, ligand

The sensor chip measures the mass increase on
the dextran layer. The chip is situated on top
to avoid sedimentation artifacts. Sediment
sinks down and thus does not falsify the signal.

Figure 2.6. Binding with BIACORE: the core of the device.

of free and bound ligand is accomplished in approximately 1 sec. With quickly dissociating binding protein/ligand complexes, the speed of the assay may outweigh its disadvantages (elaborate setup and required manual skill).

Sources
1. Hingorani, S.R., and Agnew, W. (1991). "A Rapid Ion Exchange Assay for Detergent Solubilized Inositol 1,4,5-triphosphate Receptors," *Anal. Biochem.* 194: 204–213.
2. Schneider, M., et al. (1985). "Biochemical Characterization of Two Nicotinic Receptors from the Optic Lobe of the Chick," *J. Biol. Chem.* 260: 14505–14512.

Further Binding Assays
As a last hope for the desperate researcher, there are three further binding assays that can be carried out without elaborate machinery:

1. Glatz, J., and Veerkamp, J. (1983). "A Radiochemical Procedure for the Assay of Fatty Acid Binding by Proteins," *Anal. Biochem.* 132: 89–95.
2. Li, Q., et al. (1991). "An Assay Procedure for Solubilized Thyroid Hormone Receptor: Use of Lipid Ex," *Anal. Biochem.* 192: 138–141.
3. Poellinger, L., et al. (1985). "A Hydroxylapatite Microassay for Receptor Binding of 2,3,7,8 Tetrachlo-rodibenzo-p-dioxin and 3 Methylcholanthrene in Various Target Tissues," *Anal. Biochem.* 144: 371–384.

2.2.5.2 High-tech Assays

You can bind using the BIACORE device. It is a Swedish device about the size of a photometer. In the box there is a gold plate coated with carboxymethylated dextran (sensor chip), to which the experimenter can covalently couple one of the binding partners (ligand or binding protein) (Figure 2.6). In a "flow cell," running buffer or the other partner dissolved in running buffer flows across the plate (Figure 2.7). Optics measure via the surface plasmon resonance how much mass is bound to the plate at any given point in time. This measures how much ligand/binding protein was coupled to the dextran and subsequently how much of the soluble partner has bound to the coupled partner. The device measures the mass increase, thus calculating not only protein/protein/peptide interactions but the binding of protein to DNA, DNA to DNA, and protein to sugar.

Now you are tormented by the question "What is plasmon resonance?" It is not important here! Via the plasmon resonance you can determine the refraction index of a thin layer above

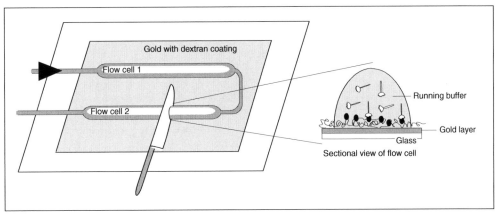

Figure 2.7. Sensor chip with two flow cells.

the gold plate. The refraction index, in turn, depends on the protein/DNA/sugar mass in the layer. The increase of (for example) the protein mass in the measuring layer can be optically determined. Thus, BIACORE devices simultaneously show the binding kinetics. You can "watch" how the binding takes place. Figure 2.8 shows the progress of such binding.

In the flow cells, the binding partner bound to the chip comes into contact with the dissolved partner (Figure 2.7). Depending on the type of device, two flow cells of 0.06 μl or four of 0.02 μl are available. The chip surface in each flow cell can have a different coating, and the binding in every flow cell is tracked by its own optics. With four flow cells, you can simultaneously measure how (for example) a hormone binds to three different receptors. One cell serves as a control and is coated (for example) with BSA or a nonbinding variant of the receptor.

BIACORE devices presumably deliver the best values for the kinetic constants k_{on} and k_{off}—and with this the K_D, the affinity constant (see Section 2.3.2).

- The device delivers the kinetics not in single points but as a continuous exact curve. The continuous injection maintains a steady ligand concentration on the chip throughout the measuring time. Depending on the model, the application of the sample is automatic or at least semiautomatic and is evaluated by computer. In short, he is happy who measures with BIACORE.
- It is desireable to work without radioactivity. Not only because there is no danger of radiation but because many ligands cannot withstand radioactive marking. Of course, the experimenter has a choice of which partner he wants to couple to the plate (provided both binding partners are purified and in solution). The other binding partner remains chemically untreated.
- BIACORE devices capture bindings of low affinity (K_D up to about 10 μM).
- You can set the measuring temperature to any value between 4 and 40° C. You can also strengthen weak bindings by lowering the temperature. You can determine the temperature dependence of the K_D and with it the van't Hoff enthalpy.
- The device gets by with small amounts of sample; 5 μl are enough. In addition the bound proteins are not lost. You can elude them again in 4 to 6 μl. This may yield substantial possibilities. Assume you have a protein X and you do not know how to classify it. What is it doing? With which proteins does X interact? In this case you bind protein X to the chip and hope that this protein does not interfere. To identify the mysterious binding partner Y, push a raw cell extract (or something adequate) across the chip. With some luck, Y will bind. Now you have two possibilities: either you place the entire chip into a MALDI-TOF (see Chapter 7) or you elude Y and analyze the eluate in the nanospray mass spectrometer. The result: you know the MW of Y. In addition, you can identify Y afterward as a peptide fingerprint in a database.

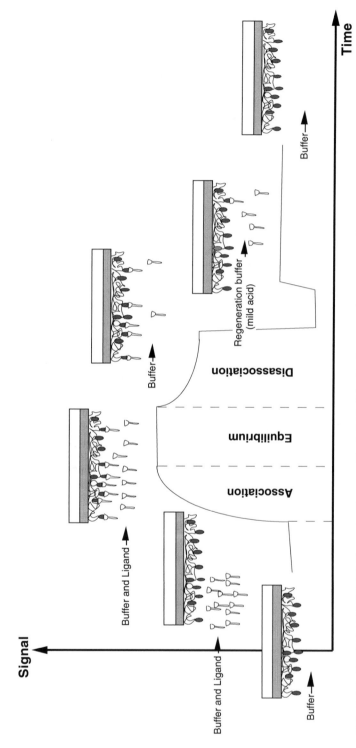

Figure 2.8. Binding with BIACORE: assay diagram. The height of the binding curve is a measure of the amount of bound protein.

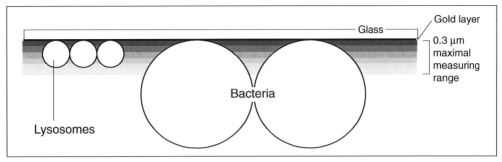

Figure 2.9. Measuring problems. Vesicles, organelles, and cells can protrude from the measuring range of the sensor chip. Round vesicles do not completely cover the surface of the chip.

BIACORE devices generally do not pose a problem when you are working with soluble proteins or DNA. This changes when you are working with vesicles or cells. It is true that viruses, bacteria, vesicles, and even eucaryotic cells can bind to the chip surface via specific proteins and that this binding is measurable. However, this is not easy. The device measures the mass increase near the chip via changes in the refractive index of a thin film above the chip surface. The emphasis is on "thin," and this means that the influence on the signal of attached molecules decreases with molecule distance from the chip surface. The attachment of molecules further than 0.3 µm has no further influence on the signal (Figure 2.9).

Eucaryotic cells have a diameter of 10 to 100 µm, bacteria 1 to 10 µm, mitochondria 1 to 2 µm, lysosomes 0.2 to 2 µm, and synaptic vesicles 0.04 to 0.05 µm. The binding of lysosomes and synaptic vesicles is also measurable. However, if vesicles/cells are larger than 1 µm the bulk extends outside the measuring range. Round vesicles/cells cover only a small part of the chip surface, which further decreases the signal (Figure 2.9).

It is even more difficult to measure the binding of dissolved proteins to coupled vesicles. This is because the mass of the proteins is small in comparison to the vesicle, because a lot of protein binds to the vesicle outside the measuring range, and because the binding density if often low. There is a further problem. The tubes of the device may be *large* enough to let cell or vesicle suspensions pass, but both cell and vesicle suspensions have the fatal tendency to aggregate. This is especially true for cells from tissues or cultures that were ripped apart by proteases or EDTA. I do not need to delve into the details of what such a fat aggregate does to the fine tubes of the device.

However, this should not keep anyone from shooting vesicles into their BIACORE device. Aggregation problems can often be solved with patience and a prefilter. Then, for small vesicles you could investigate vesicle-vesicle bindings and observe fusions. You also have the opportunity to go beyond simple binding and to ascend to function. For example, because the mass increase on the chip is measured via changes in refractive index you should be able to investigate the transport of coupled vesicles (e.g., neurotransmitter intake into synaptic vesicles). Transmitters are actively taken up and the transmitter concentration in the vesicle should, over time, increase so much in relation to the running buffer that the refractive index changes. This is probably true for all vesicles that actively take up a substance. You might even find a way to observe the ion channels at work.

It takes about one to two weeks to become familiar with the BIACORE device, and then even beginning graduates can count on experiencing success. Two users reported to me (with a sparkle in their eyes): very reproducible, great. So much for the good news. Now let's examine the problems.

Problems: The binding conformation of the coupled partners have to withstand the coupling chemistry. This is not necessarily the case, and the protein can also couple at different sites (e.g., via different lysines). If you are unlucky, this influences the binding conformation, so that a uniform binding site turns into a heterogeneous mixture.

If you want to perform several measurements with a single chip, you have to wash the bound partner from the plate after each measurement. The coupled partner has to survive this. Otherwise, the experimenter has to coat a new chip for each assay (an expensive way to have fun).

In principle, it is possible to couple unpurified protein extracts to the chip and to investigate the binding of a (purified) ligand. However, this would only rarely give you reliable measurements, due to nonspecific binding. The opposite case is said to work better, where the purified binding partner is coupled and the other binding partner swims in raw protein extract. However, highly concentrated protein solutions (e.g., serum) easily plug the extremely thin lines of the device. It is advisable to centrifuge the solutions for 1 h at 100,000 g or to filter them (sterile filter) first.

If the free binding partner swims in running buffer with a high refractive index (e.g., buffer with a lot of glycerine or DMSO), the signal decreases (compared with buffer without glycerine or DMSO). This is because glycerine or DMSO sometimes reduces the binding affinity, but also because glycerine and DMSO reduce the difference in refractive index between chip surface and running buffer. We do not want to chalk this up as a negative against the engineers. Instead, we are amazed that a sluggish pulp such as 40% glycerine is able to pass the fine channels at all and that the tiny valves can withstand the resulting pressure.

If the dissolved binding partner has an MW of under 5,000, measuring accuracy suffers because the BIACORE device measures the mass increase at the chip. Even though top-of-the-line devices detect the binding of molecules of about 200 d, it remains impossible with BIACORE devices to directly measure the binding of glycine or acetylcholine to their receptors or of Ca^{2+} to calmoduline. For such cases, you need another high-molecular binding partner that detects a change in the conformation of the protein. Another possibility would be to couple the small binding partner, but the coupling interferes with the binding capability of smaller molecules. The experimenter has to play around with coupling chemistry and spacers, which can become expensive because each trial requires a new chip. There is no guarantee of success.

In most existing publications with BIACORE devices, both binding partners were purified proteins or peptides (largely antibody/antigen). In this case, BIACORE is the method of choice (the two ecstatic users above were also working with purified binding partners). Other than that, BIACORE devices seem to have a certain "gimmick" character and are certainly expensive enough ($75,000 to $250,000) that the good old methods will survive for awhile.

Sources

1. Gruen, L., et al. (1993). "Determination of Relative Binding Affinity of Influenza Virus N9 Sialidases with the Fab Fragment of Monoclonal Antibody NC41 Using Biosensor Technology," *Eur. J. Biochem.* 217: 319–325.
2. End, P., et al. (1993). "A Biosensor Approach to Probe the Structure and Function of the P85a Subunit of the Phosphatidyl-inositol 3-kinase Complex," *J. Biol. Chem.* 268: 10066–10075.
3. Khilko, S., et al. (1993). "Direct Detection of Major Histocompatibility Complex Class I Binding to Antigenic Peptides Using Surface Plasmon Resonance," *J. Biol. Chem.* 268: 15425–15434.

Binding can also be measured via the binding heat capacity (i.e., the heat released or consumed during binding). The binding heat at steady temperature is measured by ultrasensitive isothermal titration calorimeters (ITC). These are wonderfully expensive miracle devices: a good ITC measures fractions of microcalories (the amount of heat released when you briefly rub two fingernails against each other!). It is easy to perform a binding experiment with an ITC (Figure 2.10). The binding protein is inserted into the sample cell of the calorimeter, and the ligand into the syringe. The reference cell holds the buffer. At constant temperature (isothermally), a motor presses an aliquot from the syringe into the sample cell. At the same time, the sample cell is stirred and the heat uptake or release is measured.

Sample cell and reference cell contain a volume between 0.2 and 1.4 ml, depending on the device. In the course of an experiment, 2 to 60 μl aliquots are injected 12 to 20 times, and the heat capacity is measured each time. The ligand concentration in the sample cell thereby increases in steps, whereas that of the binding protein decreases. The entire process takes about an hour and is easy to do. Once you have loaded the machine with binding partners, you enter aliquot size, binding partner concentrations, measuring temperature, stirring speed,

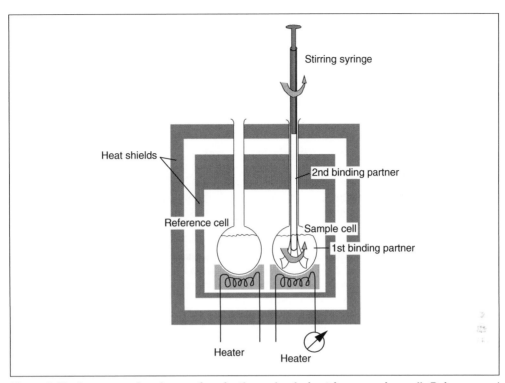

Figure 2.10. A very rough schema of an isothermal calorimetric measuring cell. Reference and sample cell are identical. The binding takes place in the sample cell. Aliquots of the second binding partner are added with a stirring syringe. The syringe is turned at the same time and the sample cell is mixed with stirring paddles at the syringe tip. During the measurement, the reference cell is electrically heated to a preset steady temperature. At the same time, the temperature difference between reference and sample cell is measured. The heating current for the sample cell is adjusted so that the temperature between both cells stays the same. If exothermal binding occurs, which releases heat in the sample cell, the device reduces the heating current for the sample cell. The current reduction is then equivalent to the heat released by the binding process.

and injection times via the attached PC and press Start. The device conducts the experiment on its own and its software reports the results.

In real life, problems occur (e.g., during the preparation of the sample). The buffer in sample cell, reference cell, and syringe must be absolutely identical (i.e., it must come from the same bottle). Again, the content of sample and reference cell may differ only with respect to the binding protein. Ideally, you dialyse the binding protein and the ligand solution against the same buffer. Banal but important: since you inject only small aliquots of the ligand solution, the ligand in the syringe must have a much higher concentration than the solution of the binding protein in the sample cell.

Also, you should ultrafilter the sample to remove dust particles. These hit the cell walls and create an unsteady baseline. After all, you are dealing with an ultrasensitive calorimeter. You should also degas the solutions. Bubbles make heat noise. You should also optimize the stirring speed. If you stir too fast, the binding protein gets sick. If you stir too slowly, the heat does not reach the surface in time, which distorts your signal. After all, the heat is measured at the surface of the cells. The optimal stirring speed is between 200 and 400 rpm. Furthermore, the time between injections has to be *long* enough for an equilibrium to return (i.e., after each injection the curve has to return to the baseline).

Another important point: heat is produced when two solutions are mixed—as evidenced by the example of boiling water and concentrated sulfuric acid. Mixing of binding protein and ligand solutions also produces heat. Obviously, this heat has nothing to do with binding and has to be subtracted. How *high* this mixing heat is you can see at the end of the run, when ligand is present in excess and does not bind anymore. Nevertheless, there is a heat peak at each ligand injection. This is the nonspecific heat: The mixing heat, the temperature differential between sample cell and syringe, and so on. If you are unlucky, or if your setup is not right, these nonspecific heat peaks can be higher than the peaks that stem from binding. Control runs are thus advisable before the binding experiment: one run with buffer only, one run with buffer and ligand only, and one run with buffer and binding protein only.

Did you run your controls? Did you prep your samples? Did you set optimal injection time and stirring speed? Then you can run your binding experiment. "You need a lot of luck," is Ilian Jelesarov's (University of Zürich) advice to heat aficionados. Many of these tricks stem from Jelesarov.

Let's assume you were lucky. Then you can map the (corrected) data in two ways. The first method is to do so differentially: the sum of the heat uptake/release at injection i divided by the ligand concentration at injection i mapped against the ligand concentration (Figure 2.11).

The integral plot, on the other hand, maps the sum of heat uptake/release at injection i against the ligand concentration. From each curve you can calculate three values: the number of binding sites n, K_D, and ΔH. ΔH is the released heat per mol of binding partner, the enthalpy. Because the values for n, K_D, and ΔH are calculated via a curve, they are not very exact. Small errors in the experimental data result in large differences in n, K_D, and ΔH. It is better to determine K_D and ΔH in separate experiments. High concentrations of binding protein at the beginning of the experiment yield exact values for ΔH (i.e., when the aliquots of the ligand still get bound completely). The K_D are best calculated on the basis of the slope of the differential curve.

The parameters from the differential and integral curves should somewhat agree. If they don't, you have a problem. Then it is advisable to determine the number of binding sites with an independent method and to enter n as a fixed value into the calculation. You get n from Scatchard plots with n = B_{max}/R_o (Figure 2.4), where B_{max} is measured and R_o must be known. It cannot hurt to repeat the measurement with different concentrations on each binding partner. If you have a 1:1 stoichiometry, you can also reverse the experiment (i.e., you put binding protein into the syringe and ligand into the sample cell and compare). In any case, if you are so inclined you can make life arbitrarily difficult for yourself.

The advantages of ITC:
- Immobilization or labeling of the binding partners is not necessary (a blessing with sensitive proteins).
- You can (theoretically) measure the binding of any ligand to any macromolecule.
- You get, in addition to the number of binding sites and the K_D, the ΔH.
- You can measure at different temperatures (between 5 and 80° C) and different pH.
- Measurements at different temperatures (T_1 and T_2) yield $\Delta H(T_1)$ and $\Delta H(T_2)$ and thus the heat capacity $\Delta Cp = (\Delta H(T_1) - \Delta H(T_2))/(T_1 - T_2)$. From the heat capacity, the expert reads off or infers various things about the molecular structure of the binding. (More about this later.)
- The device measures the heat released in a solution. Heat is generated not only by binding reactions. Why not try to use the released heat to determine the uptake of neurotransmitter into synaptic vesicles or the ATP production by mitochondria? This especially in that, theoretically, the calorimeter coolly approaches even the murkiest soup. In practice, however, murky solutions create unsteady baselines and the mitochondria do not survive the fast stirring. But there's certainly a technical solution for this.

The disadvantages:
- Modern ITC devices measure heat with stunning sensitivity (namely, ultrasensitively), but they require enormous amounts of binding partner. One experiment consumes 10 to 100 nM of protein. With an MW of 50,000, the experimenter thus goes through half a milligram! It is only worth sacrificing such amounts when you can harvest the protein in *E. coli*. The device by MicroCal, the market leader, is said to use only 50 µg of protein (Plotnikov et al. 1997). However, according to my spot checks in the literature MicroCal devices also consume milligram amounts of protein.

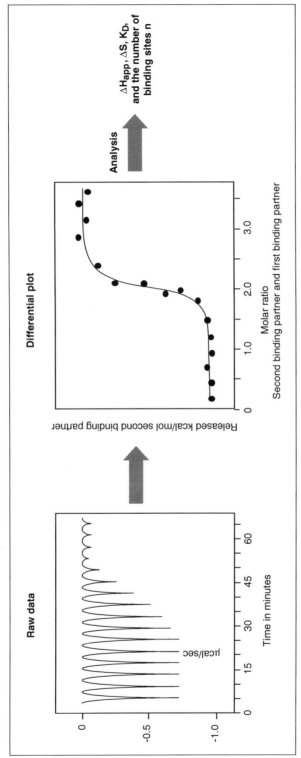

Figure 2.11. Progress of an ITC experiment. Duration is about 2 h, including preparation of cells and solutions.

- You can measure only bindings of medium affinity. This is because you can only get reasonable measurements for K_D if the concentration of binding protein is between 10 to 100 times the K_D. Thus, bindings with high affinity ($K_D < 1$ nM) are difficult to measure. Why? With a K_D of (for example) 0.1 nM, you can use a maximum of 10 nM of protein. With a sample volume of 1 ml, this means 0.01 nM, which undercuts the measuring threshold of the devices (1 nM). On the other hand, if the K_D is very low you need enormous concentrations of binding protein. The fraction of binding protein that forms complexes (and thus generates heat) is thus too small to be picked up by the device. However, at high concentrations the proteins like to aggregate. Then you measure against a backdrop of clumping and dispergation heat and only the Lord is able to figure out what the binding heat is.
- Many of the more interesting membrane proteins (e.g., neurotransmitter receptors) cannot be expressed in sufficiently high concentrations. Membrane proteins in general seem to be a problem for calometry. (I was not able to find a single calometric paper on this.)
- A good ITC costs $80,000. Moreover, if it is equipped with a differential scanning calorimeter (see material following) the price goes up to $150,000.

Researchers with a knack for bargaining can achieve substantial price reductions. However, why invest so much money in a device when you can just as well determine n and K_D with an inexpensive filter assay; Because of ΔH! With ΔH you gain access to the anatomy of the binding. This can answer questions such as the following.

- What is the proportion of hydrophobe interactions?
- Is the binding equivalent to two rigid forms snapping together like Lego pieces, or does the binding result in conformation changes (see Figure 2.12)?
- What contributes more to the binding process (to ΔG): ΔH or ΔS, the entropy?
- What individual bindings make up the overall binding (i.e., which amino acids participate in the binding)?
- Are protons of the binding protein transferred into the buffer during binding and if so, which amino acid contributes these protons? An example, including theory, is found in Baker and Murphy (1997).
- By comparing ΔH in H_2O and D_2O, you can guess at hydration effects of the binding (Chervenak and Toone 1994).

Note, however, that the insights of calometry have not been terribly clear and far-reaching so far. This is also true for the popular site-directed mutagenesis experiments that are believed to provide clues about which amino acids are involved in the binding process. The researcher switches Ala with Asp and—eureka!—ΔH increases (or decreases). But what does that tell us? My impression: often not too much. Switching an amino acid can also have indirect effects on the binding via remote conformation changes. In other words, a change in ΔH does not prove that the amino acid participates in the binding. Without additional NMR or X-ray structural data, the researcher is searching in tall grass. If he does not understand what factors he is dealing with, he is walking on quicksand. For example, the ΔH that is measured in the ITC is a compositional quantity. It is the result not only of the formation of noncovalent bindings between ligand and binding protein but of the reorganization of the solvent, of possible conformation changes of ligand and binding protein, and of changes in the protonization of solvent (buffer) and binding partners.

Strategy: It is not a bad idea to experiment for a few weeks. This allows you to get used to the device and its quirks and capabilities, and because you are working unsystematically to hit upon phenomena by accident that a systematic approach would miss. Besides, it's fun. As with all techniques, it is also advisable here to read what others are doing with it. And what you can all do! Many people like to do thermodynamic comparisons of the binding of different ligands to the same binding site, as Bradshaw et al. (1998) show. However, this presupposes the existence of different ligands. In the worst-case scenario you will make some yourself. Others measure the binding of the same ligand under different conditions. Choices are pH, buffer, ion strength, ions, and solvents. Most investigations fall into one of these two categories (i.e., different ligands or different conditions). A few authors endeavor to create colorful (i.e., really colorful) formula systems. This strategy ensures them the deference of the mathematically challenged (the *vast* majority of biologists).

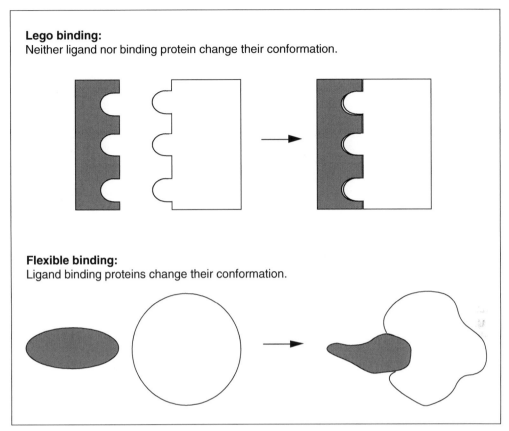

Lego binding:
Neither ligand nor binding protein change their conformation.

Flexible binding:
Ligand binding proteins change their conformation.

Figure 2.12.

Mathematically hopeless but ambitious people should stake out new areas—without a PI in their base camp. That's what science is all about. Try to apply heat measurement to unexplored processes: the fusion of vesicles, the proton uptake by lysosomes, and so on. Sometimes you also get surprising results when you measure the binding of the ligand to the binding protein under the same conditions using different methods (e.g., with ITC and plasmon resonance). The results with respect to K_D often differ in orders of magnitude. Why?

Don't forget: isothermal calometry only detects processes where heat is a factor. Purely entropy-driven processes are invisible to calometry. A lack of heat thus does not mean that there is no binding. Ferredoxin, for example, binds to ferredoxin reductase without heat.

The difficulty of calometric research is not the production of data but their processing. With modern microcalorimeters, any beginning graduate with one day of training can effortlessly shake down data like ripe pears from a tree (provided there's a continuous supply of protein). But the task of distilling the spiritual essence from these pears can lead to thermodramatic bouts of despair.

This means that we should talk about thermodynamics—but only a little bit. As you know it deals with the quantities of enthalpy (ΔH), entropy (ΔS), free energy (ΔG), heat capacity (ΔC), pressure (p), and temperature (T). Thermodynamics is characterized by a plethora of formulas and has an antiquated reputation. The latter is no surprise, in that it goes back to the nineteenth century. The concept of free energy, for example, was introduced by Josiah Willard Gibbs in 1878.

Thermodynamics! The island of incomprehensibility! Covered by a jungle of strange terminology wherein the biologist hacked through the underbrush of lush formulas for the prelims, just to paddle across the ocean of amnesia as quickly as possible (never to return). Only the development of ultrasensitive calorimeters was able to lure some people back to thermodynamics, and the island has lately turned into a hip location for adventure biologists. This shows that even the oldest rags can come back into fashion and serve as consolation to those who doggedly stick to their time-tested raiments. But now to thermodynamics! For each reaction, and for a binding reaction, it is true that $\Delta G = \Delta H - T\Delta S$. Here, ΔG is the free energy—the energy that drives the binding. ΔG results from the dissociation constant according to $\Delta G = -RT \ln K_D$. ΔH, in turn, resides the heat release or uptake of a binding reaction at a certain temperature, and hence the quantity measured by a calorimeter. If you know ΔH and ΔG, you can calculate ΔS.

The application of thermodynamic theory to isothermal calometry is presented in Jelesarov and Bosshard (1999) in a somewhat digestible form for the initiated. The book by Kensal (1998) discusses the theory in a lot of detail—unfortunately not enough to make it comprehensible for the beginner. "This is not a book for the unprepared mind," remarks one reader. I won't dare to adorn these pages with formulas. Just sit through a lecture on physical chemistry. After three months, you will know your way around. A recommendation: don't even try to visualize the thermodynamic concepts. It does not further your understanding and can lead to knots in your brain folds.

Even if you have little inclination for thermodynamic subtleties, you should not avoid measuring ΔH if you happen to have the opportunity. After all, ΔG and ΔH are independent quantities. Changes in binding behavior that go unnoticed in ΔG may perhaps show themselves in ΔH. It often happens that ΔH decreases and $-\Delta S*T$ simultaneously increases (or vice versa), so that ΔG appears constant in the end, feigning innocence and hiding with its lack of activity real conformation changes. Thus, go ahead with measuring ΔH and then call an expert for help. For the data analysis, use the following rules of thumb.

- ΔH correlates only weakly—if at all—with ΔG.
- In the temperature range of 5 to 70° C ΔC_p is generally temperature independent, which means that $\Delta H = \Delta H_O + \Delta C_p T$. ΔC_p supplies the temperature dependence of ΔH.
- Only when the binding partners behave like two rigid forms (Lego binding) and there are no protonization reactions and no changes in the hydration of the binding surface, does the measured ΔH equal the binding enthalpy of the noncovalent bindings between ligand and binding protein. You probably have a Lego binding when ΔC is small. Spolar and Record (1994) discuss this in more detail. Another clue to Lego binding is the identity of the calorimetrically measured ΔH with the change in the van't Hoff enthalpy ΔH_{vH}. The latter results from the temperature dependence of K_D (see Jelesarov and Bosshard 1999), and thus from a binding quantity. ΔH_{vH} is thus equal to the binding enthalpy and generally $\Delta H > \Delta H_{vH}$, in that ΔH captures additional processes. Unfortunately, ΔH_{vH} is difficult to determine, in that K_D is often only weaky dependent on temperature. Small measuring errors in K_D result in large errors in ΔH_{vH}.
- A positive ΔS indicates that water is pushed off the binding surface.
- A change in ΔH with the buffer (e.g., a change from Tris pH 7.0 to imidazole pH 7.0 or PIPES pH 7.0) indicates a protonization reaction.
- A large negative ΔC_p indicates a high degree of hydrophobic interaction between the binding partners. See Spolar and Record (1989) and Lin et al. (1995).

Overall, heat measurements are nice and sometimes even interesting. They can also be performed easily and quickly and always yield a result. After all, every binding must have some ΔH. If you like to calculate and play with formulas, you will find here an equally ideal and infinite range of things to do. Consider this: there are hundreds of thousands of proteins that undergo millions of bindings. If you want to characterize and understand all of these thermodynamically, you and dozens of Ph.D. students will be busy for the rest of your lives. So far, calorimetric papers have also found a home in respectable journals. Yet the important questions asked by biochemists, and especially medical researchers, aim for the function of a protein in the metabolism. The functional network of the cell will be the hit of the next decade. That's where the music will play and the grants will dance. For the determination of function, it is generally enough to know who binds whom, how strongly, and with which stoichiometry.

Exactly how this binding takes place is not necessary for the understanding of the cell function.

Sources
1. Baker, B., and Murphy, K. (1997). "Dissecting the Energetics of a Protein-protein Interaction: The Binding of Ovomucoid Third Domain to Elastase," *J. Mol. Biol.* 268: 557–569.
2. Chervenak, M., and Toone, E. (1994). A Direct Measure of the Contribution of Solvent Reorganisation to the Enthalpy of Ligand Binding," *J. Am. Chem. Soc.* 116: 10533–10539.
3. Jelesarov, I., and Bosshard, H. (1999). "Isothermal Titration Calorimetry and Differential Scanning Calorimetry as Complementary Tools to Investigate the Energetics of Biomolecular Recognition," *J. Mol. Recognit.* 12: 3–18.
4. Kensal, E., et al. (1998). *Principles of Physical Biochemistry.* Englewood Cliffs, NJ: Prentice Hall.
5. Lin, Z., et al. (1995). "The Hydrophobic Nature of GroEL-substrate Binding," *J. Biol. Chem.* 270: 1011–1014.
6. Plotnikov, V., et al. (1997). "A New Ultrasensitive Scanning Calorimeter," *Anal. Biochem.* 250: 237–244.
7. Spolar, R., and Record, M. (1994). "Coupling of Local Folding to Site-specific Binding of Proteins to DNA," *Science* 263: 777–783.
8. Spolar, R., et al. (1989). "Hydrophobic Effect in Protein Folding and Other Noncovalent Processes Involving Proteins," *PNAS* 86: 8382–8385.

You can measure the K_D of very tight bindings with an ultrasensitive differential scanning calorimeter (DSC). These devices will also give you clues regarding the conformation and conformation changes of the binding protein.

Like ITCs, DSC devices consist of two identical cells: one sample and one reference cell. The sample cell is for the solution of the binding protein or the ligand/binding protein complex, the reference cell is for the buffer. This is also true for the DSC: the buffers in reference and sample cell must be identical. The sample and reference cell are heated adiabatically (i.e., sample and reference cell are thermally isolated from each other). Depending on the different heat capacities of reference and sample cell, they heat up at different speeds. To ensure that the temperature of both cells remains equal, one of them must receive more heat input. The required power is a measure of the heat capacity ΔC_p. A typical DSC curve, a thermogram, is shown in Figure 2.13.

Usually, the ΔC_p values fall into the range of 1 to 2 $JK^{-1} g^{-1}$ and slowly rise with the temperature up to the denaturation range. There, the ΔC_p increases exponentially to a peak T_m and then falls to the ΔC_p of the denatured protein. The ΔC_p value of the denatured protein is larger than that of the native one and can be calculated from the ΔC_p of the amino acids and the peptide bindings.

What causes the peak during denaturing? First, the increasing temperature allows for more conformations. Second, the proteins more often jump from one conformation to another (fluctuation). Shape and height of the peak are thus determined by the number and energy states of the protein conformations. Pronounced conformation changes often cause multiple peaks (Blandamer et al. 1994).

Because the thermogram is determined by the number and energy states of the conformations the protein may assume, you can reversely draw conclusions from the thermogram to the protein conformation. Such "deconvolution analysis" means complicated calculations of which you can get a taste in Jelesarov and Bosshard (1999). You should only get involved in DSC work if you enjoy shooting out formulas and if a partial differential gets your heart pounding. Don't allow the theoretical exactness to dupe you into the assumption that it exactly mirrors reality. It doesn't—at least not when Mr. Theoretician applies reversible thermodynamics to irreversible processes—and the denaturing of a protein is generally irreversible.

During DSC of ligand/binding protein complexes, the thermogram often shows multiple peaks. They correspond to the denaturing of ligand, binding protein, and complex. The T_m of the complex is typically higher than the T_m of the binding protein (i.e., the ligand stabilizes the binding protein). The thermograms of binding protein and ligand/binding protein complex thus tell you whether, and to which degree, the protein is thermally stable and to what degree ligand, buffer, or buffer additions stabilize the protein (see Figure 2.14). This information can be used in purification protocols. Furthermore, you can calculate the energetic quantities of the binding reaction from the thermograms. This also holds true for tight bindings, because

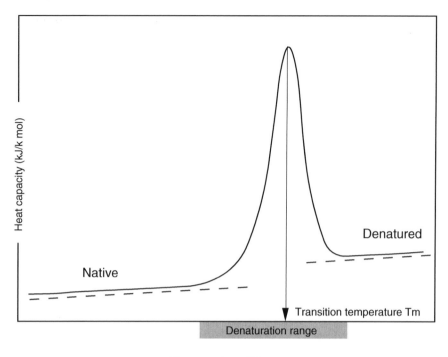

Temperature (K)

Figure 2.13. Dependence of partial molar heat capacity of a protein on the temperature (thermo-gram). The denatured protein always has a higher heat capacity than the native one. The high heat capacity at the peak stems from the increasing fluctuation in the system. In contrast to the native state, many more conformations are possible and the protein switches back and forth between different conformations.

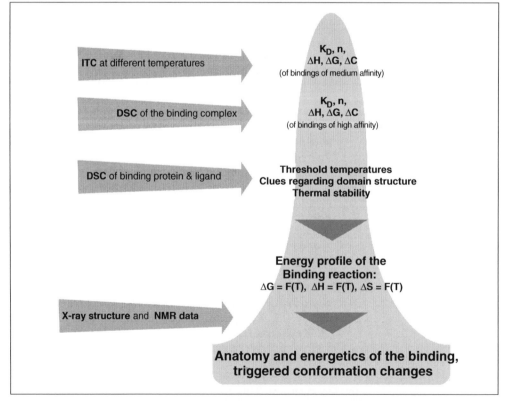

Figure 2.14. How calorimetric methods contribute to the understanding of a binding.

every binding is destroyed if you just heat it up enough. And at constant pressure and with good equipment you can heat it to 130° C.

A DSC device costs about $80,000. Typically, you purchase a complete set of equipment (i.e., ITC and DSC together). That makes sense because these devices complement each other and if you write a grant proposal you might as well write it for $150,000. The effort is the same.

The disadvantage of the method is the large amount of protein required for an experiment. This is because the binding protein is denatured afterward (i.e., unusable for most purposes). Finally, the equipment seems to lack in technical maturity. For example, the DSC device should remain powered on at all times. Otherwise, you are risking damage from it standing still.

Sources
1. Blandamer, M., et al. (1994). "Domain Structure of Escherichia Coli DNA Gyrase as Revealed by Differential Scanning Calorimetry," *Biochemistry* 33: 7510–7516.
2. Jelesarov, I., and Bosshard, H. (1999). "Isothermal Titration Calorimetry and Differential Scanning Calorimetry as Complementary Tools to Investigate the Energetics of Biomolecular Recognition," *J. Mol. Recognit.* 12: 3–18.

2.2.5.3 Developing a Binding Assay for Proteins in Solution

The choice of binding buffer is important. A physiological assay that measures the ligand's effect on an organism, organs, or cells provides clues regarding ions, ion strength, pH, and other buffer components of the binding buffer. You do not have an assay? In this case, orient yourself by the structure of the biological compartment, in which the binding protein is supposed to have an effect. Don't forget about the protease inhibitors.

The binding solution equals that used for membranes, and 4° C is a good choice for the incubation temperature. You can estimate the time until binding equilibrium from the physiological assay, and if that is not possible, 1 to 2 h at 4° C should suffice for a start. If you incubate the binding solution overnight, you risk the binding protein being digested by proteases.

As for the binding assay: if a radioactive ligand is available, try the filter assay first. It's the fastest and its applicability usually only depends on whether or not the ligand sticks to the filter. You should be able to answer this question after a few experiments. If the filter test does not work, use the column assay. If that is not possible (e.g., because ligand and binding protein are about equal in size), you can employ PEG precipitation or the methods from Petrenko et al. (1990) or Scheer and Meldolesi (1985).

If you still cannot measure any binding, or if you were not able to label the ligand, you should beg for a few days of access to a BIACORE device or an isothermal calorimeter. If that also leaves you depressed, it is time to study Section 2.2.6.

2.2.6 No Binding

> All these tempests that fall upon us are signs that fair weather is coming shortly, and that things will go well with us, for it is impossible for good or evil to last for ever.

If the radioactive ligand does not show any specific binding, this may be due to one or more of the following.
A. The affinity of the ligand is too low.
B. The affinity of the binding site decreased during membrane/extract preparation.
C. The binding assay does not work with this pair.
D. The unspecific binding is too high.
A: If the ligand's affinity for the binding site is very low ($K_D > 100$ nM), many binding assays fail. It is possible that a centrifugation assay without washing will be able to obtain some measurable binding from low-affinity ligands. For soluble binding proteins, BIACORE may be able to do something. Before you give up, consider the following.

- Often, the affinity of a ligand depends on the ion strength of the binding buffer (usually in the sense that the higher the ion strength, the higher the affinity) or its composition (e.g., divalent cations and so on). Physiological assays (e.g., the LD_{50} for toxins) give you a clue to the probable range of the ligand's affinity.
- Sometimes the ligand only appears to have a low affinity (e.g., when there are binding inhibitors in the membrane preparation or in the tissue extract). This is usually the case for ligands that also occur endogenously, such as glutamate, GABA, hormones, and so on. You remove the endogenous inhibitors from membrane preparations by washing with buffer containing 0.03% TRITON-X-100 or by repeated freezing/thawing/washing of the membranes. Washing with buffer is usually not enough, in that inhibitors that sit in the membrane vesicles are not removed. Endogenous inhibitors are removed from tissue extracts either through gel filtration, lectin chromatography, or (if the charge ratios of inhibitor and binding protein allow) ion exchange chromatography.
- During incubation with membrane or tissue extract, enzymes or chemical processes can change the ligand. For example, proteases digest peptide or protein ligands.

B: You determine or suspect that your labeled ligand works great in the physiological assay and in minuscule amounts. Nevertheless, it does not bind to membranes or membrane extracts. If this is not caused by the ligand itself, it may be due to the binding protein.

- The binding protein often loses its affinity because of effects of proteases or during extraction (e.g., during extraction of membranes with detergents). Protease inhibitors help, or 10% glycerine (stabilizes the native conformation of proteins).
- Sometimes a protein needs certain factors for retaining its binding capabilities, such as cofactors, substrates (for enzymes), ions (for ion channels), and so on, and these are lost during membrane creation.
- The pH of the binding buffer plays a role (its ion composition, ion strength, and temperature).

Dozens of parameters are at work and the only thing that helps is to try, try, try. Trial and error is an important part of research.

C: Try a different assay.

> Where one door shuts, another opens.

D: The measuring error of binding assays is about 10%. If the nonspecific binding is 80% or more of the total binding, the vacillation of measurements hides the saturable binding. The way out is clear: suppress the nonspecific binding.

- For filter assays, you may try to wash the filters better. During washing it is not so much the volume that is important but the number of washing steps. It is better to wash three times with 4 ml than one time with 12 ml.
- When the ligand is loaded, some part of the nonspecific binding is probably mediated by electrostatic interactions. To counteract, the experimenter can increase the ion strength of the binding buffer, provided this does not negatively impact the binding between ligand and binding site.
- Adding detergents (e.g., TRITON-X-100 at 0.01%) to the binding buffer suppresses the nonspecific binding. However, some proteins cannot handle TRITON-X-100, and TRITON-X-100 absorbs at 280 nm.
- Protein (e.g., BSA at 0.1 mg/ml) in the binding and/or washing buffer also helps keep the nonspecific binding down. Of course, the ligand may not interact with the BSA, and the BSA must not use up the protein binding capacity of the chosen assay (e.g., of the PEI-coated glass fiber filter).
- Often the filter contributes to the high nonspecific binding. A blind value (i.e., a binding assay without membranes or protein) is used to determine the contribution of the filter to the nonspecific binding. If it is too high, change the type of filter or coat the filter (e.g., with PEI or with BSA or with whatever else you can think of). It doesn't help? Switch to different testing methods.

• A last resort is sometimes switching the species used as the source of the binding protein. The amount of binding protein per milligram of total protein (and thus the signal-to-noise ratio) is sometimes markedly different between even closely related species.

2.3 Analysis of Binding Data

This chapter may contain too many formulas for the average taste. The formulas are supposed to:
• Point out the multitude of mechanisms and curve forms
• Urge caution in the interpretation of binding data (the facts may be much more complicated than one or another set of measurements makes you think)
• Aid in the interpretation of unusual binding data
To my knowledge, the mathematics of binding has so far only been presented piecemeal in scattered articles. For consolation: the final equations are simple. However, their derivations can be nightmarish (e.g., see Figure 2.15).

> "Look here, Sancho," said Teresa, "ever since you joined on to a knight-errant you talk in such a roundabout way that there is no understanding you."

The mathematical treatment of binding data falls into two topics.
• The binding reaction is in equilibrium.
• The binding reaction is not in equilibrium (kinetics).

2.3.1 Binding Reaction in Equilibrium

The binding reaction is in equilibrium when the concentrations of the binding variables such as binding protein/ligand complex, free ligand, and so on are not time dependent anymore (after about 3 to 4 $T_{1/2}$; see Section 2.3.2). The basis of the analysis of equilibrium binding data is the law of mass action and the retention equations.

For equilibrium binding studies, the quantities of total ligand concentration (L_0), inhibitor concentration (I_0), and so on are usually given. The concentration of the saturably bound ligand **B** is measured. From this, the desired parameters are calculated: the concentration of the free ligand **L,** the dissociation constant K_D, the IC_{50} and/or K_i of an inhibitor, the concentration of the binding protein R_0 and/or the maximum number of binding sites B_{max}, and the number of binding sites **n**.

The number of binding sites **n** is calculated quickly: if $n = B_{max}/R_0$. B_{max} is measured, R_0 must be known (e.g., by using pure binding protein).

But how are the other parameters calculated? For that you need to know some things about binding mechanisms. A ligand can bind in different ways (i.e., using different mechanisms). For each mechanism there is a set of equations: the retention equations and the laws of mass action. From these you can derive the connection of the parameters. I have done this for binding mechanisms I and II (Figures 2.4 and 2.15). The binding does not play a role in these equations because it was subtracted before.

2.3.1.1 Determining K_D and B_{max}

In the simplest case, a ligand L binds to a binding protein R according to binding mechanism I. The corresponding binding parameters K_D and B_{max} are determined via the Scatchard plot:

Two different binding proteins R_1 and R_2

$$R_1 + L \rightleftharpoons R_1L \qquad R_2 + L \rightleftharpoons R_2L$$

$K_{D1} = \dfrac{R_1 \bullet L}{(R_1L)}$	$K_{D2} = \dfrac{R_2 \bullet L}{(R_2L)}$	Law of mass action
$R_{10} = R_1 + (R_1L)$	$R_{20} = R_2 + (R_2L)$	Retention equations
$B = (R_1L) + (R_2L)$	$B_{max} = R_{10} + R_{20}$	

Derivation of the Scatchard equation (B/L as a function of B):

$$R_{10} = R_1 + (R_1L) = R_1 + \frac{R_1 \bullet L}{K_{D1}} = R_1 \bullet \left[1 + \frac{L}{K_{D1}}\right]$$

$$R_{20} = R_2 + (R_2L) = R_2 + \frac{R_2 \bullet L}{K_{D2}} = R_2 \bullet \left[1 + \frac{L}{K_{D2}}\right]$$

$$B = (R_1L) + (R_2L) = \frac{R_1 \bullet L}{K_{D1}} + \frac{R_2 \bullet L}{K_{D2}} = \frac{L}{K_{D1}} \bullet \frac{R_{10}}{\left[1 + \frac{L}{K_{D1}}\right]} + \frac{L}{K_{D2}} \bullet \frac{R_{20}}{\left[1 + \frac{L}{K_{D2}}\right]}$$

$$B = \frac{R_{10} \bullet L}{K_{D1} + L} + \frac{R_{20} \bullet L}{K_{D2} + L}$$

$$L = \frac{R_{10} \bullet K_{D2} + R_{20} \bullet K_{D1} - B \bullet [K_{D1} + K_{D2}]}{2[B - R_{10} - R_{20}]} + \sqrt{\frac{\left[B \bullet [K_{D1} + K_{D2}] - R_{10} \bullet K_{D2} - R_{20} \bullet K_{D1}\right]^2}{4[B - R_{10} - R_{20}]^2} - \frac{B \bullet K_{D1} \bullet K_{D2}}{B - R_{10} - R_{20}}}$$

$$\frac{B}{L} = \frac{B}{\dfrac{R_{10} \bullet K_{D2} + R_{20} \bullet K_{D1} - B \bullet [K_{D1} + K_{D2}]}{2[B - R_{10} - R_{20}]} + \sqrt{\dfrac{\left[B \bullet [K_{D1} + K_{D2}] - R_{10} \bullet K_{D2} - R_{20} \bullet K_{D1}\right]^2}{4[B - R_{10} - R_{20}]^2} - \dfrac{B \bullet K_{D1} \bullet K_{D2}}{B - R_{10} - R_{20}}}}$$

Axis sections of the Scatchard plot: $\qquad \left(\dfrac{B}{L}\right)_{B=0} = \dfrac{R_{10}}{K_{D1}} + \dfrac{R_{20}}{K_{D2}} \qquad \left(B\right)_{B/L=0} = R_{10} + R_{20}$

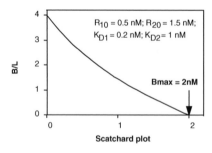

$R_{10} = 0.5$ nM; $R_{20} = 1.5$ nM;
$K_{D1} = 0.2$ nM; $K_{D2} = 1$ nM

$B_{max} = 2$ nM

Scatchard plot

Figure 2.15. Binding mechanism II.

if you add increasing concentrations of radioactive ligand (L_0) to a defined amount of binding protein and determine in each case the saturably bound ligand B, you can calculate for each L_0 the variable L and then the variable B/L. If you plot the variable B/L against the variable B, the points form a straight line B/L = a-bB with the constants a and b. The axis section a is R_0/K_D and the slope b is $1/K_D$. The line intersects with the B axis at B = B_{max} and B_{max} is equal to R_0 for mechanisms I. The plot B/L against B is called the Scatchard plot (Figure 2.4).

If you rearrange the Scatchard equation, you get different plots. For example, plotting 1/L against 1/B is also linear—this time with the axis section $-1/K_D$ and a slope of R_0/K_D. Plotting B against B/L yields a linear form with axis section R_0 and slope $-K_D$. These plots differ (when using the linear regression) in the different weighting of errors and in their beauty in the eye of the beholder and/or reviewer. The Scatchard plot gives relatively equal weighting to the points, whereas in the 1/L-against-1/B plot the values of low-ligand concentrations have an overproportional weight. The Scatchard plot is thus used most widely, although the point distribution in the 1/L-against-1/B plot looks nicer. The most uncertain points in the Scatchard plot (i.e., those with the largest error) are located at the beginning and end of the curve (i.e., are at high- and low-ligand concentrations). An aside: the error of the quantity B/L is calculated according to the error propagation law by Gauß from the errors of B and L, where the percent error of B/L is larger than that of B and of L.

Other experiments with other plots yield the same quantities. Instead of adding increasing concentrations of a radioactive ligand to a constant concentration of binding protein, the experimenter can determine K_D and B_{max} by adding different concentrations of the nonradioactive ligand Λ_0 to a constant concentration of binding protein R_0 (= B_{max}/n) and a constant concentration of the radioactive ligand L_0. For each Λ_0, he determines the concentration of the bound radioactive ligand B. In the case of L0 ≈ L and $\Lambda_0 \approx \Lambda$, plotting 1/B against Λ_0 yields a straight line $1/B = \alpha + \beta\Lambda_0$ with $\alpha = (K_D + L_0)/R_0L_0$ and $\beta = 1/R_0L_0$. The advantage of this method is that the concentration of the expensive ligand L_0 can be low.

Scatchard plots are often not linear but curved. This holds, for example, when there are often two or more binding sites with different affinity for the ligand. The high-affinity site and the low-affinity site. Suddenly there are three players (one ligand and two binding sites), and the story becomes correspondingly complicated (see equations for binding mechanisms II and III). In these cases, the constants R_{10}, R_{20}, K_{D1}, and K_{D2} are determined numerically.

The different binding sites of the ligand can be situated on different proteins (binding mechanism II in Figure 2.15) or on the same protein (e.g., binding mechanism III in Figure 2.16). Binding sites on the same protein can be dependent on each other or independent. If the binding sites are independent from each other, the binding of the ligand to one of the binding sites has no effect on the binding of the ligand to the other binding site (see binding mechanism IV, a special case, in Figure 2.17). If the binding sites are interdependent (allosterically coupled) there are two possibilities. If the binding of the ligand to the first binding site facilitates the binding to the other binding site, there is positive cooperativity (e.g., binding mechanism IV at $K_{D1} > K_{D4}$). If the binding of the ligand to the first binding site inhibits the binding to the other binding site, there is negative cooperativity ($K_{D1} < K_{D4}$ in mechanism IV, Figure 2.15).

2.3.1.2 Binding Inhibition

Molecules that decrease the binding of the ligand are called inhibitors. Inhibitors also bind to the binding protein, either via the binding site of the ligand or via another binding site. The first case is referred to as competitive inhibition and the second case as allosteric inhibition (see Figure 2.18).

Different mechanisms result in inhibition of ligand binding. The simplest inhibition mechanism (I) is competitive (i.e., ligand and inhibitor bind to the same binding site) (see Figure 2.19). A complex RLI does not exist with inhibition mechanism I.

The Scatchard equations for inhibition mechanism I and binding mechanism I have the form B/L = a − bB. For binding mechanism I, a and b are constant. For inhibition mechanism I, a and b depend on the concentration (I) of free inhibitor (Figure 2.19).

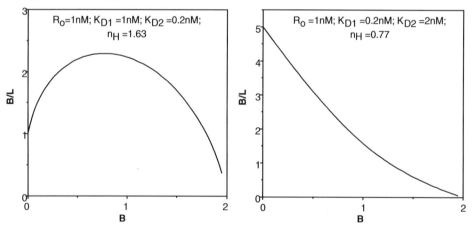

The binding protein has two binding sites (1 and 2) for ligand L:

$$1R + L \rightleftharpoons L1R2 \overset{+L}{\rightleftharpoons} L1R2L$$

$$K_{D1} = \frac{(1R) \cdot L}{(L1R2)} \qquad K_{D2} = \frac{(L1R2) \cdot L}{(L1R2L)} \qquad \text{Law of mass action}$$

$$B = (L1R2) + 2(L1R2L)$$

$$R_0 = (1R) + (L1R2) + (L1R2L) \qquad \text{Retention equations}$$

$$B_{max} = 2 R_0$$

After short calculation, this yields:

$$B = R_0 \cdot \frac{[K_{D2} + 2L] \cdot L}{K_{D1} \cdot K_{D2} + K_{D2} \cdot L + L^2}$$

After extensive calculations, this yields:

$$\frac{B}{L} = \frac{B}{\dfrac{K_{D2} \cdot [R_0 - B]}{2[B - 2R_0]} + \sqrt{\dfrac{K_{D2}^2 \cdot [B - R_0]^2}{4[B - 2R_0]^2} - \dfrac{K_{D1} \cdot K_{D2} \cdot B}{B - 2R_0}}}$$

$$\left(B\right)_{B/L=0} = 2 R_0$$

$$\left(\frac{B}{L}\right)_{B=0} = \frac{R_0}{K_{D1}}$$

Examples of Scatchard plots:

$R_0 = 1nM; K_{D1} = 1nM; K_{D2} = 0.2nM; n_H = 1.63$

$R_0 = 1nM; K_{D1} = 0.2nM; K_{D2} = 2nM; n_H = 0.77$

No thanks to a knight-errant for going mad when he has cause; the thing is to turn crazy without any provocation.

Figure 2.16. Binding mechanism III.

The binding protein has two binding sites (1 and 2) for the ligand. The binding states are in equilibrium according to:

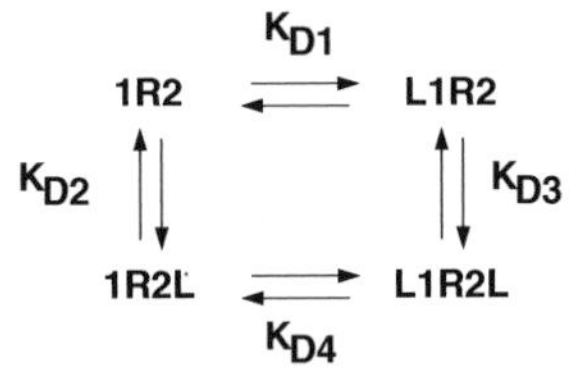

$$
\begin{array}{ccc}
 & K_{D1} & \\
1R2 & \rightleftharpoons & L1R2 \\
K_{D2} \updownarrow & & \updownarrow K_{D3} \\
1R2L & \rightleftharpoons & L1R2L \\
 & K_{D4} &
\end{array}
$$

$$
K_{D1} = \frac{(1R2) \cdot L}{(L1R2)} \qquad K_{D2} = \frac{(1R2) \cdot L}{(1R2L)}
$$

$$
K_{D3} = \frac{(L1R2) \cdot L}{(L1R2L)} \qquad K_{D4} = \frac{(1R2L) \cdot L}{(L1R2L)}
$$

Law of mass action

$$R_0 = (1R2) + (L1R2) + (1R2L) + (L1R2L)$$

$$B = (L1R2) + (1R2L) + 2(L1R2L)$$

$$B_{max} = 2R_0$$

Retention equations

From this follows:

$$
B = R_0 \cdot \frac{[K_{D3} + K_{D4} + 2L] \cdot L}{K_{D1} \cdot K_{D3} + [K_{D3} + K_{D4}] \cdot L + L^2} \qquad \text{and} \qquad K_{D1} \cdot K_{D3} = K_{D2} \cdot K_{D4}
$$

Axis sections of the Scatchard plot:

$$
\left(\frac{B}{B/L}\right)_{B/L=0} = 2R_0
$$

$$
\left(\frac{B}{L}\right)_{B=0} = R_0 \cdot \frac{K_{D3} + K_{D4}}{K_{D1} \cdot K_{D3}}
$$

Hill coefficient n_H:

$$
n_H = \frac{8 \dfrac{K_{D1} \cdot K_{D3}}{K_{D3} + K_{D4}} + 4 \sqrt{K_{D1} \cdot K_{D3}}}{4 \sqrt{K_{D1} \cdot K_{D3}} + 4 \dfrac{K_{D1} \cdot K_{D3}}{K_{D3} + K_{D4}} + K_{D3} + K_{D4}}
$$

Special case: the binding sites are independent of each other ($K_{D1} = K_{D4}$ and $K_{D2} = K_{D3}$):

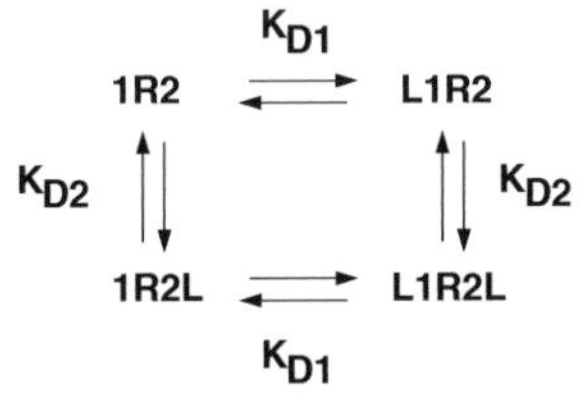

$$
\begin{array}{ccc}
 & K_{D1} & \\
1R2 & \rightleftharpoons & L1R2 \\
K_{D2} \updownarrow & & \updownarrow K_{D2} \\
1R2L & \rightleftharpoons & L1R2L \\
 & K_{D1} &
\end{array}
$$

The Hill coefficient is ≤ 1 and the Scatchard equation:

$$
\frac{B}{L} = \frac{B}{\dfrac{[R_0 - B] \cdot [K_{D2} + K_{D1}]}{4[B - 2R_0]^2} + \sqrt{\dfrac{[R_0 - B]^2 \cdot [K_{D2} + K_{D1}]^2}{4[B - 2R_0]^2} - \dfrac{B \cdot K_{D1} \cdot K_{D2}}{B - 2R_0}}}
$$

Figure 2.17. Binding mechanism IV.

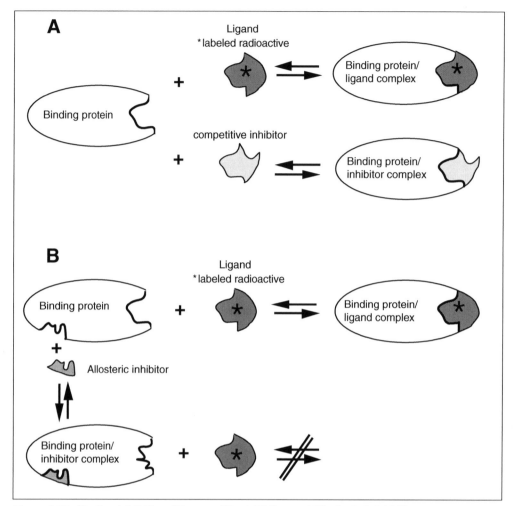

Figure 2.18. Binding inhibition. (A) competitive inhibition and (B) allosteric inhibition.

It is often true that $I >> RI$ and thus $I \approx I_0$. In this case, a and b of the Scatchard equation for inhibition mechanism I are constant and plotting B/L against B yields a straight line. Different I_0 yield different linear forms. All lines go through the point with the coordinates B/L $= 0$; $B = B_{max}$ and thus form a ray pattern (see Figure 2.20).

B_{max} is independent from I_0. The curves of more complicated mechanisms also go through the point (B/L $= 0$; $B = B_{max}$). This is the case for inhibition mechanism II, where the inhibitor occupies two binding sites on the binding protein, and for mechanism III (i.e., for allosteric inhibition). If $I >> RI$ or $I \approx I_0$ and K_D of the ligand as well as the concentration I_0 are known, the dissociation constant K_i of the inhibitor can be calculated from the slope S of the Scatchard plot (see Figure 2.21).

Scatchard plots require elaborate experiments. It is easier to determine K_i with IC_{50} curves. Into a defined concentration of ligand L_0 and binding protein R_0 you add increasingly higher concentrations of inhibitor I_0 and measure the binding B. The IC_{50} is the I_0 that reduces B to half of B_0 (binding at $I_0 \approx 0$). From the measured IC_{50} and the known K_D and L_0 of the ligand you calculate the K_i of the inhibitor according to the formula shown in Figure 2.22.

The formula in Figure 2.22 was developed by Cheng and Prusoff (1973) for enzyme kinetics. For binding experiments, this formula is only valid when $I \approx I_0$ and $L \approx L_0$. If these conditions are not given, you have to use the exact equation from Munson and Rodbard (1988). If

Inhibition mechanism I

Ligand L and inhibitor I bind to the same binding sites (1) on the binding protein (competitive inhibition):

$$R1I \; \underset{K_i}{\overset{}{\rightleftarrows}} \; R1 \; \underset{}{\overset{K_D}{\rightleftarrows}} \; R1L$$
$$+I \qquad +L$$

I Concentration of free inhibitor
L Concentration of free ligand
(R1) Concentration of free binding protein
(R1L) Concentration of binding protein/ligand complex

(R1I) Concentration of binding protein/inhibitor complex
R_o Total concentration of binding protein
I_o Sum of concentrations of free and saturably bound inhibitor, typically set to be equal to the total concentration of inhibitor

$$K_D = \frac{(R1) \cdot L}{(R1L)} \quad ; \quad K_i = \frac{(R1) \cdot I}{(R1I)} \qquad \text{Law of mass action}$$

$$R_o = (R1) + (R1L) + (R1I); \; I_o = I + (R1I);$$
$$(R1L) = B \; ; \; B_{max} = R_o \qquad \text{Retention equations}$$

The Scatchard equation is:
$$\frac{B}{L} = \frac{R_o - B}{K_D} \cdot \frac{K_i}{[K_i + I]}$$

The Scatchard plot is nonlinear, because I depends on B:

$$I = \frac{B - R_o + I_o - K_i}{2} + \sqrt{\frac{[K_i + R_o - B - I_o]^2}{4} + I_o \cdot K_i}$$

Inhibition mechanism II

The inhibitor has two binding sites (1 and 2) on the binding protein:

$$I1R2I \; \underset{K_{ii}}{\overset{}{\rightleftarrows}} \; I1R2 \; \underset{K_i}{\overset{}{\rightleftarrows}} \; 1R \; \underset{}{\overset{K_D}{\rightleftarrows}} \; L1R$$
$$+I \qquad +I \qquad +L$$

The Scatchard equation is:
$$\frac{B}{L} = \frac{R_o - B}{K_D} \cdot \frac{K_i \cdot K_{ii}}{K_i \cdot K_{ii} + K_{ii} \cdot I + I^2}$$

Inhibition mechanism III

$$1R2 \; \underset{}{\overset{K_{D1}}{\rightleftarrows}} \; L1R2$$
$$K_i \updownarrow \qquad \qquad \updownarrow K_{ii}$$
$$1R2I \; \underset{K_{D2}}{\overset{}{\rightleftarrows}} \; L1R2I$$

Ligand L binds to binding site 1, ligand I to binding site 2
With $K_{ii} > K_i$, ligand I inhibits binding of L (negative cooperation)
With $K_{ii} < K_i$, it facilitates the binding of L (positive cooperation)
B = (L1R2) + (L1R2I)
$B_{max} = R_o$

The Scatchard equation is:
$$\frac{B}{L} = \frac{R_o - B}{K_{D1}} \cdot \frac{1 + \frac{I}{K_{ii}}}{1 + \frac{I}{K_i}}$$

Figure 2.19. Inhibition mechanisms.

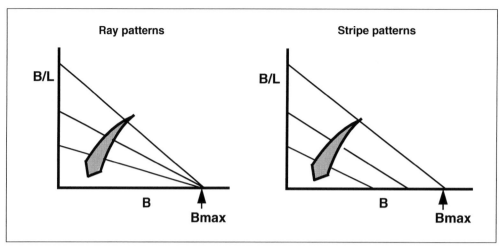

Figure 2.20. Experimental Scatchard plots in the presence of inhibitor I. Concentration of inhibitor increases in direction of arrow.

$$K_i = \frac{K_D \cdot (-S) \cdot I_0}{1 + K_D \cdot (-S)}$$

Figure 2.21. Calculation of K_i. In the presence of the inhibitor (concentration I_0), a Scatchard plot is done with ligand L. If $I \approx I_0$ for the free concentration I of the inhibitor, the K_i of the inhibitor can be calculated from the negative slope $(-S)$ of the plot, the K_D of the ligand, and I_0.

$$K_i = \frac{K_D \cdot IC_{50}}{L_0 + K_D}$$

Figure 2.22. The Cheng-Prusoff equation. The relation between the dissociation constant of an inhibitor K_i, the inhibitor's IC_{50}, and the ligand's K_D.

the saturable binding B does not reach zero even with high inhibitor concentrations, the following are possible.
- The ligand binds to several different binding sites, but the inhibitor does not cover all binding sites of the ligand.
- There is negative cooperativity. This means that for inhibition mechanism III and $K_{ii} > K_i$, B does not approach zero with increasing concentrations of inhibitor but approaches a threshold dependent on K_{ii}, K_i, K_{D1}, R_0, and L_0.

If an inhibitor concentration of several powers of 10 is needed to push the ligand binding to zero, the inhibitor presumably binds to several binding sites with different affinity. If the affinity of the binding sites is sufficiently spread out, the IC_{50} curve will show steps.

The value for K_i that is calculated from IC_{50} curves is, of course, the true K_i of the inhibitor only for simple inhibition mechanisms. For more complicated mechanisms, or when the inhibitor has two binding sites on the binding protein (e.g., inhibition mechanism II), you get only the appearance of a K_i, which consists of several constants and sometimes depends on L_0 and/or other quantities.

Determining the mechanism of an inhibition (or binding) is a work of diligence that raises a few eyebrows among experts. This is partly due to the uncertainty inherent in the proof: that the binding data match the predictions of a mechanism is no proof that the mechanism is present. Models can only be excluded, and even this requires lengthy (and boring) measurements. In addition, the average biologist fears or even detests mathematics. However, the logic behind the complicated formulas is simple and it can be a nice hobby to design hypothetical binding mechanisms and derive formulas from them.

Sources
1. Cheng, Y. C., and Prusoff, W. (1973). "Relationship Between the Inhibition Constant (K_i) and the Concentration of Inhibitor Which Causes 50% Inhibiton (I_{50}) of an Enzymatic Reaction," *Biochem. Pharmacol.* 22: 3099–3108.
2. Munson, P., and Rodbard, O. (1988). "An Exact Correction to the Cheng-Prusoff Correction," *J. Receptor Res.* 8: 533–546.

2.3.1.3 False Ideas

A widespread superstition says that Scatchard plots that with increasing I_0 show constant B_{max} but decreasing K_D are proof that ligand and inhibitor bind to the same binding site (ray pattern; see Figure 2.20). The inhibition is said to be competitive (see Figure 2.18). This is not the case. For all reversible inhibition mechanisms, even when ligand and inhibitor bind to different sites on the same binding protein, the inhibitor changes the slope of the Scatchard curves, but not the B_{max} (Tomlinson 1988; Rehm and Becker 1988). For example, inhibition mechanism III is allosteric (inhibitor and ligand bind to different binding sites on the binding protein).

Nevertheless, the Scatchard equation for III has the same form ($B/L = a - bB$) as that of mechanism I, where there is competitive inhibition. If $I = I_0$, the Scatchard plots for I and III are linear, and both go through the point ($B/L = 0$; $B = B_{max}$) for each I_0. If $I \neq I_0$, the Scatchard plots for both mechanisms are convex curves (for I, see Figure 2.23) that nevertheless extend in a ray pattern from the point ($B/L = 0$; $B = B_{max}$).

You have nonreversible inhibition when the inhibitor reacts covalently with the binding protein and thus forever blocks the binding site. Seemingly nonreversible inhibition is also caused by experimental artifacts (e.g., a contamination of the inhibition preparation with proteases that digest the binding protein).

The statement that reversible inhibition mechanisms only exhibit ray patterns seems to contradict experience. Often, the experimentally determined B_{max} seems to decrease with increasing I_0, even when the measured values are correct and the mechanism is reversible. The root cause of this phenomenon is nonlinear Scatchard plots. For the experimenter, who cannot measure all points of the curve, they create the illusion of a decreasing B_{max} (Rehm and Becker 1988). For inhibition mechanism I (competitive), the inhibitor bends the Scatchard plots downward, and the commonly performed interpolation of the measuring points creates the appearance of decreasing B_{max} with increasing I_0 (see Figure 2.23).

From the label noncompetitive inhibition for this phenomenon, many people make inferences regarding the facts, according to the following logic. The effect is called noncompetitive inhibition; thus, the inhibitor does not inhibit competitively and if it does not inhibit competitively it must inhibit allosterically. This is false. Some Scatchard plots showing seemingly smaller B_{max} with increasing concentration of inhibitor I_0 do not constitute proof of allosteric inhibition. Ray patterns such as those in Figure 2.20 instead point to nonlinear Scatchard plots and to the fact that binding is not as simple as the researcher is imagining it.

Finally, Scatchard plots are usually nonlinear in the presence of inhibitor, even for the simplest inhibition mechanism I. Strictly speaking, I is always $\neq I_0$. However, because of the measuring error the experimenter notices this only rarely.

Sources
1. Rehm, H., and Becker, C. M. (1988). "Interpreting Noncompetitive Inhibition," *TIPS* 9: 316–317.
2. Tomlinson, G. (1988). "Inhibition of Radioligand Binding to Receptors: A Competitive Business," *TIPS* 9: 159–162.

2.3.1.4 Hilly Things

The Hill plot for a ligand is log ($B/B_{max} - B$) plotted against log L. For binding mechanism I, this yields a linear form with a slope of 1. For other mechanisms (e.g., binding mechanisms

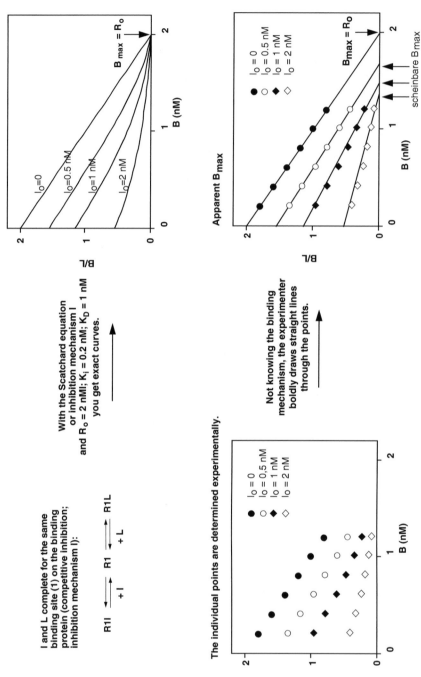

I and L complete for the same binding site (1) on the binding protein (competitive inhibition; inhibition mechanism I):

$$R1I \underset{+I}{\overset{}{\rightleftharpoons}} R1 \underset{+L}{\overset{}{\rightleftharpoons}} R1L$$

With the Scatchard equation or inhibition mechanism I and $R_o = 2$ nM; $K_i = 0.2$ nM; $K_D = 1$ nM you get exact curves.

The individual points are determined experimentally.

Not knowing the binding mechanism, the experimenter boldly draws straight lines through the points.

Figure 2.23. How correct measurements can lead to false statements.

III and IV), the Hill plot yields a sigmoid curve. The slope of the curve at half-maximal saturation (i.e., at the point $B = B_{max}/2$) is referred to as Hill coefficient n_H. In order to determine the Hill coefficient you thus use only points located near the half-maximal saturation. For the persnickety and mathematicians: the point of half-maximal saturation does not have to coincide with the turning point of the curve. For example, for binding mechanism II the Hill coefficient (≤ 1) is certainly not always the smallest slope of the curve.

The Hill coefficient is used to diagnose cooperativity. A binding protein with multiple binding sites binds cooperatively when the binding of one ligand influences the binding of the other. Positive cooperativity means that the affinity of the bond increases. Negative cooperativity, means that the affinity of the bond decreases. If $n_H > I$, the binding mechanism is positively cooperative. A Hill coefficient greater than 1 can, but does not have to, indicate negative cooperativity. Hill coefficients less than 1 also occur with noncooperative binding mechanisms, for example, for binding mechanism II (two independent binding sites on different proteins).

Examples: For binding mechanism IV (two binding sites), the formula shown in Figure 2.17 yields the following. The Hill coefficient n_H is between 0 and 2 and only depends on the dissociation constant. With $n_H \neq 1$, the Hill plot is sigmoid. With $n_H = 1$, it is linear (just like the corresponding Scatchard plot). The Hill coefficient is 1 when $K_{D3} + K_{D4} = 2\sqrt{K_{D1}, K_{D3}}$. If mechanism IV is positively cooperative, $K_{D1} > K_{D4}$ and $n_H \geq 1$. If IV is negatively cooperative, $K_{D1} < K_{D4}$ and $n_H < 1$. Finally, if $K_{D1} = K_{D4}$ and $K_{D3} = K_{D2}$ (i.e., if binding sites 1 and 2 on the binding protein are independent of each other), $n_H \leq 1$.

For binding mechanism III (simplified case of IV, two binding sites), the Hill plot is sigmoid, with $n_H \neq 1$ (forms the transition between two constraint lines with slope 1). The turning point of the Hill plot coincides with the point of half-maximal saturation [$\log (R_0/2R_0 - R_0) = 0$; $\log L = 0.5 \log (K_{D1}, K_{D2})$]. Mechanism III is positively cooperative (indepdendent of K_{D1} and K_{D2}) because the affinity of the ligand for binding site 2 is 0, as long as binding site 1 remains unoccupied. The Hill coefficient for III is dependent on K_{D1} and K_{D2} and lies between 0 and maximally 2. It is true that $n_H = 1$ for $4K_{D1} = K_{D2}$, $n_H < 1$ for $4K_{D1} < K_{D2}$, and $n_H > 1$ for $4K_{D1} > K_{D2}$.

For binding mechanism II (no cooperativity, but two binding sites), the Hill plot is sigmoid and $n_H < 1$, or for $K_{D1} = K_{D2}$, $n_H = 1$. The turning point of the Hill plot and the point of the half-maximal saturation do not necessarily coincide for II, which theoretically gives the experimenter the opportunity to distinguish between II and III ($n_H < 1$).

If $n_H = 1$, this can (but does not have to) be due to the simple mechanism I, in that (for example) mechanism III also yields $n_H = 1$ at $4K_{D1} = K_{D2}$. Similarly, mechanism IV at $K_{D3} + K_{D4} = 2\sqrt{K_{D1}, K_{D3}}$, which can be a positive cooperative binding (i.e., with $K_{D1} > K_{D4}$). The Scatchard plots for III and IV are linear at $n_H = 1$, which means that linear Scatchard plots are no proof of the existence of only one binding site. In addition, it is difficult to distinguish between $n_H = 1$ and $n_H \neq 1$, because the difference of n_H and 1 is often small and indistinguishable from the experimental error.

The Hill plot of an inhibitor is $\log (B/B_0 - B)$ plotted against $\log I$. Because I can usually not be determined without making assumptions about the mechanism, you plot $\log (B/B_0 - B)$ against $\log I_0$. B_0 is the concentration of the bound ligand B at $I_0 = 0$. The Hill plots of an inhibitor for inhibition mechanism I are sigmoid curves (with two constraint lines of slope −1). For inhibition mechanism II, the Hill plots are concave (with two constraint lines of slope −1 and −2). For inhibition mechanism III, the Hill plots are convex (with two constraint lines of slope −1 and 0). By analogy, the Hill coefficient of the inhibitor would be the slope of the curve at $B = B_0/2$. However, the concave curves of II bend near $B = B_0/2$, which makes it difficult to determine the Hill coefficient of the inhibitor. In addition, the Hill coefficient of the inhibitor depends on the inhibitor's L_0 (as for mechanism I). That is, for a specific inhibition mechanism and specific K_i and K_D it is not a fixed quantity and is of doutbful value. The Hill coefficient of the inhibitor does at least give an indication of the number of its binding sites. $n_H < −1$ is the case when the binding reaction with the inhibitor creates a complex RI_n with $n > 1$.

Furthermore, the Hill plot's form has diagnostic value, in that it provides clues regarding the inhibition mechanism. Figure 2.24 summarizes the analysis of equilibrium binding data.

Binding of a radioactive ligand

Affinity (K_D) **B_{max}** **Hill coefficient**

Experiment:

Determine the saturable binding for equal amounts of smaple with 8 to 16 different ligand concentrations (L_0). R_0 and B_{max} are thus constant and L_0 is changed. The L_0 should be 0.25 to 4 K_D and increase by a factor of 1.5 to 2. The concentration of free ligand (L) in the solution results from the total concentration of ligand in the binding solution minus the concentration of (saturably and nonspecifically) bound ligand. To determine the total concentration you count either the complete solution (^{125}I-labeled ligands) or a small portion of the binding solution (^{3}H-labeled ligands). L_0 is the sum of free ligand L and saturably bound ligand B.

Data analysis:

From the specific activity of the ligand and the saturable counts, you calculate for each ligand concentration (L_0) the concentration of bound ligand (B), free ligand (L) and the quotient B/L. B/L is a dimensionless number and is plotted against B (Scatchard plot). If the points form a straight line, you can calculate K_D (app) and B_{max}.

Calculation:

The intersection of the B axis (B_{max}) is divided by the intersection of the B/L axis. For binding mechanism I, the intersection of the B/L axis is equal to B_{max}/K_D and the division yields the K_D. For other mechanisms, you get a K_D(app), which is made up of several constants.

Calculation:

The intersection of the straight line with the B axis is equal to B_{max}. For binding mechanism I, B_{max} is the concentration of binding sites in the solution ($R_0 = B_{max}$).

Data analysis:

Determine L_{hS} = L at half-maximal saturation (at the point B = $B_{max}/2$). Plot log (B/B_{max} − B) against the log L that are located around log L_{hS}. Rule of thumb: from L at B = 0.2B_{max} until L at B = 0.8B_{max}.

Calculation:

The slope of the straight line is the Hill coefficient of the ligand.

Figure 2.24. Determining binding quantities.

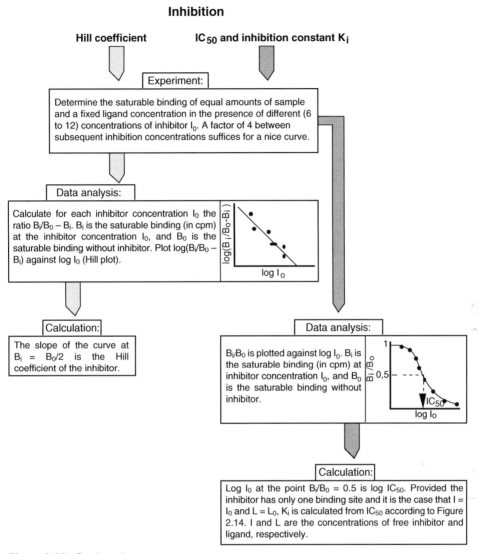

Inhibition

Hill coefficient   **IC$_{50}$ and inhibition constant K$_i$**

Experiment:

Determine the saturable binding of equal amounts of sample and a fixed ligand concentration in the presence of different (6 to 12) concentrations of inhibitor I_0. A factor of 4 between subsequent inhibition concentrations suffices for a nice curve.

Data analysis:

Calculate for each inhibitor concentration I_0 the ratio $B_i/B_0 - B_i$. B_i is the saturable binding (in cpm) at the inhibitor concentration I_0, and B_0 is the saturable binding without inhibitor. Plot $\log(B_i/B_0 - B_i)$ against $\log I_0$ (Hill plot).

Calculation:

The slope of the curve at $B_i = B_0/2$ is the Hill coefficient of the inhibitor.

Data analysis:

B_i/B_0 is plotted against $\log I_0$. B_i is the saturable binding (in cpm) at inhibitor concentration I_0, and B_0 is the saturable binding without inhibitor.

Calculation:

Log I_0 at the point $B_i/B_0 = 0.5$ is log IC$_{50}$. Provided the inhibitor has only one binding site and it is the case that $I = I_0$ and $L = L_0$, K_i is calculated from IC$_{50}$ according to Figure 2.14. I and L are the concentrations of free inhibitor and ligand, respectively.

Figure 2.24. *Continued*

Never go by arbitrary law, which is so much favoured by ignorant men who plume themselves on cleverness.

2.3.2 Kinetics

With respect to the binding kinetics of a ligand, one distinguishes between on-rate kinetics and off-rate kinetics: the ligand binding to the receptor versus the ligand dissociating from the receptor, respectively. Kinetics are often determined by hand. An experimenter may want to measure (for example) the association of ligand and binding protein at different points in time. For this she makes a binding solution with (n + 1) times more volume than for a standard assay. The binding solution here, however, is missing one of the binding partners (e.g., the radioactive ligand). At time 0, the experimenter adds (for example) radioactive ligand and mixes it in. The binding reaction begins. At n specific time points, he takes off an aliquot (each corresponding to one standard assay) and (immediately!) measures the bound ligand.

The time points are so chosen that the difference between the bound ligands at consecutive time points is significant. Most time points are thus located on the increasing slope of the B = F(t) curve, especially at its beginning. The time points also have to be far enough apart to allow for binding measurements taken between the individual aliquots. The binding solution is held at a constant temperature (water bath).

From 2 to 3 values from the shallow branch of the B = F(t) curve (i.e., at long intervals) the experimenter estimates the bound ligand at t = ∞ (i.e., at equilibrium). For the nonspecific binding, a (n + 1) solution is run in parallel with an excess of nonradioactive ligand. This soup recipe yields only one binding value per time point, but this one is exact. If you don't believe this, you are welcome to cook the soup three times. You also save yourself time and pipetting work. A similar procedure also yields the best values for dissociation. If you have fast kinetics or want to get very exact measurements, you can try BIACORE (see Section 2.2.5).

2.3.2.1 On-rate Kinetics

The on-rate kinetics measures the speed of the association of ligand and binding protein. From this, two quantities can be calculated: the half-life $T_{1/2}$ and the kinetic constant k_{on}. The half-life $T_{1/2}$ is the time it takes for the concentration of the bound ligand B to reach half its maximum value (value at binding equilibrium). $T_{1/2}$ results from the curve B (at time t) against time t, and gives an indication of how long you need to wait for binding equilibrium. $T_{1/2}$ depends on the initial concentrations of binding protein and ligand (i.e., it is not a constant). It is better to characterize the on-rate kinetics via the constant k_{on} (Figure 2.25). The value (RL) at t = ∞ for the formula in Figure 2.25 yields the kinetics, L_0 is known, and for R_0 the experimenter has to make a Scatchard.

2.3.2.2 Off-rate Kinetics

The off-rate constant k_{off} characterizes the speed of the dissociation of the ligand from the binding protein/ligand complex. The constant is determined as follows. After radioactive ligand and binding protein have reached binding equilibrium, the experimenter either reduces the concentration of the unbound ligand or its specific radioactivity. The latter is referred to as a chase. In a chase, the experimenter adds a 100- to 1,000-fold excess of nonlabeled ligand to the binding solution. However, he dilutes the binding solution—and hence the labeled ligand—20- to 40-fold with binding buffer to reduce the concentration of unbound ligand. In both cases, the on rate of the radioactive ligand does not play a role anymore, and the formula shown in Figure 2.26 is valid.

Details: In the big pot, make a (n + 1)-fold binding solution for n time points. Wait until binding equilibrium is reached. Begin at t = 0 by diluting or adding cold ligand (dilution solution and/or cold ligand and binding solution must have the same temperature!). At each time point, take an aliquot corresponding to a standard assay (e.g., with a 40-fold dilution you need of course the 40-fold volume of the standard assay) and immediately determine the amount of bound ligand.

2.3.2.3 Problems with Kinetics

The on-rate or off-rate curves (Figures 2.25 and 2.26) are often not linear but curved. This may be due to several classes of binding sites that show different on rates or off rates. The speed constant of the fastest reaction results from the slope of the tangent to the curve at the origin. Often, the phenomenon of bend curves is also caused by something trivial such as temperature changes during the experiment.

Definitions:

L   Concentration of free ligand at time **t**

L_0 $L_0 = L + (RL)$ [L_0 is independent of **t**]

R   Concentration of free binding protein at time **t**

R_0 Total concentration of binding protein [$R_0 = R + (RL)$; R_0 is independent of **t**]

(RL)   Concentration of bound ligand at time **t**

(∞RL) Concentration of bound ligand at t = ∞

Mechanism: $\qquad R + L \underset{k_{off}}{\overset{k_{on}}{\rightleftharpoons}} RL$

From the mechanism follows the differential equation: $\quad \dfrac{d(RL)}{dt} = k_{on} \cdot R \cdot L - k_{off} \cdot (RL)$

k_{on} results from:

$$k_{on} = \frac{\overset{\infty}{(RL)}}{\overset{\infty}{(RL)}{}^2 - R_0 \cdot L_0} \cdot \frac{1}{t} \cdot \left[\ln \frac{\overset{\infty}{(RL)} - (RL)}{R_0 \cdot L_0 - \overset{\infty}{(RL)} \cdot (RL)} + \ln \frac{R_0 \cdot L_0}{\overset{\infty}{(RL)}} \right]$$

Plotting the quantity $\ln \dfrac{\overset{\infty}{(RL)} - (RL)}{R_0 \cdot L_0 - \overset{\infty}{(RL)} \cdot (RL)}$ against time **t** yields a linear form whose slope S is the basis for the calculation of the constant k_{on}:

$$k_{on} = \frac{S \cdot \overset{\infty}{(RL)}}{\overset{\infty}{(RL)}{}^2 - R_0 \cdot L_0}$$

The slope S has the dimension time^{-1}, and k_{on} has the dimension time^{-1} concentration^{-1} (i.e., for example sec^{-1} M^{-1}).

Figure 2.25. Determining the on rate.

On rates and off rates are too fast to be experimentally determined without elaborate equipment. Cooling and/or lower concentrations of reactants (R_0, L_0) help. The quotient k_{off}/k_{on} should be equal to the dissociation constant K_D (following the formulas in Figure 2.4 and 2.25, at $d(RL)/dt = 0$), and normally it is. If deviations are too large (larger than factor 2), this indicates methodical errors.

2.4 Cross-linking of Ligands

While engaged in this discourse they were making their way through a wood that lay beyond the road, when suddenly, without expecting anything of the kind, Don Quixote found himself caught in some nets of green cord stretched from one tree to another; and unable to conceive what it could be, he said to Sancho, "Sancho, it strikes me this affair of these nets will prove one of the strangest adventures imaginable."

Definitions:

(RL) is the concentration of bound ligand at time **t**

(RL)$_{t=0}$ is the concentration of bound ligand at time t = 0.

Mechanism: $RL \xrightarrow{k_{off}} R + L$

Corresponding differential equation: $\dfrac{d(RL)}{dt} = -k_{off} \bullet (RL)$

Integrating yields: $\ln \dfrac{(RL)}{(RL)_{t=0}} = -k_{off} \bullet t$

Plotting **ln (RL)/(RL)**$_{t=0}$ against **t** yields a linear form with the slope $-k_{off}$.

Figure 2.26. Determining the off rate.

Many ligands can be covalently cross-linked with their binding protein. If the ligand is radio-actively labeled, the binding protein is also radioactively labeled afterward. If the ligand binds specifically, ligand and binding protein can often also be cross-linked specifically (i.e., in protein mixtures the ligand cross-links only with the binding protein). The cross-linking technique thus allows the experimenter to label binding proteins selectively and covalently, without having to purify them first. Furthermore, SDS electrophoresis and autoradiography of the cross-linked binding protein yield its MW or the MW of the ligand/binding polypeptide. Unless otherwise noted, ligand refers to radioactive ligand in the following.

2.4.1 Three-component Cross-linking (3C Cross-linking)

In 3C cross-linking, three components react with one another: binding protein, ligand, and cross-linker. Cross-linkers are molecules with two functional groups (e.g., aldehyde or imi-doester groups). Commonly used cross-linkers are shown in Figure 2.27. A number of other cross-linkers are described in the Pierce catalog, which is also a good reference for cross-linkers.

For a cross-linker to cross-link ligand and binding protein, the latter must have reactive groups that react with the functional groups of the cross-linker. Reactive groups are (for example) primary amino, carboxyl, or sulfhydryl groups. On the binding protein, the reactive group is usually a primary amino group (e.g., from lysyl residues). Cross-linking is easy and quick, but it is time consuming and laborious to find or make a suitable ligand. It is not enough that the ligand has a reactive group; it must be at the correct location. In addition, 3C cross-linking becomes an art when the ligand's K_D is in the micromolar range. The ideal ligand thus contains a primary amino group that is not involved in the binding, and it binds with nanomolar or lower K_D to the binding protein.

3C cross-linking is thus popular for protein ligands and larger peptide ligands. These are iodined and then cross-linked to the binding polypeptide via their primary amino groups. 3C

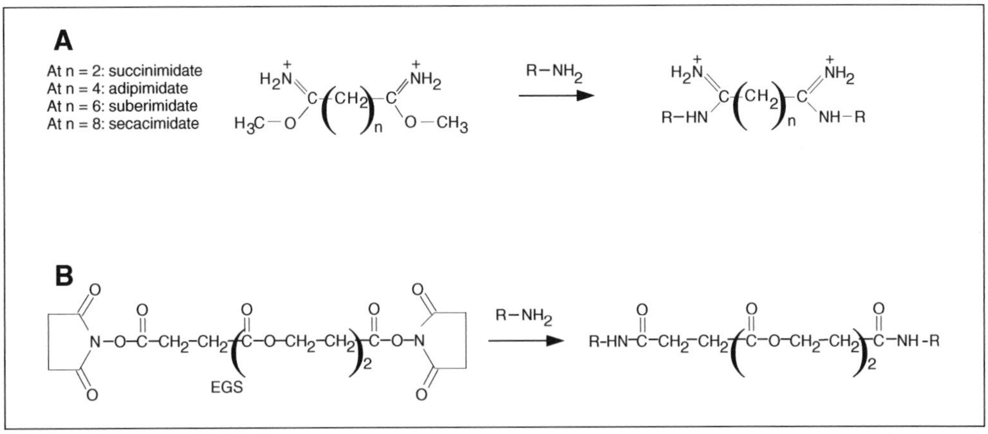

Figure 2.27. Homofunctional cross-linker. (A) At a pH between 7 and 9, bisimidates react with primary amino groups (lysyl residues of proteins preferred), creating a stable bond. The charge of the protein does not change. Bisimidates are membrane permeable, easily soluble in water, and hydrolyze with a half-life between 10 minutes and 1 h (dependent on the pH). (B) EGS (Pierce) forms stable acid amide compounds with primary amines. EGS is not very water soluble and is added as DMSO solution to the cross-linking solution. The advantage compared to bisimidates is the longer half-life of EGS in watery solutions (a few hours) and its cleavability with hydroxylamine (at pH 8.5). EGS is also available as a water-soluble sulfoanalog.

cross-linking is also possible via carboxyl groups of ligand or binding polypeptide (Schmidt and Betz 1989), but this is rarely done.

The cross-linker cross-links not only ligand and binding polypeptide but the other proteins in the reaction solution, both intermolecularly and intramolecularly. At low concentrations of cross-linkers, the chances for cross-linking ligand and binding polypeptide are low, and the signal will be low as well. At high concentrations, high-molecular adducts are created that are difficult to analyze and often not even soluble in SDS. Thus, there is an optimal concentration of cross-linker, which must be determined for each ligand.

Length is an important factor for cross-linkers without arylazide group, in that the cross-linker has to be long enough to be able to link up the reactive group of the ligand with a reactive group of the binding polypeptide. On the other hand, it cannot be too long because then it would not react with the binding protein anymore or establish intermolecular bonds.

The incorporation rate is low with 3C cross-linking: less than 1 to 2% of the bound ligand are covalently bound to the binding protein. Thus, 3C cross-linking usually only allows us to determine the apparent MW of the binding polypeptide on SDS gels, and this often does not help the experimenter much. However, if you have an appropriate ligand 3C cross-linking is not much work and is a good subject of a paper.

Sources
1. Bayley, H., and Knowles, J. R. (1977). "Photoaffinity Labeling," *Methods Enzymol.* 46: 69–114.
2. Gaffney, B. J. (1985). "Chemical and Biochemical Crosslinking of Membrane Components," *BBA* 822: 289–317.
3. Schmidt, R., and Betz, H. (1989). "Crosslinking of β-bungarotoxin to Chick Brain Membranes: Identification of Subunits of a Putative Voltage-gated K⁺ Channel," *Biochemistry* 28: 8346–8350.

2.4.1.1 Homofunctional Cross-linker

Homofunctional cross-linkers have two of the same functional groups. Popular homofunctional cross-linkers are bisimidates (e.g., dimethylsuberimidate) and the N-hydroxy-succinimidesters (e.g., EGS) (Figure 2.27). Many experimenters use dimethylsuberimidate to cross-link protein and peptide ligands.

The advantages of bisimidates lie in their specificity for primary amino group or lysyl residues (thiol, phenol, or imidazolyl groups hardly react), the stability of the resulting bonds, and the availability of a number of bisimidates of varying chain length (from dimethylsuccinimidate with n = 2 to dimethylsecacimidate with n = 8). Furthermore, retail also offers cleavable bisimidates (e.g., using a disulfide bridge), whose cross-linking is reversible. Disadvantages of bisimidates are:

- Fast hydrolysis
- The alkaline reaction requirements (facilitate, for example, disulfide exchange).

Succinimidesters are also available with different chain lengths and with a disulfide bridge in the middle. Succinimidesters are more stable in water than in bisimidates, but are often not very water soluble.

2.4.1.2 Heterofunctional Cross-linker

Heterofunctional cross-linkers have different functional groups (Figure 2.28). One group usually reacts with primary amino groups, whereas the other is activated via radiation (photolysis, Figure 2.29). The photosensitive group is usually an arylazide. Arylazides are stable in the dark, but under radiation they become aggressive nitren derivatives. Nitren derivatives have a life span of 0.1 to 1 msec and even react with C-H bonds. A reactive group on the binding polypeptide is thus not required for the nitren reaction. This is advantageous when the lysin residues of the binding protein are in an unfavorable location or the binding protein has no reactive group. Then, 3C cross-linking would fail with (for example) bisimidates, whereas a cross-linker with an arylazide group such as SANPAH could still save the experiment.

The absorption maximum of simple arylazides such as HSAB is about 265 to 275 nm, and the molar extinction coefficient is at $2 \times 10^4 \, M^{-1} \, cm^{-1}$. The substitution of the aromatic ring with iodine (a nitro or a hydroxyl group) moves the absorption maximum toward higher wavelengths (the absorption maximum of NHS-ASA is about 305 nm).

If you do not care about the state of the binding protein's amino acids tryptophane, tyrosine, and phenylalanine, you activate arylazide with shortwave absorption maximum (260 to 305 nm) using the Stratalinker (Stratagene). The device creates mostly monochromatic light with a wavelength of 254 nm and a continuum between 315 nm and 370 nm. Radiation with 0.12 Joules for 1 minute is sufficient. If you want to spare the amino acids of the binding protein, use light with longer wavelengths (310 to 360 nm). The UV lamps of cell culture hoods (280 nm) or the UV light plates or hand monitors used by molecular biologists (254/312/365 nm) can also be used to activate arylazides. However, the radiation emission of these devices is difficult to control and the exposure time needs to be 30 minutes or more.

Exposure time and distance of lamp to sample are found by trial and error (30 cm is a good start). The radiation intensity decreases with distance squared. Arylazides substituted with nitro groups can be activated in a very quick and protein-friendly way with a commercial photo flash.

2.4.1.3 3C Cross-linking Experiments

How do you correctly perform 3C cross-linking? First, you incubate the ligand with the protein sample—with and without an excess of binding inhibitor. The concentration of ligand should be in the K_D range. Once binding equilibrium is reached, remove the unbound ligand (as long as the off rate allows this). Then you add the cross-linker. You let heterofunctional cross-linkers react first with their nonphotoactivatable functional group.

The used buffers may not contain amines such as Tris or glycine. The buffers should be somewhat alkaline to ensure the availability of sufficient unprotonized amine (from ligand or protein) for the reaction. HEPES, Borat, or phosphate buffers with a pH of 7.5 to 8.5 are a good choice. To find the optimal concentration of cross-linker, the experimenter makes cross-linking solutions with different concentrations of cross-linker (for bisimidates, for example, 0.1/0.4/1.6/6.4 mM). The cross-linker is added just before the reaction to keep its hydrolysis

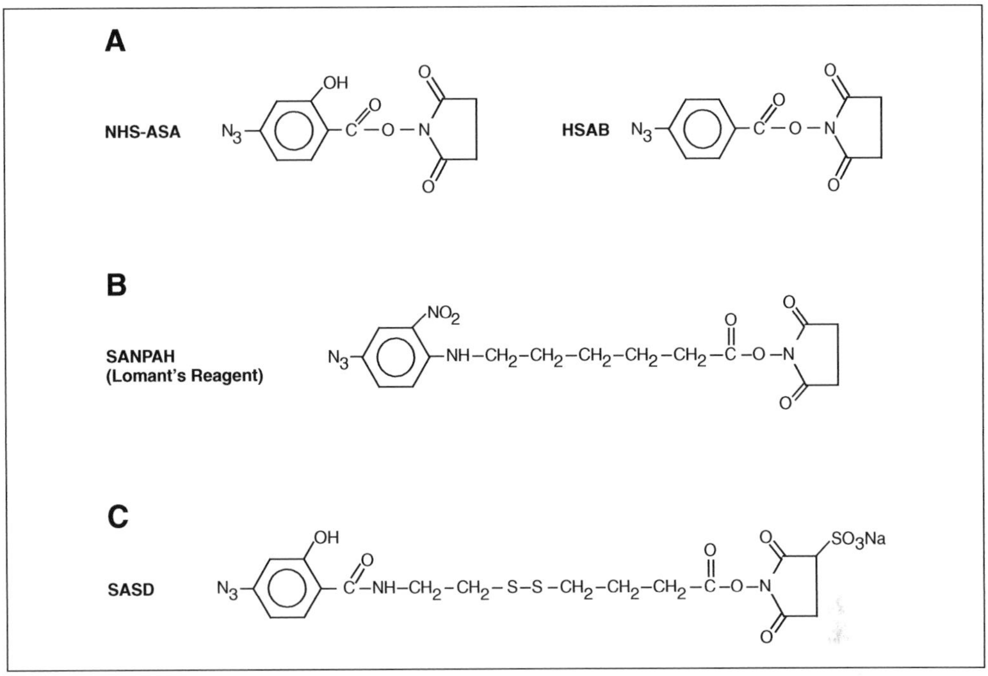

Figure 2.28. Heterofunctional cross-linker. (A) The succinimidyl group of NHS-ASA and HSAB reacts with primary amino groups to stable acid amides. The succinimidyl group hydrolyzes in water with a half-life of 2 to 10 minutes, but it is stable in organic solvents such as DMSO. The arylazide group is activated with UV light (270 to 305 nm) (photolysis). NHS-ASA can be iodined (Ji et al. 1985, and see Section 2.1.3) and can be used to create photoaffinity ligands from nonradioactive ligands with primary amino group. (B) The N-succinimidyl group of SANPAH (Pierce) reacts with primary amino groups. The nitroarylazide group is activated with long-wave UV light (320 to 350 nm). SANPAH is stable under red light. SANPAH is not very soluble in water and is added as DMSO solution to the cross-linking solution, the analog sulfo-SANPAH (sulfo group at the succinimidyl ring), on the other hand, is water soluble. (C) SASD (Pierce) can be iodined at the phenyl residue and its disulfide group can be cleaved after cross-linking. This characteristic allows for specific iodization of the binding protein. Via the sulfosuccinimidyl group, ^{125}I-SASD is converted to a photoaffinity ligand with a primary amino group of the nonradioactive ligand. The photoaffinity ligand binds to the binding protein and photolysis (270 to 305 nm) bonds the partners covalently. Afterward, the disulfide bridge is cleaved (with DTT oder mercaptoethanol). Only the iodined phenyl residue remains covalently bound to the binding protein. Phosphate buffers do not get along with SASD (SASD precipitates).

down. After incubating the cross-linker, ligand, and protein for 5 to 30 minutes, heterofunctional cross-linkers are photolyzed. Adding glycine, Tris, or ethanolamine solutions (100 mM, pH 8.0) blocks unused cross-linker. Ligand that is not bonded is washed away if this causes little effort (e.g., during cross-linking of membranes).

After cross-linking, the experimenter analyzes the solution on an SDS gel. Thick gels (3 to 1.5 mm instead of 0.75 mm) allow us to load a lot of protein per pocket and thus enhance the signal of the ligand/binding polypeptide bond in the autoradiogram or fluorogram.

In cross-linking experiments, a shotgun approach is helpful. Hoping that one solution is going to work, the experimenter makes a batch of many samples with different cross-linkers of different chain lengths in different concentrations. After the reaction, the results are analyzed in parallel with several gels and many pockets per gel. If you have the choice, you take iodined ligand, because autoradiography takes only 2 to 7 days for ^{125}I but 4 to 12 weeks for ^{3}H (see Section 1.7).

Sources

3C Cross-linking of Peptide Ligands

1. Donner, D. (1983). "Covalent Coupling of Human Growth Hormone to Its Receptor on Rat Hepatocytes," *J. Biol. Chem.* 258: 2736–2743.

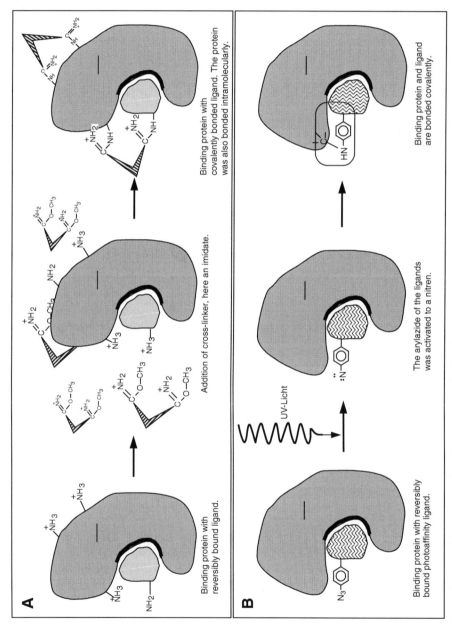

Figure 2.29. Cross-linking of ligand with binding proteins. (A) 3C Cross-linking. (B) Photoaffinity cross-linking. Not much is known about the reaction of arylnitrens with proteins, and the type of the resulting bond depends on the groups with which the nitren can react. During the reaction of arylnitren with a C-H group of the binding protein, the encircled bond is presumably created (Bayley and Knowles 1977).

2. Susini, et al. (1986). "Characterization of Covalently Cross-linked Pancreatic Somatostatin Receptors," *J. Biol. Chem.* 261: 16738–16743.
3. Wood, C., and Dorisio, M. (1985). "Covalent Cross-linking of Vasoactive Intestinal Polypeptide to Its Receptors on Intact Human Lymphoblasts," *J. Biol. Chem.* 260: 1243–1247.

3C Cross-linking of Protein Ligands
1. Rehm, H., and Betz, H. (1983). "Identification by Cross-linking of a β-bungarotoxin Binding Polypeptide in Chick Brain," *EMBO J.* 2: 1119–1122.
2. Schmidt, R., and Betz, H. (1989). "Cross-linking of β-bungarotoxin to Chick Brain Membranes: Identification of Subunits of a Putative Voltage-gated K^+ Channel," *Biochemistry* 28: 8346–8350.

2.4.2 Photoaffinity Cross-linking

2.4.2.1 Making a Photoaffinity Ligand

The experimenter can do without a cross-linker if he has a ligand that already contains a functional group. In photoaffinity ligands (PAL), this functional group is a photoactivatable one (e.g., an arylazide; see Figure 2.29). The experimenter can sythesize PALs based on assumptions (e.g., based on the structure of known ligands) and then radioactively label them. This can often take years. You are in a better situation if you have ligands whose primary amino or carboxylic group is not involved in the binding of the ligand. The experimenter can turn this group of the ligand into a PAL with help of a heterofunctional cross-linker with an arylazide group (see Figure 2.28). There are three possibilities.

- *A:* Treat the nonradioactive ligand with an excess of cross-linker. The resulting PAL is separated from the unchanged cross-linker and/or its residues. Radioactively label the PAL.
- *B:* Label the cross-linker radioactively (radioactive cross-linkers are available in retail, e.g., ^{3}H-HSAB or ^{125}I-Denny-Jaffe reagent from NEN, or other, such as NHS-ASA or SASD, are easy to iodinate). Treat the radioactive cross-linker with an excess of nonradioactive ligand. Separate the radioactive PAL from unchanged ligand.
- *C:* Treat the radioactive ligand with an excess of cross-linker. Separate the radioactive PAL from unchanged cross-linker and/or its residues.

In cases A and B, making radioactive PAL usually requires separation (typically via an HPLC run). For C, you can possibly skip the removal of the unchanged cross-linker and inactivate its photoactivatable functional group with Tris or glycine. Even, in case B, the experimenter once in a while uses the reaction mixture directly for cross-linking; namely, when the affinity of the PAL is higher than that of the nonradioactive ligand and if the excess of nonradioactive ligand is limited.

The labor-intensive separation of nonradioactive ligand is preferable but not always necessary. A simple binding assay can help determine the necessity. If radioactivity from the reaction mixture binds specifically to the protein or membrane preparation, you are dealing with the PAL, in that the radioactive cross-linker does not bind specifically. It follows further that the concentration of the nonradioactive ligand is too low to completely inhibit the binding of the PAL. Thus, you should be able to cross-link it to the binding polypeptide. Arylazide derivatives are stable in the dark, even when the benzene ring has three different groups (hydroxyl group, iodine, and azido group).

Once in a while, the ligand is *a priori* photosensitive and a derivatization is unnecessary (e.g., ^{125}I-EGF, ^{3}H-strychnine) (Graham et al. 1983). Such hopes are nurtured by the nitro group in the ligand or the label "light sensitive" on its packaging. See the "Literature" section in Section 2.8.3.3 for more on creating PALs.

2.4.2.2 Strategy of Photoaffinity Cross-linking

The advantages of photoaffinity cross-linking compared to 3C cross-linking are:
- Higher incorporation of radioactivity into the binding polypeptide (5 to 20% of the bound PAL)

- Fast reaction, avoiding collision artifacts
- Gentle treatment of binding protein

In addition, photoaffinity cross-linking often works with ligands with a K_D in the micromolar range, and there is not the extensive intermolecular and intramolecular bonding as with 3C cross-linking. The latter only pays off in subsequent attempts to solubilize the labeled binding protein. The PAL potentially also allows one to label the binding protein in the living cell and thus to investigate metabolism (destruction, proteolysis, transport of cell vesicles into the protein, and so on). The covalent bond between PAL and binding protein, however, often behaves differently from the binding protein or the binding protein/ligand complex.

The high incorporation rate of photoaffinity cross-linking sometimes allows one to isolate the covalent bond between PAL and binding protein from the PAL with the help of antibodies (immune precipitation or immunoaffinity column). The PAL generally only reacts with an amino acid near its binding site, and no intramolecular cross-linking takes place. After the photoaffinity cross-linking, the experimenter can thus cleave the binding polypeptide into peptides. The high incorporation rate allows one to detect and isolate peptides that were derivatized by the PAL. Sequencing these peptides provides clues about the PAL binding site (Winkler and Klingenberg 1992).

Source

1. Winkler, E., and Klingenberg, M. (1992). "Photoaffinity Labeling of the Nucleotide-binding Site of the Uncoupling Protein from Hamster Brown Adipose Tissue," *Eur. J. Biochem.* 203: 295–304.

2.4.2.3 Photoaffinity Cross-linking Experiments

The noblest and most difficult task during photoaffinity cross-linking is to create the derivative the researcher hopes to use as a PAL. Because this hope is often in vain, one first needs to check whether the created derivative really is a PAL. The derivative must not only be radioactively labeled and have a photoactivatable group, it also must bind to the correct protein with at least micromolar or better nanomolar K_D and must do so specifically. If this is the case, it is cross-linked.

The PAL (concentration in K_D range) is incubated with the protein sample. After reaching binding equilibrium, the experimenter removes the unbound PAL if this can be done fairly easily. Afterward, she photolyzes as with heterofunctional cross-linkers with an arylazide group (see Section 2.4.1.2). After photolysis, the unbound PAL is washed out if this can be easily done (e.g., with membranes).

The specificity of the cross-linking can be determined by adding specific binding inhibitors. At room light, arylazide derivatives without nitro group are stable for several hours. The derivatives should nevertheless be kept in the dark and never exposed to direct sunlight. The same rules apply for the analysis of the cross-linking solution with SDS gels (and for their development) as for the analysis of 3C cross-linking (see Section 2.4.1.3).

Sources

PAL with Small MW
1. Graham, O., et al. (1983). "Photoaffinity Labeling of the Glycine Receptor of Rat Spinal Cord," *Eur. J. Biochem.* 131: 519–525.
2. Lundberg, L., et al. (1984). "Photoaffinity Labeling of Mammalian α_1-adrenergic Receptors," *J. Biol. Chem.* 259: 2579–2587.
3. May, J. (1986). "Photoaffinity Labeling of Glyceraldehyde-3-phosphate Dehydrogenase by an Aryl Azide Derivative of Glucosamine in Human Erythrocytes," *J. Biol. Chem.* 261: 2542–2547.

Peptide PAL
1. Pearson, R., and Miller, L. (1987). "Affinity Labeling of a Novel Cholecystekinin-binding Protein in Rat Pancreatic Plasma Lemma Using New Short Probes for the Receptor," *J. Biol. Chem.* 262: 869–876.
2. Vandlen, R., et al. (1985). "Identification of a Receptor for Atrial Natriuretic Factor in Rabbit Aorta Membranes by Affinity Cross-linking," *J. Biol. Chem.* 260: 10889–10892.

2.4.3 Controls for Cross-linking Experiments

To check the cross-linking of ligand and binding protein, the cross-linked protein sample is applied to an SDS gel that is then autoradiographed. If the cross-linking is successful, the ligand/binding polypeptide complex appears as a band on the film. You always see one or several bands, even when the sample does not contain binding protein at all, as long as the SDS gel is kept long enough together with the film. This is due to the fact that some ligand always binds unspecifically and is thus cross-linked unspecifically. Furthermore, many protein ligands aggregate covalently to dimers, trimers, and so on during longer storage. Even if this occurs only to a small extent, the high-molecular complexes can rarely be completely separated from the sample and appear as artifact bands in the autoradiogram.

The appearance of a band is thus not sufficient evidence of specific cross-linking. You have to prove that the band has to do with the binding protein. For this, you perform the following controls.

- Unlabeled ligand (concentration in K_D range) or other specific binding inhibitors must reduce the radioactive label of the bands. The binding inhibitors are added during the first step of cross-linking; namely, the incubation of the ligand with the protein sample.
- The band may not appear with protein samples that do not contain any binding protein or whose binding protein was inactivated.
- The band may not appear in the absence of cross-linker or, in the case of PAL, without photolysis.
- The cross-linking must not change the band pattern of the protein sample in the SDS gel (run two cross-linking solutions, with and without cross-linker, on one SDS gel and stain with Coomassie).
- If you perform the cross-linking without the protein sample, no band may appear. This control is not trivial, because the radioactive band in the gel could have been the result of intermolecular cross-linking of the ligand. This danger is especially prevalent with protein ligands.

The following experiments are not totally necessary for a cross-linking paper. However, they make a good impression on the reviewers of a journal and they sometimes lead to surprising results.

- Cross-linking with different radioactive ligands.
- Cross-linking with cross-linkers of different length.
- Run the cross-linking solution under reducing and under nonreducing conditions on the SDS gel. This experiment shows whether further polypeptides are linked with the binding polypeptide via disulfide bridges.
- For protein ligands, the careful researcher checks whether the radioactivity of the band in the SDS gel truly stems from the ligand by additionally cross-linking with cleavable cross-linkers (e.g., EGS or SASD). After cross-linking and SDS gel electrophoresis, the radioactive band is cut from the gel, the gel piece is treated with a cleaving reagent in the pocket of a second SDS gel, and electrophoresis is preformed. The radioactivity now has to appear in a band whose position corresponds to the MW of the ligand.
- Membrane proteins are usually glycosylated and their apparent MW changes with the portion of acrylamide in the SDS gel. For conclusions with more general validity, the apparent MW of the ligand/binding polypeptide complex is determined with SDS gels of differing acrylamide content.

2.5 Purposes

It is nice to find the binding site of a ligand, but it would be nicer to know the purpose of the protein on which the binding site is located and to assign to it a function in the cell metabolism. The problem of knowing that a protein exists without knowing its function occurs often. If she succeeds in identifying the binding partner of a toxin via binding assays with the labeled toxin, the experimenter may know that such a binding protein exists but its function remains unknown.

During the production of monoclonal antibodies against an unpurified sample, antibodies against proteins may appear whose only known property it is to bind a monoclonal antibody (e.g., see Wiedenmann and Franke 1985). Finally, there are ever more cDNA clones that encode for proteins whose function is unknown. I venture to predict that the number of such clones will see enormous growth with the progress of the human genome project. There is no systematic strategy for investigating function, but there are starting points, (as outlined in Figure 2.30).

First, the experimenter makes tools for himself. This is an insurance against complete failure. Tools would be:

• The purified protein
• Partial amino acid sequences of the protein
• Antibodies against the protein
• The cDNA for the protein

Once in a while, the protein purification or its cloning already provides clues regarding the function (Rehm and Betz 1984; Ushkaryov et al. 1992; Schiavo et al. 1992). There is also the possibility of subjecting the purified protein to one enzyme assay after another. With the thousands of enzyme activities, however, this is an endless gamble in which the chance of success may not only be small but zero, in that not all proteins are enzymes. In addition, negative results of such assays always leave the researcher with the nagging question whether the failure may have been due to the working conditions (e.g., buffer pH or ions) or whether the activity was lost during purification. The following strategies seem more promising.

• *A:* Determine the MW, the glycosylation, and the subunit structure of the protein. Where (organs, cells) and when (ontogenetically, phylogenetically) is the protein expressed? In which cell compartments is it located? Even when these investigations do not allow one to draw definite conclusions about the protein function they are still food for thought.

• *B:* Do databases contain related protein sequences whose function is known? Does the protein have partial sequences that indicate Ca^{2+} binding, kinase activity, and so on?

• *C:* With the labeled and purified protein, the experimenter looks for (potential) binding partners (e.g., in membranes or cell extracts). If he finds a binding partner, he follows the chain of protein interactions until he encounters a known protein. Fans of quick-and-dirty experiments stain blots with the labeled and purified protein (see Section 1.7). More thorough (but not necessarily more successful) people develop a binding assay or couple the purified protein to affinity columns and try to enrich the binding partners, although it is questionable whether the binding activity can survive purification and coupling with the column. If this works and if it is also possible to solve the problem of elution from the column, the researcher has both binding partners in hand. In practice, the interpretation of such experiments is complicated by the fact that a protein affinity column always and non-specifically adsorbs proteins from concentrated protein solutions. During the necessarily nonspecific elution methods (e.g., extreme pH, 1% SDS) these nonspecifically bound proteins elude. The experimenter then helplessly faces a mixture of denatured material.

The problem of strategy C is thus proving the specificity of the binding and the nature of the binding partner. When the K_D of the binding is in the nanomolar range and a definite binding complex is created, this at the least creates interesting information. If the experimenter does not detect binding, that does not mean anything. The binding activity of the protein could have been lost during purification or labeling or an important cofactor could be missing.

• *D:* The experimenter suppresses the expression of the protein. This is done in cells, for example, by injecting anti-sense oligonucleotides against the mRNA of the protein or by creating knock-out mice. This elegant experiment can be treacherous. Its success depends on which function the experimenter can measure. It is easy to prove that the protein in question has disappeared from the cell. However, if the experimenter does not see anything else in the cell the experiment did not teach her anything. If the cell dies, nothing is won except the knowledge that the protein is important for the survival of the cell. The experimenter thus has to look for a functional area affected by the absence of the protein (e.g.,

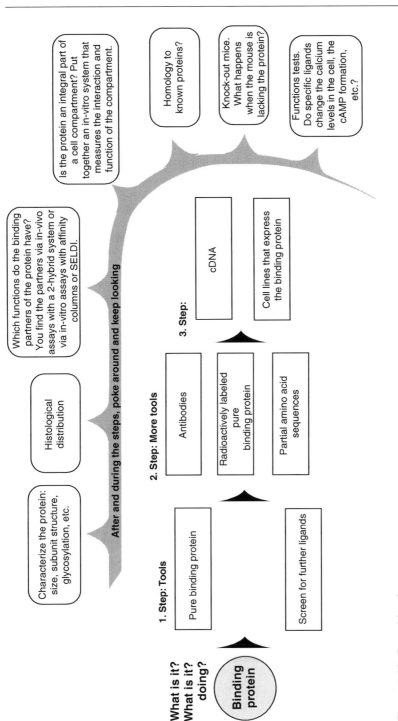

Figure 2.30. Searching for purpose.

glucolysis or neurotransmitter release). The cellular and intracellular localization is helpful here. The experimenter checks the suspicious functional area and narrows it down step by step to the trouble spot (e.g., for enzyme chains). Of course, this only works when the molecular dependencies in this functional area are largely known. Otherwise, this effort only yields the general information that the functional area does not work or works differently without the protein (e.g., see Alder et al. 1992).

- *E:* Cells without the relevant protein are permanently transfected with the gen for the protein. E is the reverse experiment of D and thus similar (opposite) conditions apply as for D. Again, you can easily show that the cells express the relevant protein and the histologist shows you where in the cell your protein is located. But what does that mean? Maybe it is only synthesized and then stored without function in a compartment, as a molecular "early retiree." Or perhaps in the new cell it assumes a different function, works in the wrong compartment, or cannot work because the right partner is missing.

The search for the function of even major proteins that are easy to isolate can take decades and burn through dozens of Ph.D. students. Examples would be the α-latrotoxin receptor or the neuronal membrane proteins synapsin and synaptophysin. The latter was discovered in 1983. Its function is as yet unknown.

Enough of this negativism. Looking for purpose may be risky and difficult, but it can lead to unexpected results (e.g., Ushkaryov et al. 1992). It takes smarts, knowledge of different methods, and a good overview of the literature—a project for ambitious Ph.D. students with nerves of steel.

Sources

1. Alder, J., et al. (1992). "Calcium-dependent Transmitter Secretion Reconstituted in *Xenopus* Oocytes: Requirement for Synaptophysin," *Science* 257: 657–661.
2. Rehm, H., and Betz, H. (1984), "Solubilization and Characterization of the β-bungarotoxin Binding Protein of Chick Brain Membranes," *J. Biol. Chem.* 259: 6865–6859.
3. Schiavo, G., et al. (1992). "Tetanus and Botulinum B Neurotoxins Block Neurotransmitter Release by Proteolytic Cleavage of Synaptobrevin," *Nature* 359: 832–835.
4. Ushkaryov, Y., et al. (1992). "Neurexins: Synaptic Cell Surface Proteins Related to the α-latrotoxin Receptor and Laminin," *Science* 257: 50–56.
5. Wiedenmann, B., and Franke, W. (1985). "Identification and Localization of Synaptophysin, an Integral Membrane Protein of Mr 38000 Characteristic of Presynaptic Vesicles," *Cell* 41: 1017–1028.

Chapter 3 Solubilization of Membrane Proteins

"Then since that may be," said Sancho, "there is nothing for it but to commend ourselves to God, and let fortune take what course it will."

If you want to purify or investigate membrane proteins, you need to get them in solution first. Integral membrane proteins (i.e., proteins with transmembrane helixes) go into solution with detergents. Peripheral membrane proteins (i.e., proteins associated with membranes) that do not have transmembrane helixes often already dissolve in buffers of high or low ion strength or high pH (Steck and Yu 1973). Proteins that anchor themselves in the membrane via glycosylphosphatidylinositol molecules (e.g., alkaline phosphatase) are solubilized through treatment with phosphatidylinositol-specific phospholipase C.

It is the experimenter's greatest wish to maintain the conformation and function of the proteins in a solubilized state. For integral membrane proteins this is often not a problem. The following pages are thus dedicated to the solubilization of integral membrane proteins and the required tools.

Source
1. Steck, T., and Yu, J. (1973). "Selective Solubilization of Proteins from Red Blood Cell Membranes by Protein Perturbants," *J. Supramol. Structure* 220–231.

3.1 Detergents

Detergents are molecular hermaphrodites. They consist of a hydrophile part and a hydrophobic one (Figure 3.1). The hydrophobic part consists of phenyl derivatives (TRITON-X-100), aliphatic chains (octylglucoside, Zwittergent 3–14, Lubrol PX), or steroid scaffolding (cholate, deoxycholate, CHAPS, BIGCHAP, deoxy-BIGCHAP, digitonin). The hydrophile detergent part consists of ionized groups (cholate, deoxycholate, CHAPS, SDS), sugar (octylglucoside, digitonin, BIGCHAP, deoxy-BIGCHAP), hydroxyl groups (cholate, deoxycholate), or polyethylene oxide (TRITON-X-100, Lubrol PX).

It is in the nature of a detergent to be torn between a love of fat and a yearning for water. It is this ambivalence that allows the detergent to bring together what does not belong together (i.e., membrane proteins and a watery solution).

3.1.1 Clean Concepts

Micelles are aggregates of detergent molecules. The hydrophobic part of the detergent molecule is located within the micelle. The hydrophile residues interact with the watery medium. The size of the micelles (i.e., the average number of detergent molecules per micelle) and their form characterize each detergent. TRITON-X-100, for example, forms large spherical micelles, whereas the steroid detergents (cholate, etc.) form liquid crystal aggregates from 2, 4, and so on monomers (Figure 3.1). Size and form of the micelles also depend on temperature, detergent concentration, salt concentration, pH, presence of phospholipids, and so on (Figures 3.2 and 3.3).

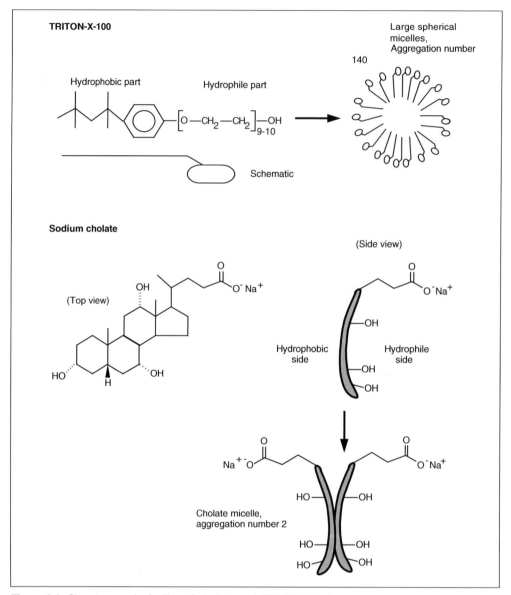

Figure 3.1. Structure and micelles of cholate and TRITON-X-100.

The critical micelle concentration (CMC) is the detergent concentration above which micelles form. Below the CMC, the detergent molecules are in monomer solution. The CMC depends on the temperature, the ion strength, the buffer pH, and the concentration of nonionic substances such as urea or alcohols.

The cloud point is the temperature at which the detergent precipitates in watery solution. A number of further concepts that are less important to daily work—such as the critical micelle temperature, the Krafft point, and the hydrophile/lipophile equilibrium—are described in the overview article by Helenius and Simons (1975), which is the best piece of writing about detergents in spite of its age.

Source
1. Helenius, A., and Simons, K. (1975). "Solubilization of Membranes by Detergents," *BBA* 415: 29–79.

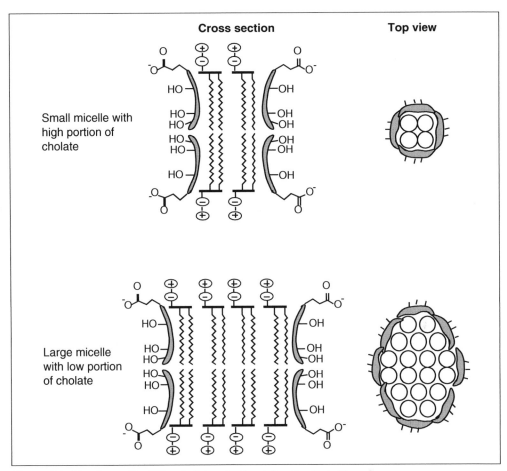

Figure 3.2. Structure of lecithin-cholate micelles (after Small et al.).

3.1.2 Handling Detergents

3.1.2.1 General

Nonionic detergents with high CMC are easy to dialyze. Detergents with low CMC, on the other hand, are almost impossible to remove via dialysis. The CMC of an ionic detergent decreases when the buffer's ion strength is increased. The CMC of nonionic detergents, on the other hand, for the most part does not depend on ion strength.

Temperature has almost no effect on the CMC of ionic detergents, but the CMC of nonionic detergents significantly decreases with increasing temperature. The CMC of a mixture of two detergents (with different CMC) is close to the lower CMC, even when the solution contains little of the lower CMC detergent. Many nonionic detergents can be autoclaved, but they precipitate in the process (cloud point) and have to be mixed again after autoclaving.

3.1.2.2 Specifics

TRITON-X-100 and its analogs (such as TRITON-X-114) are mixtures of p-t-octylphenyl-polyoxyethylenes, where the length of the polyoxyethylene chains in each case are near an average value. TRITON molecules with short polyoxyethylene chains establish especially

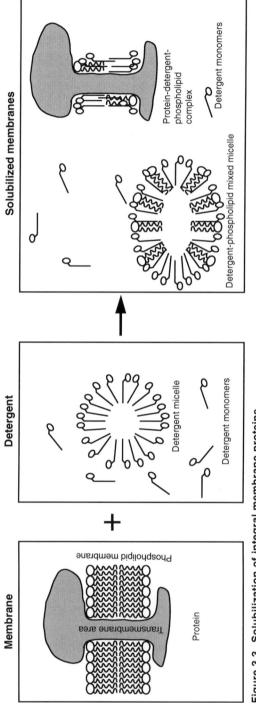

Figure 3.3. Solubilization of integral membrane proteins.

strong bonds with hydrophobic surfaces. The phenyl group in the TRITON molecule causes strong absorbtion at 280 nm.

Because of the glutinous, honey-like consistency of pure TRITON-X-100, you better prepare a 20% (w/v) stock solution. Na-EDTA pH 7.4 (0.1 to 1 mM) in the stock solution prevents fungal growth and inhibits the development of oxidants (Chang and Bock 1980).

The micelle size of TRITON increases exponentially with the temperature until at the cloud point the TRITON precipitates. The phase separation after Bordier (1981) takes advantage of this effect: membranes are dissolved in TRITON-X-114 at 4° C and the solution is then heated to 30° C. This causes the TRITON-X-114 to fall out of the solution and centrifuging yields a TRITON-X-114-rich phase and a TRITON-X-114-poor phase. According to Bordier, the TRITON-X-114-rich phase contains the integral membrane proteins, whereas the soluble proteins swim in the detergent-poor phase. In my experience, this separation is far from being as clean as Bordier claims.

If you can, you should avoid digitonin. Digitonin preparations available via retail are not only expensive but contaminated with toxic substances (up to 50% and above), whose exact chemical nature is unknown. Furthermore, digitonin preparations usually dissolve only with heat (max. concentration 4% w/v). Over the days, material falls out of the solutions at 4° C, which plugs up the columns and is a source of endless trouble. The digitonin preparations of different companies, and sometimes even of the same company, are of varying quality and often unsuitable for solubilization. I had the best experience with digitonin products from the companies Wako, Fluka, and Sigma. Kun et al. (1979) describe the purification of digitonin.

Of the detergents listed in Table 3.1, doxycholate solubilizes best (after SDS). About 70 to 80% of the protein in a membrane preparation goes into solution with deoxycholate. Still, deoxycholate is not the ideal detergent, in that its solubilization characteristics, CMC, and micelle sisze are strongly dependent on pH and ion strength. With acidic pH, deoxycholate forms a gel, and it precipitates with divalent cations. Furthermore, it cannot be used for purification that takes advantage of charge differences, and it changes the conformation of many proteins.

Sources
1. Bordier, C. (1981). "Phase Separation of Integral Membrane Proteins in TRITON-X-114 Solution," *J. Biol. Chem.* 256: 1604–1607.
2. Chang, H., and Bock, E. (1980). "Pitfalls in the Use of Commercial Nonionic Detergents for the Solubilization of Integral Membrane Proteins: Sulfhydryl Oxidizing Contaminants and Their Elimination," *Anal. Biochem.* 104: 112–117.
3. Kun, E., et al. (1979). "Stabilization of Mitochondrial Functions with Digitonin," *Methods Enzymol.* 55: 115–118.

3.2 Solubilization

During solubilization, the hydrophobic residues of detergent molecules dock to the hydrophobic sites (e.g., transmembrane areas) of the protein and partially push out the phospholipids (see Figure 3.3) (Roth et al. 1989). If sufficient detergent molecules attach, the membrane protein goes into solution. Solubilized membrane proteins are thus complexes from detergent, phospholipid, and protein, where the proportion of the individual components depends on the composition of the utilized buffer. The proportion of detergent and phospholipid molecules in the complex is typically about 10 to 50%. During solubilization of integral membrane proteins, two problems occur.

• The protein does not go into solution but aggregates.
• The protein goes into solution and denatures.

Each membrane protein requires different conditions to go into solution in native conformation. The detergent and its concentration are always critical, but the buffer, the pH, the ions, the ion strength, and the presence or absence of certain ligands can also play a role. The same is true for the type and amount of phospholipids and for the ratio of protein to detergent. Each of these variables can be crucial, but none has to be.

Table 3.1. Detergents.

Name	MW Monomer	MW Micelle	CMC % (w/v)/(mM)σ*	Dialyzability	Special Characteristics
BIGCHAP	862	6,900, 13,800	0.12/(1.4)	0.60+	
CHAPS	615	6,150	0.25/(4)	0.81+	
Sodium cholate	431	1,700	0.36/(8)	0.771+	Forms unsoluble complexes with divalent cations. Precipitates when pH < 6.5.
Deoxy-BIGCHAP	862	7,000	0.12/(1.4)	0.63+	+
Na-deoxycholate	415	700–9,000	0.11/(1–2.7)	0.778	Forms giant micelles when pH < 7.8. Precipitates when pH < 6.9. CMC decreases with increasing salt concentration. Precipitates with Ca^{2+}, Mg^{2+}.
Taurodeoxycholate	522	4,200–30,000	0.05–0.2/(1–4)	0.76	CMC decreases with increasing salt concentration, whereas micelle size increases.
Digitonin	1,229	70,000	0.031/(0.25)	–	Toxic, dissolves only with heat.
Octylglucoside	292	8,000	0.73 (25)	0.858+	
Lubrol PX	582	64,000	0.006/(0.1)	0.958–	
SDS	288	18,000	0.24/(8.2)	0.870+	Insoluble with K^+ and divalent cations.
Dodecylmaltoside	511	50,000	0.008/(0.16)	–	
MEGA-8	321		1.9/(58)	+	
MEGA-10	350		0.22/(6.2)	+	
HECAMEG	335		0.65/(19.5)	+	In high concentrations interferes with BCA assay and Lowry. Incompatible with Bradford. Adsorbs at 280 nm. Cloud point at 64° C.
TRITON-X-100	650	90,000	0.02/(0.24)	0.908	
TRITON-X-114	536		0.009/(0.20)		Adsorbs at 280 nm. Cloud point at 22° C.
Zwittergent 3-14	364	30,000	0.011/(0.3)		
Nonidet P 40	606		0.015/(0.25)		Cloud point at 65° C
Tween 20	1,228		0.0074/(0.06)		Cloud point at 76° C

* Partial specific volume.

Detergent: Because detergents dock to the protein with their hydrophobic residues, the detergent residue influences the conformation of the solubilized protein. According to Table 3.1, there are three basic forms of hydrophobic residues: steroid scaffolding, aliphatic chains, and phenyl derivatives. During the screening of different detergents, each basic form should be present. Detergents with nonionic hydrophile parts (e.g., octylglucoside) are gentle with the membrane protein and typically leave it in its native conformation. On the other hand, ionic detergents prevent aggregation due to ionostatic repulsion. However, you cannot count on this, not even with deoxycholate.

For solubilization, the detergent concentration has to be higher than the CMC, because the membrane lipids must be able to incorporate into micelles. The upper threshold for the detergent concentration is at 2 to 3% (w/v). More is of questionable utility, because of the inactivation of many proteins and the viscosity of the solution.

Solubilization buffer: HEPES, MOPS, and Tris buffers are a good choice, if compatible with the assay. The concentration should be about 20 mM to ensure sufficient buffer capacity. A high ion strength (add 0.1 to 0.4 M NaCl or KCl) in the solubilization buffer is rarely bad. Especially with the ionized detergents deoxycholate and cholate, high concentrations of NaCl (up to 1 M) facilitate the solubilization. Adding phospholipids is not necessary for solubilization (the membranes supply enough endogenous phospholipids).

Phosphate buffer (0.1 to 0.2 M), the chaotropic rhodanide salts (0.2–0.4 M), and urea (2 to 6 M) reinforce the solubilization power of detergents, where urea forms complexes with nonionic detergents. CHAPS combined with guanidine-chloride is said to be especially effective for the solubilization of aggregated proteins, but nonionic detergents precipitate with high concentrations of guanidine-chloride. The denaturing additives rhodanide, urea, and guanidine-chloride are used by experimenters only in emergencies if they value the native conformation of their proteins.

Popular stabilizers in the solubilization buffer are 10 to 20% (w/v) glycerine, 1 mM DTT, and 0.1 to 1 mM EDTA. The protease inhibitors PMSF, bacitracine, trypsin inhibitor, leupeptin, benzamide, and benzamidine are also a blessing, in that binding activities in solution are more sensitive against protease than in the membrane (see Table 5.1). PMSF irreversibly inhibits serin proteases via covalent derivatization of the active center. A one-time application is thus sufficient, especially because PMSF falls apart in watery solution (half-life of a few hours).

Detergent/protein ratio: During solubilization, not only the concentration of the detergent is important but the mass ratio of detergent/protein (DPR). Figure 3.4 shows how dependent the solubilizaton of a membrane protein is on the DPR. Systematic investigations are amiss, but there seems to be an optimal DPR for each detergent/membrane protein pair. For CHAPS, the optimal DPR is typically between 0.5 and 1, for cholate between 0.5 and 2.5, for TRITON-X-100 between 2 and 4, and for digitonin between 3 and 6. Thus, the solubilization rule of thumb: detergent concentration 0.5 to 2% (w/v); protein concentration 2 to 5 mg/ml (DPR 1 to 10).

Storage: At 4° C and in buffers with stabilizers, most solubilized membrane proteins keep for a few days. If you want to keep detergent extract for a longer period, you should freeze it in liquid N_2 and store it at −80° C. Solubilized membrane proteins survive repeated freezing and thawing, provided the freezing is done in liquid N_2. Freezing and storing at −20° C is in my experience not any better than storage at 4° C.

How do you solubilize? You incubate the finely suspended membranes (protein concentration 2 to 5 mg/ml) at 4° C in a suitable buffer with 0.5 to 2% (w/v) of detergent. Some stir with a Teflon spatula; others vortex every 10 minutes; and others use an orbital shaker. However, the pro makes waves only on congresses and sonication can also denature the proteins. After about 1 h, the extract is centrifuged (1 h, 100,000 g or more) and the pellet, after repeated washing, is resuspended in the same volume of solubilization buffer (without detergent). Examine supernatant and resuspended pellet for the desired activity!

First, conduct 2 to 3 pilot experiment with TRITON-X-l00, cholate, octylglucoside, and the buffer in which you usually measure their activity. If the pilot experiments fail, you should consider whether you should might better continue with another project—especially when (for example) the binding activity disappears during the extraction and is neither to be found in the pellet nor in the supernatant. Did it solubilize, but the binding assay does not work? Did it not solubilize and the protein sits in the pellet aggregated and denatured? Does the assay work and the protein was solubilized in nonbinding state? You can struggle with these questions for years. Because there is no rational recipe for success for solubilization, you have to screen: detergents, ions, stabilizers, and so on (see Figure 3.5).

If the protein is solubilized in native state, you lower the detergent concentration for subsequent investigations (target value 0.05 to 0.2% (w/v) for detergent, 0.01 to 0.04% (w/v) for phospholipid), for economic reasons, and because so many proteins dislike a high detergent

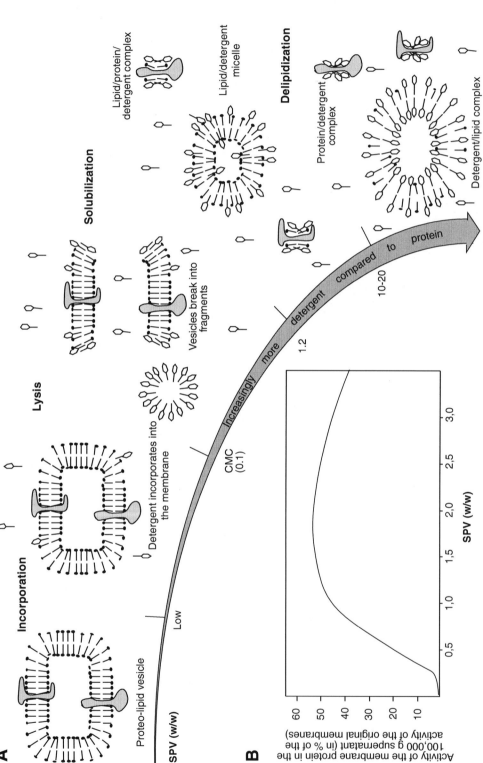

Figure 3.4. The influence of the detergent/protein ratio (DPR) on the solubilization of membrane proteins. (A) SPV and membrane solubilization schematic (after Hjelmeland 1990). (B) Dependence of the solubilized activity on the DPR. Aliquots of a membrane suspension with a certain protein concentration were solubilized with increasing detergent concentrations. Then the experimenter determined the activity of the solubilized activity on the DPR. Aliquots of a membrane protein (e.g., ligand binding, enzyme activity) in 100,000 g supernatants and in the original membrane suspension (100% value). Generally, the percentage of solubilized activity depends on the DPR in an S-shaped pattern. However, the utilized detergent, buffer, and so on influence the position of the curve, the height of its peak, and the magnitude of the inactivation of the activity (at high detergent concentrations, the right arm of the curve). In addition, the curve often moves to the left if you raise the concentration of total protein, so that (for example) under otherwise identical conditions a DPR of 0.5 solubilizes 10% activity with a membrane protein concentration of 3 mg/ml but 40% with a concentration of 6 mg/ml.

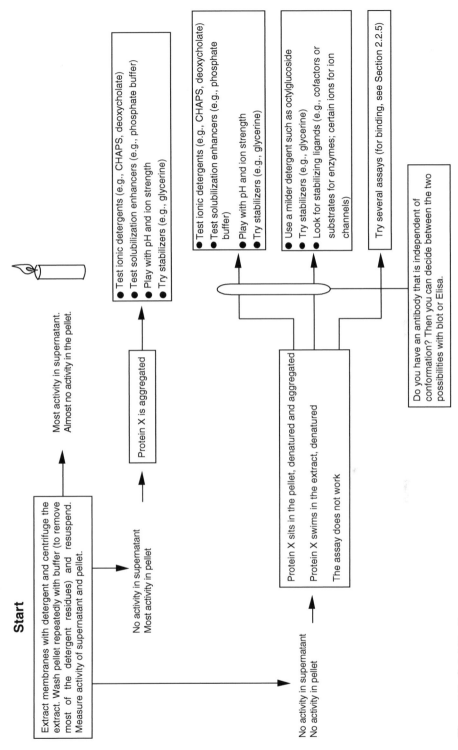

Figure 3.5. Solubilization.

concentration in the long run. By the way, many solubilizers believe that membrane proteins remain in solution only above the CMC of a detergent. In my experience, this is not true for all detergents and membrane proteins. However, all buffers with which the solubilized protein comes into contact should contain detergent and phospholipid. Of course, you check whether your protein falls out under the new conditions (let it stand for a few hours, centrifuge, test supernatant and pellet, and compare). You notice: solubilization of membrane proteins requires little intelligence, but diligence and perseverance.

> All the while he rode so slowly and the sun mounted so rapidly and with such fervour that it was enough to melt his brains if he had any.

Sources

Overview Articles
1. Hjelmeland, L. (1990). "Solubilization of Native Membrane Proteins," *Methods Enzymol.* 182: 253–264.
2. Roth, M., et al. (1989). "Detergent Structure in Crystals of a Bacterial Photosynthetic Reaction Center," *Nature* 340: 659–662.

Neurotransmitter Receptors
1. Bristow, D., and Martin, I. (1987). "Solubilization of the γ-aminobutyric Acid/Benzodiazepine Recentor from Rat Cerebellum: Optimal Preservation of the Modulatory Responses by Natural Brain Lipids," *J. Neurochem.* 49: 1386–1393.
2. Hooper, R. (1986). "Optimization of Conditions for Solubilization of the Bovine Dopamine D2 Receptor," *J. Neurochem.* 47: 1080–1085.

Ion Channels
1. Rehm, H., and Betz, H. (1984). "Solubilization and Characterization of the β-bungarotoxin Binding Protein of Chick Brain Membranes," *J. Biol. Chem.* 259: 6865–6859.
2. Seagar, M., et al. (1987). "Solubilization of the Apamin Receptor Associated with a Calcium-activated Potassium Channel from Rat Brain," *J. Neuroscience* 7: 565–570.

Transporters
1. Yamada, K., et al. (1988). "Solubilization and Characterization of a ^{3}H-hemicholinium-3 Binding Site in Rat Brain," *J. Neurochem.* 50: 1759–1764.

Hormone Receptors
1. Dufau, M., et al. (1973). "Characteristics of a Soluble Gonadotropin Receptor from the Rat Testis," *J. Biol. Chem.* 248: 6973–6982.
2. Johnson, W., et al. (1984). "Solubilization and Characterization of Thyrotropin-releasing Hormone Receptors from Rat Brain," *Proc. Natl. Acad. Sci. USA* 81: 4227–4231.

Other Receptors
1. Bogonez, E., et al. (1985). "Solubilization of a Vectorial Transmembrane Receptor in Functional Form: Aspartate Receptor of Chemotaxis," *Proc. Natl. Acad. Sci. USA* 82: 4891–4895.
2. Perdue, J., et al. (1983). "The Biochemical Characterization of Detergent-solubilized Insulin-like Growth Factor II Receptor from Rat Placenta, *J. Biol. Chem.* 258: 7800–7811.

3.2.1 Solubilization Criteria

If a protein is still in the supernatant after one hour of centrifugation with 100,000 g, it is generally assumed to be solubilized. This criterion fails if the solubilization buffer contains glycerine or other components that raise the density of the solution and decrease the sedimentation speed. For large centrifugation volumes, and thus long sedimentation distances, 100,000 g × h are also not sufficient for sedimenting all unsoluble particles. A gel filtration of the detergent extract to Sepharose CL 6B provides better evidence: if the protein is solubilized, it must appear in the inclusion volume of the column.

Density gradient centrifugations (e.g., in 5 to 20% sucrose gradients) are another solubilization criterion. Nonsolubilized protein forms pellets at the bottom of the centrifuge tube. A solubilized protein, on the other hand, has a peak in the gradient, whose position depends on its sedimentation constant. Density gradients barely dilute the sample and allow one to examine several samples at the same time under different conditions. Via accompanying marker proteins, the density gradient centrifugation measures the apparent sedimentation coefficient.

Part of a flawless proof of solubilization is gel filtration and the density gradient. The proof is by no means trivial. At the end of the 1980s, a half-dozen papers were published by different laboratories that all claimed they had solubilized the neurotransmitter receptor for NMDA in native form with deoxycholate, TRITON, or cholate. The centrifugation criterion was given as the only proof. However, the NMDA receptor does not even go in solution with deoxycholate.

3.2.2 Physical Parameters of Solubilized Membrane Proteins

From solubilized membrane proteins you can—without major equipment—determine MW, Stokes' radius, sedimentation coefficient, proportion of bound detergent and phospholipid, isoelectric point, and the apparent frictional coefficient. Knowledge of Stokes' radius, MW, and isoelectric point comes in handy during planning of a purification or drafting of a detection assay (the other measures are good for the library).

A gel filtration (e.g., to Sepharose 6B or CL 6B) yields the Stokes' radius of the protein-detergent-phospholipid complex, provided you also run marker proteins of known Stokes' radius (e.g., β-galactosidase 6.9 nm; apoferritin 6.1 nm; catalase 5.2 nm; BSA 3.55 nm). The method does *not* determine the protein MW (because of detergent and phospholipid in the complex).

Sedimentation coefficient, MW, apparent frictional coefficient, and the proportion of bound detergent and phosholipid can be determined via two density gradient centrifugations in H_2O and D_2O. The prerequisite is that the partial specific volumes of utilized detergent and protein (assumed to be 0.73 cm^3/g) are different and the Stokes' radius of the protein-phospholipid-detergent complex is known (Clarke 1975). Lubrol PX, TRITON-X-100, and octylglucoside are suited for the H_2O/D_2O centrifugation (deoxycholate is not). The relevant calculations are complicated, and some parameters (e.g., the partial specific volume of the protein proportion) have to be based on assumption. Nevertheless, the calculated MW often agrees well with reality (i.e., with the results of other determination methods).

With radiation inactivation, MW of membrane proteins can also be determined (Harmon et al. 1985). The experimenter irradiates freeze-dried or frozen sample with different dosages of high-energetic γ- or β-radiation (> 1 MeV). In an "everything or nothing" process, the radiation destroys the molecular structure of the protein and with it its biological activity. The experimenter determines the remaining enzymatic activity (V_{max}) or the number of the binding sites (B_{max}) for every dose. From this she calculates the likelihood of a hit and the MW of the "functional unit." The method requires a high-energetic radiation source. In addition, the sample must allow freeze-drying and rehydration or freeze/thawing without a large loss of activity.

Chromatofocusing is suited for determining the isoelectric point of membrane proteins. It also provides substantial enrichment. For chromatofocusing, the protein must also be stable with low ion strength (salt interferes with focusing), and you may not use any ionic detergents such as deoxycholate, cholate, or SDS.

The isoelectric point of integral membrane proteins cannot be determined via isoelectric focusing in polyacrylamide gels. Integral membrane proteins do not focus in IEF gel, but smudge across several pH units (see Section 1.3.3). The reason for this is presumably aggregation/adsorption problems that cannot even be prevented by adding nonionic detergents such as TRITON-X-l00. For the undeterred: Dockham et al. (1986) claim that membrane proteins can be focused if you dissolve them in lysine, 9 M urea, and 2% TRITON-X-100.

Sources

MW Determination for Membrane Protein
1. Clarke, S. (1975). "The Size and Detergent Binding of Membrane Proteins," *J. Biol. Chem.* 250: 5459–5469.
2. Haga, T., et al. (1977). "Hydrodynamic Properties of the β-adrenergic Receptor and Adenylate Cyclase from Wild Type and Variant S49 Lymphoma Cells," *J. Biol. Chem.* 252: 5776–5782.

3. Harmon, J., et al. (1985). "Molecular Weight Determinations from Radiation Inactivation," *Methods Enzymol.* 117: 65–94.
4. Patthi, S., et al. (1987). "Hydrodynamic Characterization of Vasoactive Intestinal Peptide Receptor Extracted from Rat Lung Membranes in TRITON-X-l00 and n-octyl-β-D-glucopyranoside," *J. Biol. Chem.* 262: 15740–15745.
5. Tanford, C., et al. (1974). "Molecular Characterization of Proteins in Detergent Solution," *Biochemistry* 13: 2369–2376.

Isoelectric Focusing and Chromatofocusing
1. Dockham, P., et al. (1986). "An Isoelectric Focusing Procedure for Erythrocyte Membrane Proteins and Its Use for Two-dimensional Electrophoresis," *Anal. Biochem.* 153: 102–115.
2. Siemens, I., et al. (1991). "Solubilization and Partial Characterization of Angiotensin II Receptors from Rat Brain," *J. Neurochem.* 57: 690–700.
3. Thomas, L., et al. (1988). "Identification of Synaptophysin as a Hexameric Channel Protein of the Synaptic Vesicle Membrane," *Science* 242: 1050–1053.

Solubilization has a certain poetic flair. The experimenter floats like an eagle in windy heights well above the steady ground and searches for the little mouse that does not want to show itself. Lonely and relentlessly, he circles between making membranes, pipetting, stirring, centrifuging, suspending, and the activity assay. Some call this mind-numbing, but don't we all yearn for the soothing ritual and the impression of doing something for its own sake? Thankfully, only in exceptional cases does it take longer than a year to have success with solubilization. Afterward, the road is clear for more interesting tasks such as purification, subunit structure, and so on.

When Don Quixote saw himself in open country, he felt at his ease, and in fresh spirits to take up the pursuit of chivalry once more; and turning to Sancho he said, "Freedom, Sancho, is one of the most precious gifts that heaven has bestowed upon men."

Chapter 4 Protein Detection via Functional Measurements

4.1 Translocators

The principles and details of enzyme assays are discussed in Bergmeyer (1983) and Suelter (1990). The object of this chapter is thus not proteins that change molecules (enzymes) but proteins that control the spatial distribution of molecules or ions: transporters and ion channels. Transporters and ion channels are also called translocators.

An open (active) ion channel is a water-filled protein pore that allows certain ions to pass through. The ion stream is driven by the ions' concentration gradient and the electric potential above the membrane. However, ion channels are no molecular water fountains, through which the ions flow in a steady stream. A channel molecule opens and closes whenever it feels like it (i.e., stochastically), and the likelihood of it opening depends on certain parameters such as membrane potential or ligand concentration. Channels thus rather resemble a damaged toilet tank that delivers water in bursts as long as the lever is depressed. One distinguishes voltage-dependent ion channels (e.g., voltage-dependent Na^+ channel) and ligand-controlled ion channels (e.g., acetylcholine receptor). There are ion channels that let only one specific ion through (e.g., Na^+) and channels that are permeable for several ions (e.g., Na^+ and Ca^{2+}). Some ion channels are oligomer proteins with four subunits (K^+ channels) or five subunits (ligand-controlled channels), others consist of a large protein (Na^+ channels, Ca^{2+} channels) with four domains that are homologous. Ion channel proteins are integral membrane proteins with several transmembrane helixes.

Transporters mediate the admission or delivery of lipid-unsoluble materials (sugars, amino acids, ions) via the phospholipid membranes of the cell. Transporters do not have a porous structure, but transport their freight in three steps: binding of the substrate to one side of the membrane, conformation change or position change of the transporter-substrate complex, and dissociation of the substrate from the transporter on the other side of the membrane. Often, not only one molecule is being transported but several at the same time, either in the same direction (symport; e.g., H^+ and lactose) or in opposite direction (antiport; e.g., Na^+ versus Ca^{2+}). Primary active transport processes (pumps) are powered by ATPases (e.g., Na^+/K^+ATPase), while secondary active transport processes are coupled to H^+ or Na^+ gradients or to the membrane potential.

The currently known transporter proteins have up to a dozen transmembrane helixes, some of which are amphipatic and the rest hydrophobic. The amphipatic helixes of the transporter presumably form an inner circle that is held together by an outer circle of hydrophobic helixes in the membrane.

Transporters have binding sites for their substrate. Transport processes, just like the kinetic quantities of transport speed and exchange rate, are characterized by the affinity of the substrate to the transporter protein and the number of its binding sites. The transport speed (e.g., in μM substrate/min) reaches a maximum value with increasing substrate concentration, the maximum transport speed (Vmax). The exchange rate (in sec^{-1}) measures the number of substrate molecules transported by a transporter molecule per second.

The transport speed of the transporter is equivalent to the ion flux of the ion channel. However, the ion flux of channels does not exhibit saturation with an increase of the ion concentration and it is larger by orders of magnitude than the transport speed of transporters. The ion flux is directly proportional to the ion stream. The basis of the function of translocators is membrane-enclosed compartments: cells, organelles, vesicles, or proteoliposomes (prolis). Prolis and liposomes play a role in purification and reconstitution of translocators. The following sections are dedicated to them.

Table 4.1. Characteristics of liposomes.

Vesicle Type	Creation Method	Size Range (Ø in nm)	Capacity (l/mol lipid)	Advantages	Disadvantages
MLV	Dispersion of the lipid in water	—	1–4	Easy and quick to make	Low capacity and heterogeneous; badly characterized vesicle mixture
Oligolamellar vesicles and LUV	Homogenization of MLV with liposome extruder and polycarbonate membranes of defined pore size	—	—	—	Heterogeneous vesicle mixture; requires specialized equipment
SUV	Ultrasonication of MLV	206,020–60	0.2–1.5	Uniform size; easy to make	At low capacity the liposomes do not take up any molecules whose MW is higher than 40 kd; the SUV are unstable and form aggregates; sonication degrades lipids
SUV/LUV	Dialysis of phospholipid/detergent micelles; detergents: cholate, deoxycholate, octylglucoside	20–150	0.2–7	Uniform size; good reproducibility; vesicle size can be changed by different parameters (detergents, dialysis, speed), residues of cholate, and deoxycholate stabilize the vesicle	Detergent residues in the vesicles?
SUV	Adsorption of detergents from phospholipid-detergent micelles to hydrophobic matrices	—	—	—	—
SUV	Injection of phospholipid mixtures in ethanol into watery solutions	30–60	0.4–1.5	Easy to make	Heterogeneous vesicle mixture; the liposome suspension contains ethanol

Sources
1. Bergmeyer, U. (1983). *Methods of Enzymatic Analysis.* Weinheim: Verlag Chemie.
2. Suelter, C. H. (1990). *Experimentelle Enzymologie.* Stuttgart: Gustav Fischer Verlag.

4.1.1 Liposomes

Liposomes are vesicles consisting of phospholipid membranes. Among liposomes, one distinguishes multilamellar vesicles (MLVs), small unilamellar vesicles (SUVs), and large unilamellar vesicles (LUVs) (Table 4.1). MLVs consist of many liposomes that are folded around each other like the layers of an onion. They form during contact of phospholipids with watery solutions.

SUVs form when MLV suspensions are sonicated, during dialysis, during gel filtration of detergent-phospholipid mixed micelles, during injection of an ethanolic solution of phospholipids into a watery solution, and while pushing MLV suspension through filters with specific pore sizes.

The diameter of an SUV is about 20 to 60 nm. An SUV of diameter 20 nm contains 33,300 water molecules and on average about 0.6 molecules from a substance of concentration 1 mM. An SUV of diameter 60 nm, in contrast, contains approximately 2.5 million water molecules and on average 45 molecules of a substance with concentration 1 mM. SUVs are unstable and over time they merge into aggregates, MLV and LUV. In contrast to their bigger companions, SUVs are insensitive against fluctuations in osmolarity. LUVs have a diameter of 100 nm to 10 μm. Smaller LUVs of diameter 100 to 240 nm originate during the dialysis of detergent-phospholipid solutions and the size of the vesicles depends on the molar detergent/lipid ratio and on the type of detergent.

The properties of liposomes (e.g., their permeability) depend on the phospholipids of which they are made. The basis is largely phosphatidylcholine. The often-used phospholipids from egg yolk or soy beans (asolectin) are mixtures of different phospholipids with unknown substances. The products from Avanti polar satisfy higher demands regarding purity. Cholesterol (e.g., phosphatidylcholine:cholesterol 2:1) increases the stability and decreases the permeability of the liposomes, and negatively charged phospholipids such as phosphatidylserin prevent aggregation because of the electrostatic repulsion.

The properties of the liposomes also depend on the temperature, because the phospholipid membrane changes its structure at the phase change point. Below this temperature the phospholipids are in ordered state with their acyl chains in all-trans conformation. Above the phase change point the state is more liquid, and the liposomes become permeable. The phase change point of the liposomes depends on type and length of the acyl chains of their phospholipids. Phosphatidylcholines with long saturated acyl chains form especially stable liposomes.

Store liposomes in neutral pH and under nitrogen, because extreme pH hydrolyze phospholipids and the resulting lysolecithins make the liposome membrane permeable. Finally, longer contact with oxygen leads to lipid oxidation.

Sources

1. Saito, Y., et al. (1988). "Giant Liposomes Prepared by Freezing-thawing Without Use of Detergent: Reconstitution of Biomembranes Usually Inaccessible to Patch-clamp Pipette Microelectrode," *BBRC* 154: 85–90.
2. Woodle, M., and Papahadjopoulos, D. (1989). "Liposome Preparation and Size Characterization," *Methods Enzymol.* 171: 193–217.

4.1.2 Proteoliposomes

Liposomes with built-in membrane proteins are called proteoliposomes (prolis). Working with prolis has the advantage that the internal and external media, the phospholipid structure of the membrane, and the type and amount of the proteins stored in the membrane are controlled by the experimenter. She can, for example, build up arbitrary ion gradients above the proli membrane. Accordingly, prolis are used for characterizing the properties of a purified translocator in a defined environment, but also for the detection of a translocator (e.g., in a purification protocol). The measured quantity is usually the flux of the translocator substrate into or out of the prolis (see Section 4.3). If the proli properties change due to the substrate flux, this can occasionally be used for the purification of the translocator (Barzilai et al. 1984).

The properties of prolis resemble those of liposomes. They also occur as MLV, SUV, and LUV, and they have the same permeability characteristics as liposomes. However, whereas SUV and LUV liposomes pass through sterile filters the corresponding prolis are adsorbed. MLV prolis are not suited for flux assays because the phospholipid membranes of the external and the next inner vesicle are so close to each other that the intervolume is too small. In addition, the proteins of the inner onion skins are not reachable by many assays. SUV prolis can be transformed into smaller LUV prolis by freezing/thawing and subsequently passing them through filters of defined pore size (e.g., polycarbonate filter from Nuclepore) and liposome extruders. Because of their large inner volume, LUVs are well suited for flux measurements. The application of prolis for flux assays is not free from pitfalls.

- The small inner volume of SUVs limits the intake capacity of the prolis and thus the signal size of the assay. In addition, the importance of macroscopic quantities such as concentration or pH is dubious inside such small vesicles.
- The surface of SUVs and LUVs is large in comparison to their inner volume. The proli surface is also charged, depending on the protein and phospholipid composition, and it can adsorb charged materials such as Ca^{2+}, proteins, and peptides. Hence, it is sometimes difficult to make a distinction between the intake of material into the inner proli volume and its binding to the proli surface.
- Prolis have or develop leaks (i.e., they become unspecifically permeable for ions or sugar) because (for example) SUVs are not static entities but undergo fusion, change position, and so on. Larger prolis are osmometers that swell up and finally burst when the osmolarity of the buffer drops. Leaking prolis are suited for assays that do not measure flux and require neither membrane potential nor an ion gradient across the membrane—and, of course, for assays for which both the inner and outer surface of the membrane protein must be accessible to hydrophile substrates. Stability and density of prolis depend on their protein and phospholipid composition, the production method, the temperature, the ion composition (above all, Ca^{2+} and Mg^{2+}), and the pH of the surrounding buffer. Accordingly, wildly different values are given for the stability of concentration gradients across proli membranes (from minutes to days). If other detergents than deoxycholate or cholate were used in the production of the prolis, these must be removed completely. Even the minutest residues of (for example) TRITON-X-100 make prolis permeable.

4.2 Reconstitution

To reconstitute a membrane protein means to incorporate the protein into a phospholipid membrane while retaining its function. With most methods, this creates prolis. It is the goal of the experimenter to reconstitute all functions and properties of the translocator. With larger proteins, with their huge number of ligand binding sites and regulation possibilities, this is an ambitious task. The most important function to reconstitute is the translocation of the substrate. Other properties of the translocator (such as substrate or ligand binding, enzyme activities, and so on) serve as guidance in the search for optimum reconstitution conditions because they are easier to detect. Prolis are created by:

- Incorporating membrane proteins into premade liposomes
- Removing the detergent from a detergent-phospholipid-protein solution

4.2.1 Reconstitution from a Solution

If the experimenter reduces the detergent concentration of a detergent-phospholipid-membrane protein solution, prolis are created. A portion of the membrane protein is incorporated into the phospholipid membrane. Another portion aggregates. The incorporated membrane proteins can be incorporated functionally as active or inactive and in two different directions (outside out or inside out). Three variables determine the relative proportions:

- The detergent that keeps protein and phospholipids in solution, and the method of its removal
- Type and amount of phospholipids and the ratio of protein to phospholipid
- Composition, ions, ion strength, and pH of the buffer

The easiest method for lowering the detergent concentration is to dilute the solution of phospholipids, detergent, and membrane protein in a suitable buffer. The created prolis can immediatly be used in the flux assay or first concentrated or washed (e.g., by centrifuging). This method works well with cholate or deoxycholate solutions. The created SUVs are stable and tight because the residues of the bile salt stabilize the vesicle membrane similar to cholesterol, and with their negative charge they keep the vesicles from clumping. If the membrane proteins are in other detergents, the experimenter adds cholate before diluting (Ambesi et al. 1991).

Prolis are also created during dialysis of the detergent-phospholipid-membrane protein solution, during their passage over a gel filtration column (e.g., Sephadex G-50) and when an ultrafiltrate of the solution is diluted with detergent-free buffer. For all three methods, the detergent must be dialyzable or have a high CMC (octylglucoside, cholate, and so on; see Table 3.1). Gel filtration has the disadvantage that the column adsorbs part of the prolis, and during ultrafiltration the filter membrane can become plugged. Depending on the detergent, SUVs (cholate, deoxycholate) or smaller LUVs (octylglucoside) are created. To a certain extent the size of the prolis can be controlled via the ratio of detergent to phospholipids.

Detergents with low CMC (such as digitonin, TRITON-X-100, or Lubrol PX) cannot be removed by dialysis (see Section 3.1.2). The experimenter either switches them with a dialyzable detergent (e.g., via ion exchange or lectin chromatography) or adds a dialyzable detergent. Nondialyzable detergents form mixed micelles with dialyzable detergents, and this detergent mixture is dialyzable provided the concentration of nondialyzable detergent is substantially lower than that of dialyzable detergent (e.g., 0.05% TRITON-X-100 with 1% octylglucoside; Scheuring et al. 1986; Rehm et al. 1989). The addition of a small quantity of ^{3}H-TRITON-X-100 allows one to determine whether and how much TRITON-X-100 was removed from the solution.

Hydrophobic matrices (e.g., SM-2 Biobeads from Bio-Rad, Extracti-gel from Pierce) adsorb detergents from watery solutions (either in the batch or during the passage through a column). This technique also removes detergents with low CMC (such as TRITON-X-100 or Lubrol). The adsorption matrices are moderately efficient. Switching the nondialyzable detergent with a dialyzable detergent followed by dialysis is often more advantageous. The adsorption method partially leads to the formation of MLV and aggregates that are not suitable for translocator assays and must be post-treated. Nevertheless, in the presence of 1 mM EDTA and in absence of divalent cations SUVs can also develop.

Regarding the choice of phospholipids: most receptors, ion channels, and transporters work in membranes made from very different phospholipids, albeit with different degrees of cheerfulness. Phospholipids from yolk, brain, or soy beans with addition of cholesterol make good prolis. The experimenter who would like to fuse the vesicles with other membranes afterward (e.g., for electrophysiological investigations), needs special phospholipids (Rehm et al. 1989).

The buffer in which the prolis were produced is the buffer that is later within the prolis. The experimenter can easily and quickly change the ion composition of the outer medium (e.g., while she pours the prolis over a gel filtration column equilibrated with another buffer). This yields pH gradients and ion gradients that due to the different permeability of the ions lead to the formation of electric potentials across the proli membrane.

Sources
1. Levitzki, A. (1985). "Reconstitution of Membrane Receptor Systems," *BBA* 822: 127–153.

Reconstitution by Diluting
1. Ambesi, A., et al. (1991). "Sequential Use of Detergents for Solubilization and Reconstitution of a Membrane Ion Transporter," *Anal. Biochem.* 198: 312–317.
2. Hell, J., et al. (1990). "Energy Dependence and Functional Reconstitution of the γ-Aminobutyric Acid Carrier from Synaptic Vesicles," *J. Biol. Chem.* 265: 2111–2117.
3. Huganir, R., and Racker, E. (1982). "Properties of Proteoliposomes Reconstituted with Acetylcholine Receptor from *Torpedo californica*," *J. Biol. Chem.* 257: 9372–9378.
4. Newman, M., and Wilson, T. (1980). "Solubilization and Reconstitution of the Lactose Transport System from *E. coli*," *J. Biol. Chem.* 255: 10583–10586.

Reconstitution by Dialysis
1. Barzilai, A., et al. (1984). "Isolation, Purification and Reconstitution of the Na^+ Gradient-dependent Ca^{2+} Transporter (Na^+-Ca^{2+} Exchanger) from Brain Synaptic Plasma Membranes," *Proc. Natl. Acad. Sci. USA* 81: 6521–6525.
2. Cook, N., et al. (1986). "Solubilization and Functional Reconstitution of the cGMP-dependent Cation Channel from Bovine Rod Outer Segments," *J. Biol. Chem.* 261: 17033–17039.
3. Epstein, M., and Racker, E. (1978). "Reconstitution of Carbamylcholine-dependent Sodium Ion Flux and Desensitization of the Acetylcholine Receptor from *Torpedo Californica*," *J. Biol. Chem.* 253: 6660–6662.
4. Rehm, H., et al. (1989). "Dendrotoxin-binding Membrane Protein Displays a K^+ Channel Activity That Is Stimulated by Both cAMP-dependent and Endogenous Phosphorylations," *Biochemistry* 28: 6455–6460.

5. Scheuring, U., et al. (1986). "A New Method for the Reconstitution of the Anion Transport System of the Human Erythrocyte Membrane," *J. Membrane Biol.* 90: 123–135.

Reconstitution by Adsorption to Hydrophobic Matrices
1. Hanke, W., et al. (1984). "Reconstitution of Highly Purified Saxitoxin-sensitive Na⁺ Channels into Planar Lipid Bilayers," *EMBO J.* 3: 509–515.
2. Home, W., et al. (1986). "Rapid Incorporation of the Solubilized Dihydropyridine Receptor into Phospholipid Vesicles," *BBA* 863: 205–212.
3. Talvenheimo, J., et al. (1982). "Reconstitution of Neurotoxin Stimulated Sodium Transport by the Voltage-sensitive Sodium Channel Purified from Rat Brain," *J. Biol. Chem.* 257: 11868–11871.

Reconstitution by Gel Filtration
1. Garcia-Calvo, M., et al. (1989). "Functional Reconstitution of the Glycine Receptor," *Biochemistry* 28: 6405–6409.
2. Haga, K., et al. (1985). "Functional Reconstitution of Purified Muscarinic Receptors and Inhibitory Guanine Nucleotide Regulatory Protein," *Nature* 316: 731–732.

4.2.2 Reconstitution in Preformed Liposomes

It is possible to incorporate membrane proteins in liposomes without using detergents (or in the presence of low detergent concentrations). In particular, you can let membrane proteins move directly from the native membrane into liposomes, without inserting a solubilization step. This technique usually requires specialized and highly purified phospholipids and a precise knowledge of their properties.

Sources
1. Dencher, N. (1989). "Gentle and Fast Transmembrane Reconstitution of Membrane Proteins," *Methods Enzymol.* 171: 265–274.
2. Zakin, D., and Scotto, A. (1989). "Spontaneous Insertion of Integral Membrane Proteins into Preformed Unilamellar Vesicles," *Methods Enzymol.* 171: 253–264.

4.3 Flux Assay

Developing an assay that measures the translocation or the flux of a substrate is more difficult the smaller the compartments. The difficulty peaks when it comes to detecting transport or ion channel function of a translocator in prolis. Some of the trouble results from the fact that with prolis the researcher does not know at which step the experiment failed. Was the translocator not incorporated into prolis in the correct conformation and orientation or was the flux protocol not suitable? In addition, ion channels involve quick fluxes and rare membrane proteins (see Table 4.2). Often only few of the vesicles contain an ion channel. The experimenter has to distinguish the flux in this small volume from the nonspecific flux in the large remainder of the vesicle population. It helps a little bit to check whether other properties of the translocator were reconstituted (e.g., enzyme activity or binding sites). However, there is not necessarily a connection: often (for example) a binding site of the translocator is reconstituted but not its translocation function. The question remains open whether the reconstitution failed or the flux assay was unsuitable. In the development of a flux assay, the following controls should be carried out independently of the type of compartment and translocator.
- Does the vesicle preparation still contain endogenous substances which influence the translocator (e.g., glycine with the glycine receptor, glutamate with glutamate receptors)?
- Does the translocation exhibit pharmacological specificity? How high is the unspecific flux (i.e., the flux that does not run across the translocator, for example, in the presence of an inhibitor of the translocator)? Then, the translocator flux is the difference between total flux and nonspecific flux. The nonspecific flux is a measure for the quality of the compartments and the presence of other translocators for the substrate. It can constitute more than 90% of the total flux.

Table 4.2. Ion flux assays.

Ion	Assay Molecule	Measurement	Remarks	References
Na+	^{22}Na+	Direct (radioactivity)	Strong γ radiation source: centimeter-thick lead brick needed as shield.	Epstein and Racker 1978; Rosenberg et al. 1984[*]
K+	^{86}Rb+	Direct (radioactivity)	Strong γ and β radiation source; half-life 19 d; centimeter-thick lead bricks needed as shield.	Arner and Stallcup 1981[*]
Ca2+	^{45}Ca^2	Direct (radioactivity)	Measures the ^{45}Ca2+ of the entire cell. The specific activities of the Ca2+ in cell and medium are typically different.	Barzilai et al. 1984; Akerman and Nicholls 1981[*] Combettes et al. 1990[*] Borte et al. 1975[*] Barrit et al. 1981[*]
—	Fura-2	Indirect (fluorescence)	Measures cytoplasmatic Ca2+. Fura-2 is more selective for Ca2+, works with longer wavelengths, and has a larger fluorescence intensity than Quin-2. Fura-2 is the Ca2+ indicator of choice.	Pandiella and Metdolesi 1989[*] Morgan et al. 1987[*] Grynkiewicz et al. 1985[*]
—	Indo-1	Indirect (fluorescence)		Grynkiewicz et al. 1985[*]
—	Quin-2	Indirect (fluorescence)		Arslan et al. 1985[*]
—	Arsenazo III	Indirect (adsorption)		Koch and Kaupp 1985
—	Equorin	Indirect (light emission)	The protein equorin binds Ca2+ under light emission. The light emission does not depend linearly on Ca2+. Equorin is injected or transfixed. Depending on the transfixion technique, it appears in the cytoplasma or in the mitochondria. Ag+ interferes.	Olsen et al. 1988[*] Rizzuto et al. 1992
H+	Pyranin	Indirect (fluorescence)	—	Page et al. 1988
—	4-methyl-umbelliferon	Indirect (fluorescence)	—	Page et al. 1988
Cl−	^{36}Cl−	Direct (radioactivity)	—	Thympy and Barnes 1984
—	I− and MSQ[*]	Indirect (fluorescence)	—	Garcia-Calvo et al. 1989
Anions (HCO$^-$$_3$ etc.)	^{35}SO^{2-}$_4$	Direct (radioactivity)	—	Scheuring et al. 1986

[*] MSQ, 6-methoxy-N-(3-sulfopropyl)quinolinium.
[†] See "sources" section following.

- No translocation without driving gradients or enzyme activities!
- Do compartments that contain no translocator still take up substrate? No translocation without translocator!

Was the substrate really taken up or did it only bind on the outside to the membrane of the compartments? Lysis of the compartments or substances that make the membrane permeable for the substrate (e.g., Ca^{2+} release after addition of A-23187) should set the substrate free that is associated with the compartments.

4.3.1 Influx Assay

With the direct method of the influx assays, the experimenter incubates the radioactive substrate (e.g., $^{22}Na^+$-, ^3H, amino acids) with the compartments in a suitable buffer. At the desired point, he interrupts the translocation (e.g., by addition of a specific blocker) and separates the compartments from the medium. The separation is achieved using size, adsorption, or charge differences between compartment and substrate (i.e., filtration through glass fiber filters (such as a Whatman GF/C or nitrocellulose filters, such as Millipore, GS-TF 0,22 μm), IEC in Dowex 50 W, gel filtration, or (for cell culture) simply washing. For cells or larger vesicles, you could also try centrifugation through a sucrose or mineral oil gradient. In this case, the experimenter also determines the portion of extracellular space in the cell pellet or vesicle pellet with a nonmembrane-permeable substance such as ^3H sucrose.

Note with filter assays: glass fiber filters with adsorbed ^3H or ^{45}Ca must be incubated with scintillator for at least 18 h, with occasional shaking. Only then the cpm/vial will have stabilized. Preincubation of the filters with a solution of 10 mM $CaCl_2$ should inhibit the nonspecific binding of $^{45}Ca^{2+}$ to the filters.

For the prolis to survive a filtration without losing their content, the temperature is lowered below the phase change point of the phospholipids being used. The filtration procedure resembles those for membrane binding assays (see Section 2.2.3). However, they are different insofar as you need to keep an eye on the stability and tightness of the compartments (and perhaps on ion gradients between compartment and media). The compartments must not be subjected to osmotic stress.

Measuring fluxes through ion channels in small compartments is an art, because of the speed of the process, the small inner volume of the compartments, and the rarity of the channel proteins. Gaily et al. (1983) developed a method that is as simple as ingenious (see Figure 4.1). They measure the influx of the radioactive ion against a chemical gradient of the corresponding nonradioactive ion. Paradoxical? The compartments contain a high concentration of the ion for which the channel is permeable, whereas the outer media contain the identical concentration of an ion for which the channel is impermeable. Because of the unequal distribution of the ion for which the channel is permeable, a diffusion potential develops (negative with cations inside) across the membrane of those compartments that contain the channel. The other compartments are not permeable for the ion, and hence no diffusion potential develops there (ideal case). Now the experimenter mixes into the outer medium small quantities of the radioactive ion, for which the channel is permeable. The radioactive ion distributes in accordance with the membrane potential and hence selectively accumulates (up to 100-fold) in those vesicles that contain the channel. It mimics, so to speak, the distribution of its mighty relative. This procedure delivers a signal that is higher by orders of magnitude than ion channel fluxes that measure only a balance. In addition, the process runs slowly (in the minute range) and can be measured comfortably.

In indirect methods of influx measurement, a color reaction, fluorescence reaction, enzyme reaction, or density change of the compartments indicate the translocation of the substrate. The indicator molecules are generally inside the compartments, but not in the outer medium. If the substrate enters the vesicles, it reacts with the indicator molecule, the indicator molecule changes its properties, and this change is measured. The reaction with the indicator generally happens quickly, so that with adsorption or fluorescence measurements the substrate intake or loss can be traced continuously.

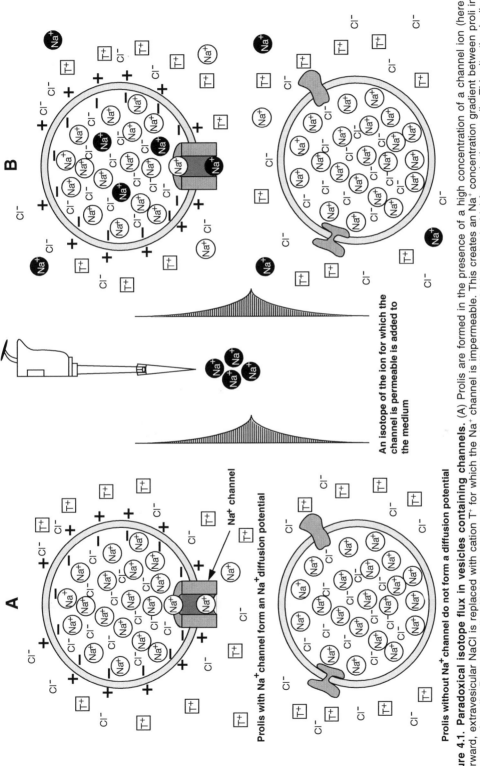

Figure 4.1. Paradoxical isotope flux in vesicles containing channels. (A) Prolis are formed in the presence of a high concentration of a channel ion (here Na$^+$). Afterward, extravesicular NaCl is replaced with cation T$^+$ for which the Na$^+$ channel is impermeable. This creates an Na$^+$ concentration gradient between proli internal and external media. Because prolis with Na$^+$ channels are permeable for Na$^+$, an Na$^+$ diffusion potential (negative inside) forms in these prolis. This situation is displayed in A. Now radioactive Na$^+$ is added. The following happens (B): outside there is only radioactive Na$^+$, but the channel lets Na$^+$ pass in both directions. Thus, radioactive Na$^+$ migrates through the channel into the vesicle. The radioactive Na$^+$ cannot easily come out of the vesicles. Reason? If you throw a black ball (= radioactive Na$^+$) into a box with many white balls (= Na$^+$), shake the box, and take out a ball again, the likelihood that you get a white ball is high. However, the channel-containing vesicles contain a lot of Na$^+$ ions (white balls), because the inside negative diffusion potential keeps the Na$^+$ ions in the vesicles. The radioactive Na$^+$ thus fills the vesicles that have Na$^+$ channels and contain a lot of Na$^+$.

Of course, the properties of the indicator molecule may only change with the substrate. Thus, membrane-permeating ligands of the translocator—whose effect on the translocator should be ascertained—can (for example) quench the fluorescence of the indicator and thus create the appearance of a change in the substrate translocation.

For the indirect influx assay, the compartments must be tight, and the indicator must be brought into the compartment beforehand. An elegant solution to the latter problem is lipid-soluble indicator precursors that easily penetrate the membrane and are then transformed within the compartment (e.g., by esterases into the lipid-unsoluble hydrophile indicator). The experimenter thus measures substrate influx only in the compartments that contain the suitable esterases. The living cell thus hydrolyzes fura-2 acetoxymethylester to fura-2 only in the cytoplasm. Fura-2 is a fluorescent stain that selectively forms complexes with Ca^{2+} and then changes its excitation spectrum. With fura-2, the experimenter can measure the change of the Ca^{2+} level in the cytoplasm, but not (for example) in the endoplasmic reticulum or in the mitochondria. Rizzuto et al. (1992) were able to do the latter. Rizzuto transfixed cells with a cDNA for aequorin. Aequorin is a protein that binds Ca^{2+} under light emission. Because aequorin cDNA was fused with a mitochondrial label sequence, the cell expressed them selectively in the mitochondria. This allowed Rizzuto to also selectively measure the changes in the Ca^{2+} content of these organelles. Finally, influx measurements with $^{45}Ca^{2+}$ (a direct method) registers the total $^{45}Ca^{2+}$ taken up by the cell.

Sources

Influx into Celis
1. Thampy, K., and Barnes, E. (1984). "γ-aminobutyric Acid-gated Chloride Channels in Cultured Cerebral Neurons," *J. Biol. Chem.* 259: 1753–1757.
2. Thayer, S., et al. (1988). "Measurement of Neuronal Ca^{2+} Transients Using Simultaneous Microfluorymetry and Electrophysiology," *Pflügers Arch.* 412: 216–223.

Influx into Vesicles
1. Gaily, H., et al. (1983). "A Simple and Sensitive Procedure for Measuring Isotope Fluxes Through Ion-specific Channels in Heterogenous Populations of Membrane Vesicles," *J. Biol. Chem.* 258: 13094–13099.
2. Kish, P., and Ueda, T. (1989). "Glutamate Accumulation into Synaptic Vesicles," *Methods Enzymol.* 174: 9–25.
3. Rizzuto, R., et al. (1992). "Rapid Changes of Mitochondrial Ca^{2+} Revealed by Specifically Targeted Recombinant Aequorin," *Nature* 358: 325–327.

Influx into Prolis
1. Barzilai, A., et al. (1984). "Isolation, Purification and Reconstitution of the Na^+ Gradient-dependent Ca^{2+} Transporter (Na^+-Ca^{2+} Exchanger) from Brain Synaptic Plasma Membranes," *Proc. Natl. Acad. Sci. USA* 81: 6521–6525.
2. Claassen, D., and Spooner, B. (1988). "Reconstitution of Cardiac Gap Junction Channeling Activity into Liposomes: A Functional Assay for Gap Junctions," *BBRC* 154: 194–198.
3. Epstein, M., and Racker, E. (1978). "Reconstitution of Carbamylcholine-dependent Sodium Ion Flux and Desensitization of the Acetylcholine Receptor from *Torpedo Californica*," *J. Biol. Chem.* 253: 6660–6662.
4. Garcia-Calvo, M., et al. (1989). "Functional Reconstitution of the Glycine Receptor," *Biochemistry* 28: 6405–6409.
5. Kasahara, M., and Hinkle, P. (1977). "Reconstitution and Purification of the D-glucose Transporter from Human Erythrocytes," *J. Biol. Chem.* 252: 7384–7390.
6. Page, M., et al. (1988). "The Effects of pH on Proton Sugar Symport Activity of the Lactose Permease Purified from *Escherichia coli*," *J. Biol. Chem.* 263: 15897–15905.
7. Radian, R., et al. (1986). "Purification and Identification of the Functional Sodium- and Chloride-coupled γ-aminobutyric Acid Transport Glycoprotein from Rat Brain," *J. Biol. Chem.* 261: 15437–15441.

Theoretical Analysis
1. Kotyk, A. (1989). "Kinetic Studies of Uptake in Yeast," *Methods Enzymol.* 174: 567–591.
2. Stein, W. (1989). "Kinetics of Transport: Analyzing, Testing and Characterizing Models Using Kinetic Approaches," *Methods Enzymol.* 171: 23–62.

4.3.2 Efflux Assay

Instead of measuring the translocation of a substrate into a compartment, it is sometimes more useful to determine the translocation out of the compartment. Because you can only get something out of a container if something is in it, the compartments are first loaded with the substrate for the efflux assay. How do you load? You can incubate the compartments for hours with high concentrations of substrate until the inner and outer concentrations have balanced out. This loading method is lengthy but reliable and gentle (Scheuring et al. 1986; Hunter and Nathanson 1985). If you are working with prolis, you can form these in the presence of substrate or you can subject your prolis to a freeze/thaw cycle in the presence of substrate.

Once the compartments are loaded, the experimenter removes the substrate that was not taken up (by dialysis, centrifuging, gel filtration, and so on) and uses the loaded compartments in the assay as quickly as possible. The assay typically consists of exposing the loaded compartments to an agent for a certain time and then separating them from the surrounding medium. The amount of released substrate is measured (directly or indirectly). Alternatively, the amount of substrate remaining in the compartments is measured.

Sources

Efflux from Cells
1. Hunter, D., and Nathanson, N. (1985). "Assay of Muscarinic Acetylcholine Receptor Function in Cultured Cardiac Cells by Stimulation of $^{86}Rb^+$ Efflux," *Anal. Biochem.* 149: 392–398.
2. Lukas, R., and Cullen, M. (1988). "An Isotopic Rubidium Ion Efflux Assay for the Functional Characterization of Nicotinic Acetylcholine Receptors on Clonal Cell Lines," *Anal. Biochem.* 175: 212–218.

Efflux from Vesicles
1. Koch, K., and Kaupp, B. (1985). "Cyclic GMP Directly Regulates a Cation Conductance in Membranes of Bovine Rods by a Cooperative Mechanism," *J. Biol. Chem.* 260: 6788–6800.

Efflux from Prolis
1. Scheuring, U., et al. (1986). "A New Method for the Reconstitution of the Anion Transport System of the Human Erythrocyte Membrane," *J. Membrane Biol.* 90: 123–135.

4.4 Constructive Thoughts

Reconstitutions are technically simple, require little training, and teach you how to handle detergents, phospholipids, membranes, and membrane proteins. Once the reconstitution has succeeded, you can observe your protein in a defined environment, free of influences from the cell metabolism. Afterward, you can construct in vitro entire metabolic chains or processes and thus build your own organelle. Sometimes reconstitution of the translocator function and flux assay is a prerequisite to the purification of the translocator protein. Figure 4.2 shows the possibilities of this field.

With larger compartments, such as cells or organelles, flux assays are easy to develop. Even the inexperienced have mastered the assay within a few months and can proceed with more worthwhile tasks. Assays for the detection of translocation in prolis, possibly even with an indirect method, are more difficult. If the assay does not work right away, all types of things could be to blame: prolis that are too small or not tight enough, their phospholipid composition, the protein/phospholipid ratio, the substrate, the translocator protein that was either not incorporated at all or in the wrong way (or which denatured during the incorporation), the flux assay, and so on. The researcher aimlessly gropes about in the darkness, until he gets lucky. Often, more than half a decade passes between the purification of a protein and its reconstitution (e.g., acetylcholine receptor, glycine receptor). In addition, the know-how that is so laboriously gained with reconstitutions is not particularly sought after, and the reconstitution of many proteins is a dead end that does not open any new experimental perspectives (Figure 4.2). Thus, the reconstitution of ion channels and receptor proteins has lost appeal, because the questions for which a reconstitution of the protein used to be necessary can be answered today more quickly and better with molecular biological methods (e.g., expression of cRNA in Xenopus oocytes). Do not be discouraged by this. You should have learned patience during reconstitution.

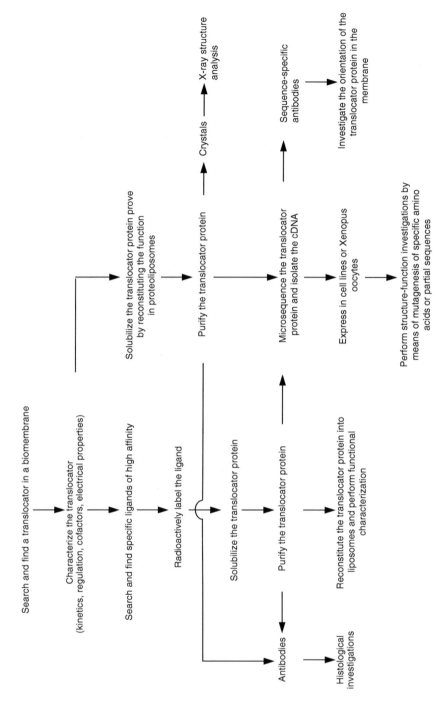

Figure 4.2. Translocator research.

"In the king's name!" exclaimed Sancho, "what have squires got to do with the adventures of their masters? Are they to have the fame of such as they go through, and we the labour? Body o' me! if the historians would only say, 'Such and such a knight finished such and such an adventure, but with the help of so and so, his squire, without which it would have been impossible for him to accomplish it'; but they write curtly, 'Don Paralipomenon of the Three Stars accomplished the adventure of the six monsters'; without mentioning such a person as his squire, who was there all the time, just as if there was no such being."

Chapter 5 Cleaning and Purifying

> But I may tell you this much by the way, that there is nothing in the world more delightful than to be a person of consideration, squire to a knight-errant, and a seeker of adventures. To be sure most of those one finds do not end as pleasantly as one could wish, for out of a hundred, ninety-nine will turn out cross and contrary. I know it by experience, for out of some I came blanketed, and out of others belaboured. Still, for all that, it is a fine thing to be on the look-out for what may happen, crossing mountains, searching woods, climbing rocks, visiting castles, putting up at inns, all at free quarters, and devil take the maravedi to pay.

5.1 Pure Fun

Uninteresting proteins are in principle just as difficult to purify as interesting ones. Hence, the protein purifier without long-term employment searches for a protein whose purification catches the attention of wider circles. This is often possible. The first purification of a protein makes an impression, above all when you can present the purification briefly and clearly, with a purification table and a band pattern on the SDS gel.

A clean protein is the key to a huge number of further findings (see Section 5.5). You can produce specific antibodies against it, find partial amino acid sequences for cloning, craft crystals for the X-ray structural analysis, and determine the subunit composition and stochiometry. That is, you could do so if after several years of purification your employment contract had not run out and you would not have to change laboratory and subject. Now, a good employee is still proud of his protein purification, just like a tailor's apprentice is proud of his first suit.

The purification of a protein crucially depends on two things: the proof assay and the material from which the protein is to be isolated. The assay is typically taken from the predecessor or from the literature. "I have fine-tuned it for years; the assay is optimized," assures the predecessor. Do not pay attention. Try to make the assay faster and simpler, even at the expense of accuracy.

The protein purifier has two requirements for the source material: high concentration of the sought-after protein and easy access. It is worthwhile to systematically test several species, different organs, and different developmental stages for their concentration of the sought-after protein before the first purification is attempted. Even if this preliminary work brings no advantages for the purification, you have acquainted yourself with the subject matter.

The progress of a purification is reflected in the increase of the specific binding or activity of the protein (e.g., enrichment in fM binding sites/mg protein) and the reduction of the number of bands in the SDS gel. For both purity indicators it is helpful to know the MW of the protein to be purified. If suitable ligands are available, you should determine the MW of the ligand/binding subunit of your protein with cross-linking assays—during or before the purification (see Section 2.4).

According to which criteria do you select the purification steps? How do you combine them? My recommendation: perform at most four purification steps and keep an eye on enrichment, yield, and work required for every purification step. If possible, perform no step with a purification factor of under 5, no step with a yield of less than 30%, and no step that takes longer than a day and a night. Concentrating purification steps saves a lot of trouble. The clever purifier also combines the steps in such a way that the enriched fractions can be loaded directly into the next step, without labor-intensive intermediate treatments such as

Table 5.1. Protease inhibitors.

Protease Inhibitor	Protease Class	Working Concentration	Special Characteristics
• PMSF	Serine proteases	1–10 µM	Disintegrates in watery solution; soluble in ethanol and isopropanol; stable for years in 100% isopropanol
Benzamidine	Serine proteases	~1 mM	
Aprotinin	Serine proteases	5 µg/ml	Inactivated through repeated freeze/thawing and alkaline pH (> 10)
Antithrombin III	Serine proteases	1 U/ml	Forms irreversible 1:1 complexes with serin proteases
Trypsin inhibitors	Serine proteases	10–100 µg/ml	Trypsin inhibitor from soy beans is instable in alkaline solutions
• Pepstatin A	Acidic proteases	1 µg/ml	Soluble in methanol (1 mg/ml)
• Leupeptin	Thiol proteases	1 µg/ml	
Antipain	Thiol proteases	1 µg/ml	
Cystatin	Thiol proteases	250 µg/ml	
E64	Thiol proteases	5 µg/ml	Soluble in 1:1 ethanol/water
• EDTA	Metal proteases	0.1–1 mM	
Phorphoramidon	Metal proteases	100 µg/ml	

dialysis, buffer exchange, and so on. Lengthy and boring steps such as column washing are done overnight.

Countless are the circumstances that cause a protein purification to fail: the column draws air, the pump tube is worn through, the wrong buffer was used, inlet or outlet tubes are leaking, the column is blocked, the dialysis tube has a hole, the cooling cell fails, and fraction collectors have a tendency to quit just when the precious protein eludes in the last purification step. Lucky is he who owns a pump, has a reliable fraction collector (e.g., from Gilson) and good columns (e.g., from Pharmacia), and does not have to lend these tools to colleagues.

You save yourself some aggravation if you connect column, fraction collector, and pumps with short tubes, each made of one piece. This measure provides for short waiting periods, involves small dead volume, and reduces the chance of leakage.

When a large buffer reservoir sits on the column material and if the elution buffer has a higher density than the column buffer, a mixing chamber develops. The mixing chamber transforms step gradients into linear gradients. The consequence is wide peaks and diluted eluate.

Successive buffers must be compatible. For example, Ca^{2+} solutions should not follow buffers that contain phosphate or deoxycholate. Stoll and Blanchard (1990) provide an overview of the most important buffer systems according to pH ranges, temperature dependence, and production (see Section 1.1).

Protein purifiers often liven up their coffee breaks with complaints about the disappearing activity of their protein. The activity disappears because of irreversible adsorption of protein to the column materials, proteolytic digestion, and/or unstable conformation. Cofactors are treacherous—of whose existence the experimenter has no idea but that are nevertheless important for the stability of the protein and that get lost during the purification. The existence of a low molecular cofactor may be suspected if the protein is stable in the raw cell or membrane extract but loses—activity during dialysis against extraction buffer.

Many things help against proteases, but nothing helps against all proteases. Therefore, it is customary to add a cocktail of 5 to 6 protease inhibitors to the protein extract (Table 5.1). The most important components of the cocktail are EDTA and PMSF.

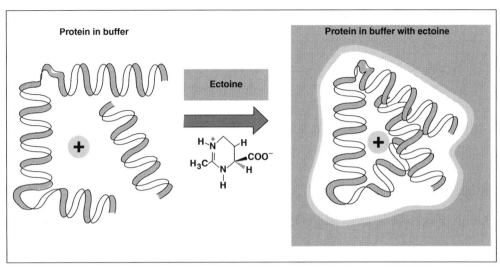

Figure 5.1. With ectoines you will have a ball. In ectoine solutions, gradients form between protein surface and water phase. As a result, the proteins take a round shape and become more resistant to proteases and heat.

The protease inhibitors in Table 5.1 marked with a large bullet suffice for most protein purifications. If you want to be extra careful, you also add the inhibitors marked with a small dot to your buffers. If the activity is still unstable, the experimenter strengthens the cocktail with the remaining protease inhibitors or searches for other reasons for the instability.

Digestion of protein can be prevented not only by inhibiting proteases but by stabilizing the protein. Salt-loving bacteria, for example, produce low molecular compounds called ectoines. These substances protect the bacteria against the extreme osmotic conditions of their environment. At the same time, ectoines also stabilize proteins against proteases and denaturing. How do ectoines manage to do this? They do so by taking advantage of the fact that water has a different structure near proteins than in a salt solution.

The structure of liquid water resembles the construction of ice (Ice I, to be exact). In ice, every water molecule is connected with four other water molecules via hydrogen bridges. The average distance of oxygen to oxygen amounts to 0.276 nm. In liquid water at 0° C, every water molecule is connected with 3.6 water molecules on average, and the O to O distance is about 0.280 nm. Thus, no big difference: liquid water is similar to ice. Near proteins, however, the hydrophile amino acid residues disrupt the water structure. Here, the water no longer resembles ice. And now here come the ectoines! They preferentially bind to ice-like water. Thus, the ectoines are repelled by the water film on the protein surface. Between the surface of each protein and its solution, an ectoine (mini) concentration gradient develops. The gradient causes the protein to reduce its surface and to become round (Figure 5.1). Simplified explanation: the smaller the surface the smaller the gradient surface and the smaller the balance disruption (and thermodynamics does not like disruptions). Thus, the protein becomes round because (as a ball) it has the smallest surface. The compact ball is more difficult for proteases to attack and more stable under heat.

Ectoines appear to protect enzymes as well as antibodies—at least some of them. Add to this that ectoines are stable and nontoxic (e.g., in contrast to EDTA!) and do not stop cell metabolism, even in high concentrations. However, you need high concentrations (between 0.4 and 2 M), and this costs. In addition, ectoine and hydroxyectoine are nitrogen-containing zwitterions (see Fig. 5.1) and thus could distrupt some protein assays and probably also purification procedures that rely on the protein charge. The conformation of many ligands can also be stabilized with glycerine (10%) and/or ligands (see Section 2.2.5.2 DSC).

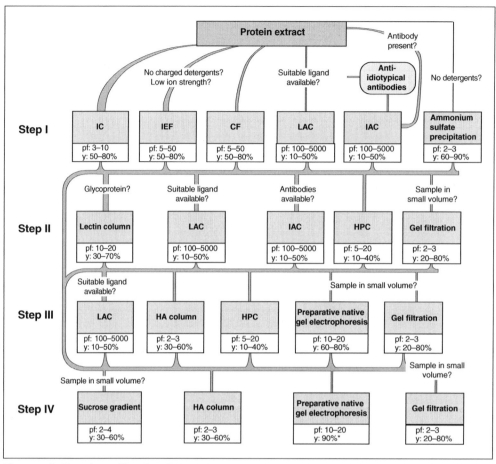

Figure 5.2. Protein purification. This chart should help in the planning of a protein purification. The purification factors (pf) and the yield (y) are provided for rough estimates and do not apply to every protein. The farther on the left a method box appears, the sooner it should be performed during the relevant step. It rarely makes sense to repeat a method over the course of a purification protocol. LAC: ligand affinity chromatography; IAC: immuno-affinity chromatography; HPC: hydrophobic chromatography. *Refers to protein, not activity.

Protect full columns you do not need for a longer period of time against microorganisms with azid (0.02%) and label the device (content, your initials). Thus, the matrix material remains free from suspect gray fungal spots and the person responsible for the cold room will hesitate to remove the column. With some luck, it will still be there years after your departure and be a monument to your work. Figure 5.2 provides an overview of popular purification methods.

Source
1. Stoll, V., and Blanchard, I. (1990). "Buffers: Principles and Practice," *Methods Enzymol.* 182: 24–38.

> With this, Sancho wheeled about and gave Dapple the stick, and Don Quixote remained behind, seated on his horse, resting in his stirrups and leaning on the end of his lance, filled with sad and troubled forebodings.

5.2 Conventional Purification Methods

Conventional purification methods are relatively easy to perform and yield valuable additional information (e.g., regarding the stability of the sought-after protein and its charge or size). Therefore, begin with a conventional method if you are likely dealing with a multistep purification. Here, the number of variables is limited, and after two weeks (at most) you see whether the method is usable. You rarely have to try more than three methods before you find a procedure that exhibits a satisfactory purification factor and yield. The experience of success right at the beginning gives to you the momentum you need to tackle the subsequent steps.

> Don Quixote pursued his journey in the high spirits, satisfaction, and self-complacency already described, fancying himself the most valorous knight-errant of the age in the world because of his late victory.

5.2.1 The Column Technique

With almost all adsorption materials, the column chromatography is superior to the batch method. Commercial columns (Pharmacia, Bio-Rad) are preferable to makeshift contraptions made from syringes, pipettes, and rubber plugs. The required time and the chance of misfortune increases with homemade devices. And in the end, it is not the doctoral candidate's money that is spent on proper columns but her own time that is saved.

For many protein purifications, inexpensive low-pressure columns suffice, which are run with tube pumps. However, low-pressure separation methods are sometimes too inefficient and/or too slow (e.g., with proteins that must be purified about 1,000-fold and for which no affinity column exists).

In these cases, it pays to do a few trials with HPLC (high-performance liquid chromatography). It offers high reproducibility and—with peptides and small stable proteins—high yields.

Reversed-phase HPLC with water/acetonitrile gradients is the method of choice for peptide purification (e.g., from tryptic digestion). However, it can also efficiently separate small stable proteins (e.g., toxins). Reversed-phase HPLC columns use picomolar amounts of sample. You can immediately use the peaks for Edman sequencing (see Chapter 7) or for mass determination via MALDI-TOF (see Section 7.5). A typical run takes an hour.

The column material of the reversed-phase HPLC consists of porous silica particles that are generally coated with n-alkyl chains. Peptides and/or proteins adsorb to the hydrophobic surface. The chains are 4-, 8-, or 18-C long and their length unpredictably changes the separating properties of the HPLC. Two peptides that show two separate peaks on C_{18} columns may exhibit only one peak on C_4 columns (or vice versa). The conformation of proteins is also influenced by the length of the n-alkyl chain (largely in the direction of denaturing). C_4 chains often have a higher protein yield than C_{18} chains.

For peptides, you use silica particles with pore dimensions of 100 to 300 Å. For proteins with 300 to 4,000 Å, the pore diameter should be 10 times bigger than the peptide or protein. The diameter of the silica particles lies between 3 and 5 μm. Generally, the smaller the diameter the better the separation. The sample is normally eluded with a rising acetonitrile gradient. Methanol or 2-propanol also serve well, the latter being the stronger eluens.

The quality of the separation depends, among other things, on the steepness of the gradient and the temperature. The temperature is in play because peptides can maintain a secondary structure (α-helix, β-fold) under the conditions of the reversed-phase HPLC, which influences the adsorption. High temperatures prevent secondary structures. Because it is more comfortable, people predominantly work at RT. Regarding the column dimensions: the separation of peptides and smaller proteins improves with longer columns. For tryptic digestion, people typically use 20-cm columns with a 4.6-mm standard diameter. Larger proteins are separated on shorter columns because the yield otherwise becomes too low. Larger proteins like to denature under reversed-phase conditions. Because of the danger of denaturation, reversed-phase HPLC is rarely used to separate larger proteins.

However, the reversed-phase principle is not the only separation principle of HPLC. You can also run an HPLC with other column materials and watery salt solutions. But then you must use the corrosion-proof HPLC devices with titanium, steel, or Teflon parts. With these devices you can also do ion exchange chromatography (IEC) or gel filtration. Pharmacia/LKB or Merck offers especially developed matrix materials (monobeads or tentacle gel) that are packed in glass, stainless steel, or polyether etherketone columns that permit high flow rates at medium pressure (5 to 10 megabar). These columns are available for gel filtration, IEC, chromatofocusing, hydrophobic chromatography, and affinity chromatography to proteins A and G.

Problems: For affinity chromatography, HPLC offers no fundamental advantage compared to low-pressure chromatography. For the separation of membrane proteins with IEC or chromatofocusing, low-pressure chromatography even delivers better results. Glycoproteins generally cause trouble with HPLC. The booklet *FPLC Ion Exchange and Chromatofocusing* from Pharmacia describes some tricks for removing these, such as adding betaine or taurine to the buffer. The experimenter has to sterile-filter the protein solutions for HPLC before use (because of particles). The same is true for homemade buffer solutions. These also have to be degassed.

Pharmacia/LKB developed a complete solution (FPLC, fast protein liquid chromatography). FPLC is suited to projects that frequently call for isolation of larger nonmembrane proteins (e.g., isolation of monoclonal antibodies from hybridoma supernatants). With an FPLC, an IEC or a gel filtration is done within 30 to 60 minutes. In comparison to low-pressure chromatography, the separation results are often substantially better (except for membrane proteins). The individual runs are very well reproducible, and the device can accept arbitrary gradients, is easily programmable, and impresses every lay person with its professional appearance. However, the supply of ion exchanger matrices is limited to a single type of cation exchanger and a single type of anion exchanger, the device reaches only medium pressure, and a reversed-phase chromatography is not possible. Finally, FPLC needs a lot of room and is expensive (about $25,000).

Sources
1. Aguilar, M., and Heam, M. (1996). "High-resolution Reversed-phase High-performance Liquid Chromatography of Peptides and Proteins," *Methods Enzymol.* 270: 3–26.
2. Chicz, R., and Regnier, F. (1990). "High-performance Liquid Chromatography: Effective Protein Purification by Various Chromatographic Modes," *Methods Enzymol.* 182: 392–421.
3. Meyer, V. (1992). *Praxis der Hochleistungsflüssigchromatographie.* Verlag Salle & Sauerländer.
4. Unger, K. K. (Hrsg.) (1989). *Handbuch der HPLC: Leitfaden für Anfänger und Praktiker.* GLT Verlag, Darmstadt. "FPLC Ion Exchange and Chromatofocusing," available from Pharmacia/LKB (Tel.+49-0761/490 30). Good overview of different HPLC systems in (1992) *Messages from Chemistry* 40: Ml–M32.

5.2.2 Purification Based on Size Differences

5.2.2.1 Gel Filtration

Gel filtration or size exclusion chromatography (SEC) is based on the different distribution of molecules between a gel compartment (Table 5.2) and the surrounding media. Although it is widely used, low-pressure SEC is not suitable for protein purification, with exceptions. The reasons are: pouring and running the column is a lot of work, the resolution is bad (purification factors between 3 and 6), the sample volume is limited, and the chromatography takes a long time because the flow speed is limited. Finally, the sample is diluted by at least a factor of 3.

If you do not heed this warning, or if you do not have any other recourse, you should pay attention to the following during the SEC. Maintain high ion strength (at least 0.1 M salt) in the column buffer; plumb the column (use a level); pour the column at RT and degas the gel

Table 5.2. Gel compartments.

Name	Supplier	Matrix Type	Stability		Grain Ø (µm)	Separation Range (in kd)	Typical Flow Speed (cm/h)
			pH	T(°C)			
Biogel A-0.5 m	Biorad	Agarose	4–13	2–30	40–300	10–500	7–25
Biogel A-5 m			—	—	—	10–5000	7–25
Biogel P-2		Polyacrylamide	2–10	1–80	<45–90	0.1–1.8	5–10
Biogel P-100			—	—	45–180	5–100	3–6
Sephacryl S-100 HR	Pharmacia	Dextran cross-linked with bisacrylamide	2–11	1–120	25–75	1–100	15
Sephacryl S-500 HR			—	—	—	40–20,000	15
Sephadex G-10		Dextran cross-linked with epichlorohydrin	2–10	1–120		<0.7	5
Sephadex G-200			—	—		5–600	2
Sepharose CL-6B		Cross-linked agarose	4–10	1–120	45–165	10–4000	18
Sepharose CL-2B			—	—	60–200	7–40,000	18
Ultrogel A 6	IBF	Cross-linked agarose	3–10	2–36	60–140	25–2400	5
Ultrogel A 4			—	—	—	55–9000	4
Ultrogel AcA 54		Cross-linked agarose/ polyacrylamide	3–10	2–36	60–140	5–70	4.5
Ultrogel AcA 22			—	—	—	100–1200	2.5
Protein Pak 60	Waters	Silica	2–8	1–90	10	1–20	
Protein Pak 300			—	—	—	10–500	
Shodex WS 802.5	Showa Denko	Silica	3–7.25	10–45	9	4–150	
Shodex WS 80e			—	—	—	10–2000	
Superose 12 HR 10/30	Pharmacia	Cross-linked agarose	2–12	4–40	8–12	1–300	
Superose 6 HR 10/30			—	—	11–15	5–5000	
SynChropak 60	SynChrom	Silica	2–7	1–60	5–10	5–30	
SynChropak 500			—	—	—	30–2000	
TSK G2000SW	Toyo-Soda	Silica	3–7.5	1–45	10–13	5–60	
G4000SW			—	—	—	5–1000	
Zorbax GF-250	DuPont	Silica	3–8.5	1–100	4–6	4–500	
Zorbax GF-450			—	—	—	5–900	

HPLC

suspension before pouring to avoid the formation of air bubbles. To avoid the formation of layers in the gel, pour the column evenly in one go (use a large container). Apply the sample carefully. The sample volume should not be more than 5% of the column volume, and the gel surface should be even and smooth. You get best results if you load the sample onto the column with a Peleus ball or a Pipetus-Akku. The sample should have a higher specific weight than the column buffer and should slowly and evenly run down along the edge of the column approximately 0.5 cm above the gel surface. The buffer reservoir above the gel surface should be 2 to 3 cm high (i.e., the sample slides underneath).

Distorted bands are the result of bad sample application, adsorption of protein to the column material, or a balance of protein between different polymerization states. A preliminary run with BSA and/or adding TRITON-X-100 or salt to the column buffer helps against the adsorption to the column material. Ideally, you apply colored markers (e.g., a mixture of cytochrome c, dextran blue, potassium dichromate) to the column before the run to check for run properties and band distortion.

As a precaution against blocked columns, the sample should be centrifuged at 100,000 g for 1 h before application. For the same reason, you rinse the column with buffer containing sodium azid or 1 mM EDTA right after the run, because the sugary nature of many SEC gels leads to bacterial infestation within a few days with neutral or slightly alkaline column buffers. If the column still blocks, rinsing with 1% SDS in 10 mM NaOH may help.

The resolution of SEC can be improved with longer columns, slower throughput, finer matrix material, and an FPLC device (see Section 5.2.1). Recommendation for aggregation problems: isopronanol in concentrations up to 15% should prevent hydrophobic interaction between proteins.

Source
1. Stellwagen, E. (1990). "Gel Filtration," *Methods Enzymol.* 182: 317–328.

5.2.2.2 Preparative Gel Electrophoresis

The preparative SDS gel electrophoresis (see Section 1.3.1) is suitable for the purification of proteins, where the experimenter does not care about biological activity. The preparative SDS gel electrophoresis separates according to size. The protein's MW, which is known from (for example) cross-linking experiments (see Section 2.4), is used for the detection of the sought-after protein. After electrophoresis, the sought-after protein is available in the form of a denatured protein/SDS complex. Hence, the method makes sense as the last purification step (i.e., after purification steps that depend on charge differences or on activity have already been performed).

The preparative SDS gel electrophoresis differs from the analytic one by using thicker spacers (3 mm) and a wide gel pocket. A gel with 3-mm-thick spacers and a 10-cm-wide pocket can take a maximum of 20 mg of protein. For the protection of sensitive amino acids such as tryptophane and methionine, the cathode buffer contains 0.1 mM sodiumthioglycolate. The purification factor of preparative gel electrophoresis depends on the length of the gel and its structure (linear or exponential gradient gels purify better than simple gels). Gradient gels are preferable to long gels (> 10 cm) because the latter's better purification comes at the expense of yield. During SDS gel electrophoresis, small amounts of protein remain stuck in the run (sensitive protein stains make the comet tail visible, which follows the protein band in the gel).

Preparative native gel electrophoresis (see Section 1.3.2) separates according to charge and size and exhibits substantial purification factors with good yield. The biological activity of the sought-after protein is often preserved, and the protein can be detected in the gel after electrophoresis via its activity. Native gel electrophoresis is not suitable for membrane proteins (see, however, Section 1.3.2, special cases). Regarding the methodology, it is similar to preparative SDS gel electrophoresis. Eby (1991) offers a comparison of different preparative electrophoresis methods (SDS, native, IEF, and more) with regard to price, capacity, company, and so on.

After the run it is advisable to stain the proteins with a careful method; for example, with sodium acetate (see Section 1.4.2). You then cut out the interesting bands with a razor blade. Different devices for electroelution of proteins from gel pieces are available in retail. If you often use preparative electrophoreses, you should try the prep cell from Bio-Rad. This device allows you to preparatively electrophorese and elude proteins in one step.

Sources
1. Eby, M. (1991). "Prep-phoresis: A Wealth of Novel Possibilities," *Bio/Technology* 9: 528–530.
2. Hunkapiller, M., et al. (1983). "Isolation of Microgram Quantities of Proteins from Polyacrylamide Gels for Amino Acid Sequence Analysis," *Methods Enzymol.* 91: 227–236.

5.2.3 Purification Based on Charge Differences

5.2.3.1 Ammonium Sulfate Precipitation

Ammonium sulfate precipitation, which was traditionally applied as the first step, has fallen out of fashion—for good reasons. The losses are big, the purification factor is unsatisfactory (about 2 to 3), and the time involved is substantial. Better results are achieved if the protein sample is completely precipitated with high ammonium sulfate concentrations onto an inert carrier (e.g., silica gel), the carrier suspension then poured with the precipitated protein into a column, and the sought-after protein subsequently eluded again with a descending gradient of ammonium sulfate. TRITON-X-100s and other neutral detergents salt out as oily masses in high concentrations of ammonium sulfate.

Source
1. Englard, S., and Seifter, S. (1990). "Precipitation Techniques," *Methods Enzymol.* 182: 285–300.

5.2.3.2 Ion Exchanger

In ion exchanger chromatography (IEC), proteins bind to a matrix via electrostatic interactions. The matrix carries positively charged groups (anion exchanger) or negatively charged groups (cation exchanger). Extent and strength of the binding of a protein to the ion exchanger depends on the pH and ion strength of the buffer, the isoelectric point of the protein, and the density of the charges on the matrix.

As far as the technical aspects are concerned, the IEC is easier than the SEC. The IEC does not require perfectly poured columns, and the sample volume can amount to a multiple of the column volume. In addition, the IEC exhibits better purification factors than the SEC (between 3 and 15, depending on protein and conditions) and concentrates the sample. The yield is also often better (between 50 and 80%). With multistep purifications, it is a good idea to have the IEC at the beginning. Afterward, you have the sample in a small volume, free of endogenous ligands or other disruptive extract components.

Experimental things: First, the experimenter checks under which pH and salt conditions the sought-after protein is stable. Then, she loads the sample on the column which was equilibrated with a buffer of low ion strength (e.g., 20-mM salt). Unbound proteins are washed out, and the sought-after protein is eluded through increasing the salt concentration or changing the pH. As a precaution, the sought-after protein is also measured when passing through the column.

It is easiest to chromatograph at a steady pH. The sought-after protein is adsorbed or eluted through stepwise changes in ion strength. With bigger proteins and integral membrane proteins, it is fruitless to try to improve the IEC with a salt or pH gradient. The concentration effect of the IEC gets lost, and the purification factors are increased only slightly. It is more efficient and reproducible to elude the protein with a well-balanced step gradient. For preparative IEC, gradients are only used for estimating the elution conditions. A linear or exponential gradient often separates mixtures of smaller proteins and peptides into a spectacular panorama of peaks (e.g., in Harvey and Karlsson 1980).

The best purification factors are attained if the column is run at its capacity limit for the sought-after protein. The separating capacity of IEC also depends on the flow rate (the slower the better). The matrix material, its substitution degree with ionized groups, and the type of these ionized groups influence the separation result. It is worthwhile to try out different ion exchangers, possibly eliminating the historism, which is not always only CM or DEAE Sephadex! As for column dimensions, the small thick ones are the best. With IEC of membrane proteins, charged detergents such as cholate or deoxycholate are to be avoided.

Sources
1. Choudhary, G., and Horvath, C. (1996). "Ion-exchange Chromatography," *Methods Enzymol.* 270: 47–82.
2. Harvey, A., and Karlsson, E. (1980). "Dendrotoxin from the Venom of the Green Mamba, *Dendroaspis angusticeps*," *Naunyn-Schmiedebergs Arch. Pharmacol.* 312: 1–6.
3. Rossomando, E. (1990). "Ion-exchange Chromatography," *Methods Enzymol.* 182: 309–317.

5.2.3.3 Isoelectric Focusing

A mixture of zwitterionic compounds (ampholyte) distributes over an electric field according to the compounds' isoelectric points. A pH gradient forms (low pH at the anode, higher pH at the cathode). A protein added to this system migrates in the electric field to the position in the pH gradient at which its net charge is zero (Figure 5.3). The electric field concentrates the protein at its isoelectric point, which is why the process is called isoelectric focusing (IEF). For the pH gradient to remain stable, you have to stabilize the electrolyte system against convection (e.g., with a 5% polyacrylamide gel, analytic IEF) or by suspension of granular gel globules (preparative IEF).

For preparative IEF, the Rotofor cell (Bio-Rad) can be used. Here, a column is used for focusing. Convection streams are inhibited by a row of inserted filters and the rotation of the column around the horizontal axis. The Rotofor cell was developed by Egen et al. 1984.

The IEF with soluble ampholines suffers from the fact that convection stabilization alone yields only moderate results. This is because with longer focusing time the ampholines happily migrate into an electrode and disappear there together with the proteins. Electroendosmotic and other effects are at fault. This happens especially with basic gradients. What to do? Görg et al. (1994) have solved the problem by immobilizing the ampholytes in the gel. The immobiline system (Pharmacia) co-polymerizes buffering acrylamide derivatives (immobilines) with acrylamide and Bis. Two acrylamide/immobiline solutions of different pHs and a gradient mixer provide a covalently fixed (immobilized) pH gradient. The result: the pH gradient does not migrate anymore and you get—in comparison to IEF with ampholines—a 10-fold better resolution. If you use gels with a narrow pH range (e.g., pH 5.5 to pH 6.5), you can still separate proteins whose isoelectric point differs only by 0.01 pH unit. Among other reasons, this is because with immobilines you can focus longer and at a higher voltage. The results are also more reproducible. Finally, it is technically easier to use because immobiline IEF gels are available in retail poured on foil, dried, and cut into 3- to 4-mm-wide strips (see Section 7.3). Table 5.3 summarizes the IEF procedure.

Problems: IEF can only be carried out at low ion strength, because even low salt concentrations lead to band distortions. However, many proteins need a certain ion strength or bivalent cations such as Ca^{2+} or Mg^{2+} for their native conformation. Membrane proteins smudge in the polyacrylamide gels of the analytic IEF. Overall, applying the sample application and keeping it in solution in the IEF gel are obstacles that already many a doctoral candidate failed to surmount. If compatible with your setup, you can dissolve your proteins in 8 M urea, 4% (w/v) CHAPS, 40 mM Tris, and 65 mM dithiotreitol (see Section 7.3). This generally has a negative effect on the native conformation of the protein, but it puts and keeps a lot—but not everything—in solution. The required immobiline strips are soaked in 8 M urea, 2% (w/v) CHAPS, 10 mM dithiotreitol, and 2% (v/v) resolyte of the suitable pH range.

If you focus with soluble ampholytes, the ampholytes (ampholines) interfere with many applications (e.g., protein determination). In addition, the pH gradients remain stable only for a limited time (approximately 3 h). Extremely basic proteins are not separated very well. If you take immobilines, you avoid these problems, but the focusing can take 2 to 3 days.

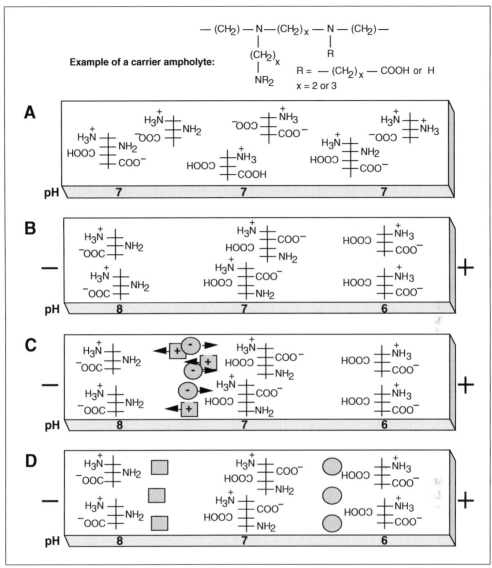

Figure 5.3. Isoelectric Focusing. (A) Mixture of carrier ampholytes (schematic) in the nonfocused IEF gel. (B) Apply voltage. The carrier ampholytes migrate to their isoelectric points, and the pH gradient forms. (C) Apply the sample (this can also be done before the pH gradient forms). (D) Focus the sample.

Advantages: IEF can process large amounts of material, concentrate the sample, provide high purification factors (10 to 100) with good yield, and determine the isoelectric point. Provided the desired protein is stable in ampholine solutions or does not aggregate under the conditions of the immobiline IEF, IEF is a better and faster alternative to IEC.

Sources
1. Corbett, J., et al. (1994). "Positional Reproducibility of Protein Spots in Two-dimensional Polyacrylamide Gel Electrophoresis Using Immobilised pH Gradient Isoelectric Focusing in the First Dimension: An Interlaboratory Comparison," *Electrophoresis* 15: 1205–1211.
2. Egen, N., et al. (1984). In *Electrophoresis 83*, H. Hirai (ed.), p. 547, Berlin: de Gruyter.
3. Grafin, D. (1990). "Isoelectric Focusing," *Methods Enzymol.* 182: 459–477.
4. Righetti, P., et al. (1988). "Immobilized pH Gradients," *TIBS* Sep; 13(9): 335–338.

Table 5.3. Isoelectric focusing.

Applications:
- Determining the isoelectric point.
- Checking for microheterogeneity (e.g., because of phosphorylation, sulfation, etc.).
- Purity proof (analytically).
- Purification of a protein mixture according to charge (preparative).

Recommendations for the gel:
- If your protein sample is only soluble with detergents, you can add CHAPS, CHAPSO, octylglucoside (1 to 2%), or TRITON-X-100 (0.1%) to the gel. Urea (3 to 8 M) does not disturb the focusing either. If you want to capture as many proteins as possible (e.g., cell extracts, plasma, etc.), it is recommended to dissolve the sample in 8 M urea, 4% (w/v) CHAPS, 40 mM Tris, and 65 mM DTT. Immobiline gels are similarly soaked in 8 M urea, 2% (v/v) CHAPS, 10 mM DTT, and 2% (v/v) resolyte of the suitable pH range.
- You fear that the cysteine and methionine residues of your proteins oxidize to cysteine acid and methionine sulfoxide? Then gas the IEF solutions beforehand with nitrogen.
- The thinner the gel the better the heat dissipation, the lower the diffusion, and the better the resolution.

Recommendations for sample application:
- Do not apply it too close to the electrodes (protein denatures).
- The sample has to be free of salt and precipitate.
- In raw extracts, DNA can interfere (e.g., by distorting bands). Remedy: pretreatment of the sample with DNAse.

Recommendations for focusing:
- During long focusing (> 3 h) the pH gradient migrates into the cathode (cathode drift). Remedy: immobilines.
- Oligosaccharide residues of glycoproteins are modified in unfavorable pH ranges.
- The metal ions in metalloproteins can be anodically oxidized or cathodically reduced. Metalloproteins can also lose their metal ion.

Control:
Repeat the focusing and apply the sample once in the basic and once in the acidic range.

5.2.3.4 Chromatofocusing

Chromatofocusing (CF) separates proteins according to their isoelectric points. It takes advantage of the pH gradient that forms when an ion exchanger column equilibrated in buffer pH A (start buffer) is eluded with a buffer pH B (elution buffer) (Table 5.4). That is, when A > B.

The proteins are first bound to the column in start buffer, and then the pH gradient separates the bound proteins in the order of their isoelectric points. At the same time, focusing occurs, because the migration speed of a protein in the column depends on the pH (Figure 5.4). CF supposedly still separates proteins whose isoelectric points differ only by 0.05 pH units. My best result was 0.1 pH units.

A CF column must be poured like a gel filtration column, evenly and free of gas bubbles. The art of even pouring is described in Section 5.2.2.1. Because the HCO_3^- pH gradients interfere, the start and elution buffer must not contain carbon dioxide. The experimenter thus has to degas the buffers. Finally, the protein sample should have a low ion strength. Otherwise, CF is as easy as IEC and does not require power sources or gradient mixers. If need be, you can even do chromatofocusing in the presence of dissociating agents (urea and DMSO) or nonionic detergents (NP-40, TRITON-X-l00, octylglucoside).

Once the column is prepared, chromatography is done within a few hours. You can chromatofocus even faster and with somewhat better results with an HPLC or FPLC device. CF often gives better purification factors than IEC, shows the isoelectric point, and can handle large amounts of protein. Hence, CF is an alternative to IEC if the sought-after protein is stable in the elution buffer (polybuffer).

Recommendations: Check the quality of your CF column with a cytochrome C solution (Giri 1990).

Clean sample application is as important with CF as with SEC (see Section 5.2.2.1). In particular, the gel surface must be even. Giri (1990) recommends applying a layer (1 to 2 cm) of Sephadex G-25 coarse onto the PBE gel bed. The Sephadex layer should facilitate an even sample application and serve as a mixing chamber. It also prevents the surface of the ion exchanger gel from being disturbed. It is common practice to equilibrate the sample before

Table 5.4. Buffers and gels for chromatofocusing (after Giri 1990).

pH Range	Gel	Start Buffer	Elution Buffer	Dilution of the Elution Buffer	Dead	Gradient	Total
					Volumes (in Column Volumes)		
10.5–8	PBE 118	pH 11; 0.025 M triethylamine HCl	pH 8; Pharmalyte (pH 8–10.5) HCl	1:45	1.5	11.5	13.0
10.5–7	PBE 118	pH 11; 0.025 M triethylamine HCl	pH 7; Pharmalyte (pH 8–10.5) HCl	1:45	2.0	11.5	13.5
9–8	PBE 94	pH 9.4; 0.025 M ethanolamine HCl	pH 8; Pharmalyte (pH 8–10.5) HCl	1:45	1.5	10.5	12.0
9–7	PBE 94	pH 9.4; 0.025 M ethanolamine HCl	pH 7; polybuffer 96 HCl	1:10	2.0	12.0	14.0
9–6	PBE 94	pH 9.4; 0.025 M ehanolamine CH_3COOH	pH 6; polybuffer 96 CH_3COOH	1:10	1.5	10.5	12.0
8–7	PBE 94	pH 8.3; 0.025 M Tris-HCl	pH 7; polybuffer 96 HCl	1:13	1.5	9.0	10.5
8–6	PBE 94	pH 8.3; 0.025 M Tris-CH_3COOH	pH 6; polybuffer 96 CH_3COOH	1:13	3.0	9.0	12.0
8–5	PBE 94	pH 8.3; 0.025 M Tris-CH_3000H	pH 5; polybuffer 96 (30%) + polybuffer 74 (70%) CH_3COOH	1:10	2.0	8.5	10.5
7–6	PBE 94	pH 7.4; 0.025 M imidazol CH_3000H	pH 6; polybuffer 96 CH_3000H	1:13	3.0	7.0	10.0
7–5	PBE 94	pH 7.4; 0.025 M imidazol HCl	pH 5; polybuffer 74 HCl	1:8	2.5	11.5	14.0
7–4	PBE 94	pH 7.4; 0.025 M imidazol HCl	pH 4; polybuffer 74 HCl	1:8	2.5	11.5	14.0
6–5	PBE 94	pH 6.2; 0.025 M histidine HCl	pH 5; polybuffer 74 HCl	1:10	2.0	7.0	9.0
6–4	PBE 94	pH 6.2; 0.025 M histidine HCl	pH 4; polybuffer 74 HCl	1:8	2.0	7.0	9.0
5–4	PBE 94	pH 5.5; 0.025 M piperazine HCl	pH 4; polybuffer 74-HCl	1:10	3.0	9.0	12.0

applying it to the start or elution buffer, and the sample volume should be less than half of the column volume. You pour a few ml elution buffer over the column, and then over the sample, and finally over the elution buffer. The flow speed is not critical for CF and normally lies between 20 and 40 cm/h.

Problems: Protein sample and elution buffer must not contain any salt. However, some proteins need a certain salt concentration for the preservation of their native conformation. With membrane proteins or with level pH gradients, the focusing effect is weak and the sample is diluted.

The eluate contains polybuffer/ampholines. This interferes with some assays (e.g., with the BCA assay; see Section 1.2.1). You remove them using one of the following techniques.

- Precipitating the protein with ammonium sulfate (80% saturation; see Section 5.2.3.1 and [Giri 1990])
- Ion exchange chromatography (binding, washing, eluding only; no gradient!)
- Lectin chromatography (in the case of glycoprotein)
- GC to Sephadex G-75 (only in emergencies)

Sources
1. Giri, L. (1990). "Chromatofocusing," *Methods Enzymol.* 182: 380–392.
2. Siemens, I., et al. (1991). "Solubilization and Partial Characterization of Angiotensin II Receptors from Rat Brain," *J. Neurochem.* 57: 690–700.
3. Sluyterman, L. (1982). "Chromatofocusing: A Preparative Protein Separation Method," *TIBS* 7: 168–170.

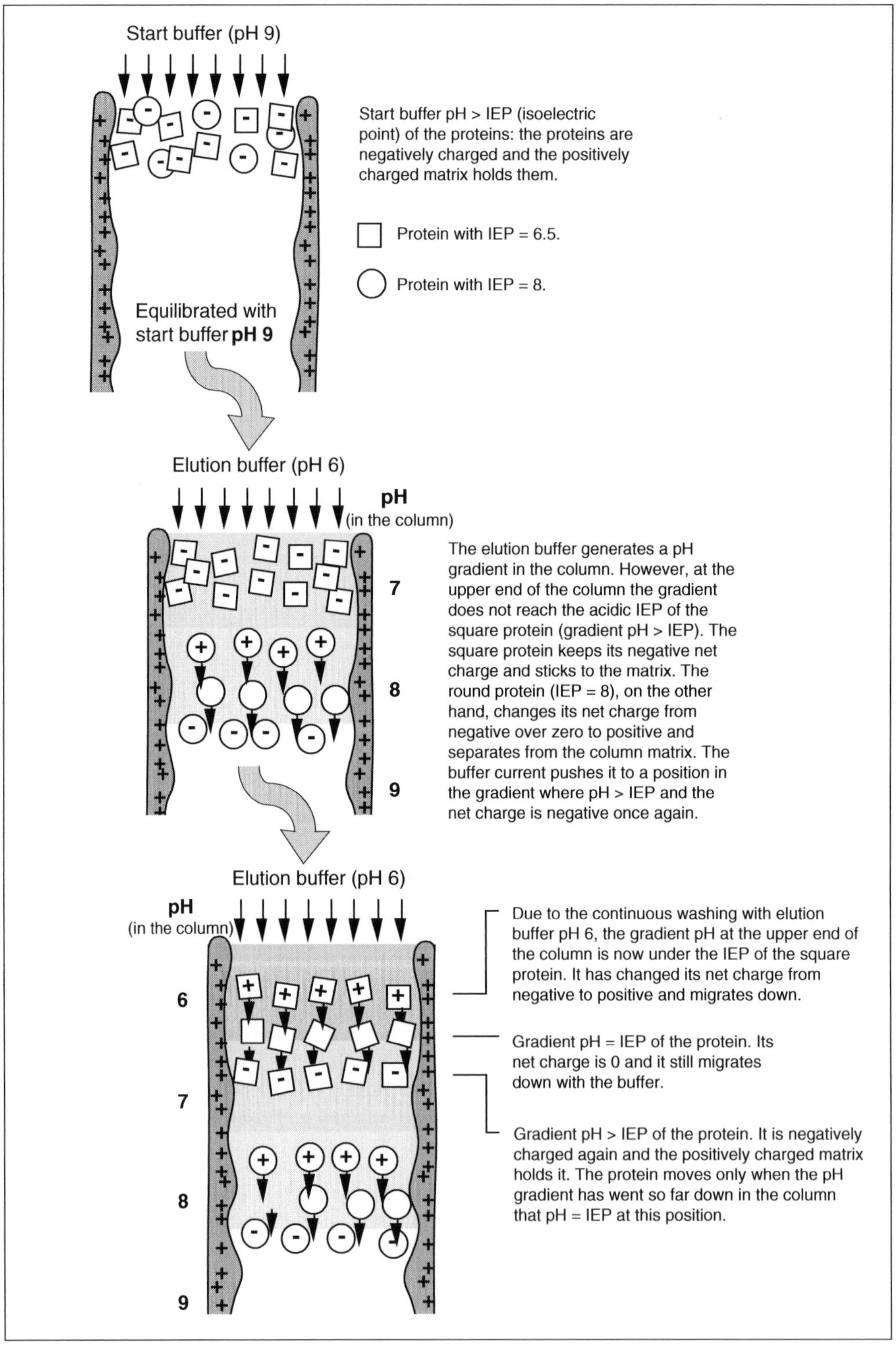

Start buffer (pH 9)

Start buffer pH > IEP (isoelectric point) of the proteins: the proteins are negatively charged and the positively charged matrix holds them.

☐ Protein with IEP = 6.5.

◯ Protein with IEP = 8.

Equilibrated with start buffer **pH 9**

Elution buffer (pH 6)

pH
(in the column)

The elution buffer generates a pH gradient in the column. However, at the upper end of the column the gradient does not reach the acidic IEP of the square protein (gradient pH > IEP). The square protein keeps its negative net charge and sticks to the matrix. The round protein (IEP = 8), on the other hand, changes its net charge from negative over zero to positive and separates from the column matrix. The buffer current pushes it to a position in the gradient where pH > IEP and the net charge is negative once again.

Elution buffer (pH 6)

pH
(in the column)

Due to the continuous washing with elution buffer pH 6, the gradient pH at the upper end of the column is now under the IEP of the square protein. It has changed its net charge from negative to positive and migrates down.

Gradient pH = IEP of the protein. Its net charge is 0 and it still migrates down with the buffer.

Gradient pH > IEP of the protein. It is negatively charged again and the positively charged matrix holds it. The protein moves only when the pH gradient has went so far down in the column that pH = IEP at this position.

Figure 5.4. Chromatofocusing.

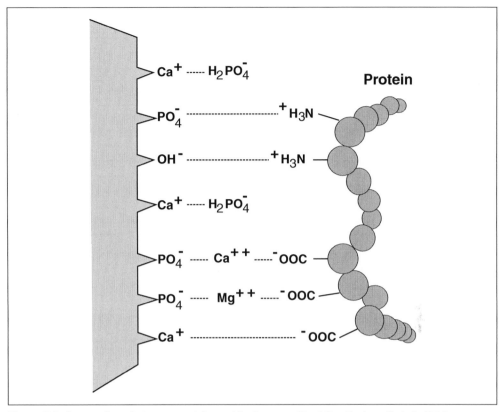

Figure 5.5. Interactions between protein and hydroxyapatite (after Gorbunoff et al. 1984).

5.2.3.5 Hydroxyapatite

Hydroxyapatite (HA) chromatography is a typical "I didn't know what else to do" method and thus most purifiers use it as the last step. HA chromatography is an inexpensive and easy method that can process large amounts of protein. However, the purification factors are often bad (2 to 5) and the yields moderate (40 to 60%). With acidic proteins, HA sometimes exhibits stunning purification effects. This is especially true for preparations that are prepurified via IEC, because HA separates acidic proteins by principles other than an ion exchanger. HA is a calcium phosphate mineral that binds proteins via two mechanisms (Figure 5.5).

- Basic proteins bind via their amino groups to the negative surface charge of the mineral (i.e., via electrostatic interactions).
- Acidic proteins form complex bindings via their carboxyl groups with the Ca^{2+} of the mineral.

Denatured proteins and substances of low MW (e.g., amino acids) do not bind to HA. Whether peptides bind depends on the number of their charged amino acids.

When you are loading the column with protein and eluting the bound protein, you must pay attention to pH and ion strength of the elution buffer and check whether its ions enter specific interactions with the HA mineral. This is the case with F^-, phosphate, and Ca^{2+}.

Basic proteins elude from HA either with solutions of 0.1 to 0.3 M of univalent cations (anion: Cl^-, F^-, SCN^-, or phosphate) or with low concentrations (1 to 3 mM) of divalent cations (Ca^{2+}, Mg^{2+}). In the first case, HA works like a cation exchanger. In the second case, the divalent cations bind to the negative phosphate of HA, edging out the amino groups of the proteins in the process.

Acidic proteins elude with F^- and phosphate salt solutions (0.05 to 0.15 M), but not with Cl^- or divalent cations. This is because F^- and phosphate edge out the carboxyl groups of the protein from the mineral's calcium. Divalent cations enhance the binding of acidic protein to HA.

In practice, the HA column is equilibrated with a phosphate buffer of low concentration (50 mM or less). Then the protein sample is loaded and the sought-after protein—if it was bound by the column—is eluded again with phosphate buffer. The optimal elution concentration is determined either with a gradient or via eluding in several successive steps (e.g., 1 to 2 column volumes of 100 mM of phosphates followed by 1 to 2 column volumes of 150 mM of phosphates).

Crystalline HA is unstable (i.e., HA columns do not have a long life span). However, Bio-Rad offers a ceramic (sintered) HA said to be chemically and mechanically stable that allows for high flow rates.

Sources
1. Gorbunoff, M. (1984). "The Interaction of Proteins with Hydroxyapatite: Role of Protein Charge and Structure," *Anal. Biochem.* 136: 425–432. (See also the follow-up papers in the same issue to page 445.)
2. Hartshorne, R., and Catterall, W. (1984). "The Sodium Channel from Rat Brain: Purification and Subunit Composition," *J. Biol. Chem.* 259: 1667–1675.
3. John, M., and Schmidt, J. (1984). "High-resolution Hydroxyapatite Chromatography of Proteins," *Anal. Biochem.* 141: 466–471.

5.2.4 Hydrophobic Chromatography

Many soluble proteins have hydrophobic areas on their surface. In watery solution, these hydrophobic areas associate with hydrophobic surfaces. The inclination toward association depends on the structure of the water, and this in turn from the salt dissolved in it.

High concentrations of certain ions increase the hydrophobic interactions, whereas chaotropic salt disrupts the water structure and thereby reduces the inclination for hydrophobic interactions (Figure 5.6). Hydrophobic chromatography takes advantage of this circumstance. It is compatible with ammonium sulfate precipitation (see Section 5.2.3.1).

The protein sample is loaded with high ion strength (largely ammonium sulfate) onto a hydrophobic matrix. The protein binds, the column washed, and the sought-after protein is eluded again with a decreasing salt gradient or a lower salt concentration (step). For the elution, people often use detergents that cover the hydrophobic areas of the proteins and thereby loosen them from the matrix.

Matrices for hydrophobic chromatography (Pharmacia, Spectrum) are derivatized either with phenyl residues (e.g., phenylsepharose) or octyl residues (e.g., octylsepharose). If the hydrophobicity of the sought-after protein is unknown, it is best to start with phenylsepharose because it is less hydrophobic than octylsepharose, and the proteins already elude under gentle conditions. Octyl-substituted matrices are suited for weakly hydrophobic proteins and membrane proteins.

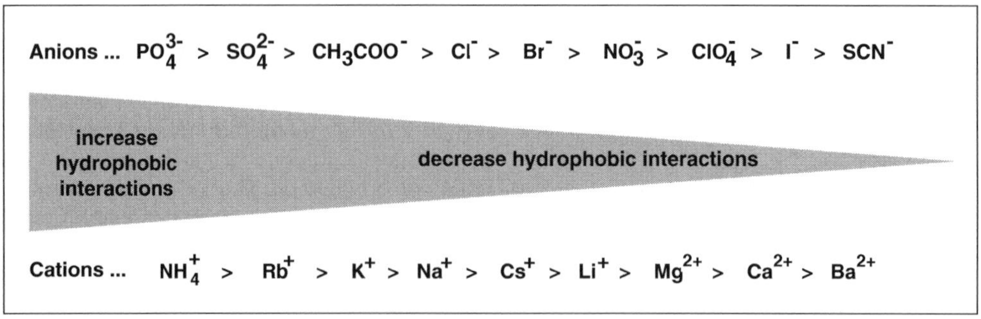

Anions ... PO_4^{3-} > SO_4^{2-} > CH_3COO^- > Cl^- > Br^- > NO_3^- > ClO_4^- > I^- > SCN^-

increase hydrophobic interactions   decrease hydrophobic interactions

Cations ... NH_4^+ > Rb^+ > K^+ > Na^+ > Cs^+ > Li^+ > Mg^{2+} > Ca^{2+} > Ba^{2+}

Figure 5.6. Ions for hydrophobic chromatography.

Sources
1. Kennedy, R. (1990). "Hydrophobic Chromatography," *Methods Enzymol.* 182: 339–343.
2. Sarnesto, A., et al. (1990). "Purification of H Gene Encoded β-galactoside α1←2-fucosyltransferase from Human Serum," *J. Biol. Chem.* 265: 15067–15075.
3. Wu, S., and Karger, B. (1996). "Hydrophobic Interaction Chromatography of Proteins," *Methods Enzymol.* 270: 27–47.

5.2.5 The Blue Gel

"Blue gels" are matrices that contain the stain Cibacron Blue 3GA as ligand. Agarose, sepharose, or silica globules serve as matrices. Cibacron Blue 3GA is a textile stain—a triazine that can bind to proteins with dinucleotide pockets. In addition, Cibacron Blue 3GA has negative charges (it is hydrophobic and deep blue). This versatility of the ligand makes blue gel chromatography a gamble in that it is not predictable how a protein behaves on a blue gel column. Proteins without nucleotide pockets can also bind to it, because of the ion exchange properties, the hydrophobicity, or other properties of this complicated stain. As with all games of chance, the following rule applies to blue gel chromatography: you largely lose.

Hence, my love of the blue gel is based more on the nice color than on success with it. I feel that the blue looks pretty among all the white in the column forest. In the United States, at a VA medical center, I once saw a Bavarian colleague cry in front of his white-blue column grove. I do not know whether it was because of homesickness (white and blue are the state colors of Bavaria) or because he had lost his sample. After all, the purification effect is quite low. If you achieve an enrichment of 5 with a blue gel and only lose half your sought-after protein you may pat yourself on the back. Recommendation: use small columns (a volume of 1 to 1.5 ml or less is enough).

The blue gel is recommended if you want to rid media at BSA. Depending on matrix and substitution degree, BSA binds in amounts between 8 and 15 mg/ml in the blue gel. Even if you know or suspect that your sought-after protein has a dinucleotide pocket, you can do chromatography on blue gel. This, in any case, is true when nothing better occurs to you. I would not risk the entire preparation. It is best to test the purification effect first with an aliquot. It happens occasionally that a protein completely disappears in the blue depths.

Depending on the protein, the blue gel is eluded with high ion strengths (e.g., 1.5 M KCl) or with detergents. You can also try using NADH for eluding proteins with dinucleotide pockets.

Source
1. Kolossov, V., and Rebeiz, C. (2001). "Chloroplast Biogenesis 84: Solubilization and Partial Purification of Membrane-bound (4-vinyl) Chlorophyllide, a Reductase from Etiolated Barley Leaves," *Anal. Biochem.* 295: 214–219.

5.3 Affinity Chromatography

5.3.1 Lectin Chromatography

Lectins are proteins that reversibly bind monosaccharides, polysaccharides, or sugar residues from glycoproteins. Lectins differ in their sugar specificity, their construction, and in the cofactors necessary for binding the sugar. Lectins that are covalently coupled to matrices are popular affinity materials and are available in retail (Table 5.5). Their specificity is shown in Table 9.1.

The requirement for lectin chromatography is a lectin matrix that binds the sought-after protein. To find such a matrix, the experimenter orders lectin matrices with different sugar specificity (WGA, Con A, and three to five others) and checks the adsorption of the sought-after protein via batch method. Lectin matrices with sugar specificity found in many

Table 5.5. Matrices for lectin affinity chromatography.

Lectin	Cofactors	Matrix (Supplier)	Elution Sugar	Special Characteristics
Wheat germ agglutinin (WGA)	—	4 % Agarose (Sigma) Sepharose 6 MB (Sigma) SpectraGel* (Spectrum)	0.02–0.2 M N-acetyl-D-glucosamine	Stable in 0.07% SDS and 1% deoxycholate
Concanavalin A (Con A)	Mn^{2+}; Ca^{2+}	4% Agarose (Sigma) SpectraGel* (Spectrum) Sepharose (Pharmacia)	0.1–0.2 M α-D-methylmannoside, 10 mM α-D-methylglucoside	Buffer must not contain EDTA
Bandeiraea simplicifolia lectin (BS-I)	Mn^{2+}; Ca^{2+}	Sephadex A25 (not available in retail)	0.05–0.1 M melibiose	—
Bandeiraea simplicifolia lectin (BS-II)	—	4% Agarose (Sigma)	N-acetyl-D-glucosamine	—
Lentil lectin (LCA)	Mn^{2+}; Ca^{2+}	4% Agarose (Sigma) Sepharose 4B (Pharmacia) SpectraGel* (Spectrum)	0.1–0.2 M methyl-α-D-mannoside	Stable in 1% deoxycholate; buffer must not contain EDTA
Ricinus communis agglutinin (RCA-120)	—	4% cross-linked Agarose (Sigma)	0.3 M β-methyl-galactopyranoside	—
Peanut lectin (PNA)	—	4% cross-linked Agarose (Sigma)	—	—
Soy bean lectin (SBA)	—	4% cross-linked Agarose (Sigma)	N-acetyl-D-galactosamine	—
Jacalin	—	4% Agarose (Vector Lab.) 6% Agarose (Peirce)	25–100 mM melibiose or 10 mM α-methyl-galactopyranoside	Binds O-glycosylated proteins
Ulex europaeus I (UEAI)	—	4% Agarose (Sigma)	0.1–0.3 M α-L-(-)Fucose	—
Helix pomatia (HPA)	—	4% cross-linked Agarose (Sigma)	N-acetyl-α-D-galactosamine	—

* Matrix material not known.

glycoproteins bind the sought-after glycoprotein with high probability, but provide only moderate purification factors. Lectins with specificity for rare sugars bind the sought-after protein with low probability, but if they do they promise pleasant surprises (e.g., Barbry et al. 1987).

Some glycoproteins bind the lectin only after their sialic acid residues are split off with neuraminidase. A double chromatography takes advantage of this fact: the lectin-binding proteins are removed in the first step, but the sought-after protein is allowed to pass. The column pass is treated with neuraminidase and once again poured over the regenerated lectin column. This time the sought-after protein binds.

Running two different lectin columns in sequence gives good results every now and then. The first column binds the sought-after protein together with other glycoproteins. The bound glycoproteins are eluded and the eluate is loaded onto the second column. This likewise binds many glycoproteins, but allows the sought-after protein to pass.

To suppress ion exchanger effects (lectins carry charges as proteins), lectin chromatography uses buffer of high salinity (0.2 to 0.5 M NaCl). Lectins need several hours to bind their ligands. Thus, the protein sample is often incubated with the lectin gel overnight and the loaded gel is poured in a column only for washing and eluding. After several runs, the capacity of a lectin column decreases. To regenerate, wash it three times alternately with 0.1 M sodium acetate pH 4.0 and 0.1 M Tris Cl buffer pH 8.5, where the pH changes should occur abruptly.

WGA (wheat germ agglutinin) and Con A (concanavalin A) are the favorites of the protein purifiers. WGA is often used for the purification of membrane proteins. The lectin binds ligands in the absence of bivalent cations and withstands 0.07% SDS, 1% deoxycholate, or 1 mM EDTA. WGA binds N-acetylglucose and N-acteylglucosamine residues and, with low affinity, sialic acid residues. Weakly binding proteins elude from WGA with 3 to 30 mM N-acetylglucosamine. Tightly bound ligands require 100 to 300 mM N-acteylglucosamine. The elution is also possible with high salt (0.5 $MgCl_2$). The purification factors of WGA chromatography are at 10 to 20 for raw protein extracts, the yield at 30 to 70%.

Binding and elution from Con A columns is most effective at room temperature. With WGA, the temperature does not play a role. Binding of glycoproteins to soy bean lectin or peanut lectin is optimal at 4° C.

Sources

Overview Article
1. Gerard, C. (1990). "Purification of Glycoproteins," *Methods Enzymol.* 182: 529–539.

Bandeiraea simplicifolia *Lectin*
1. Barbry, P., et al. (1987). "Purification and Subunit Structure of the ³H-phenamil Receptor Associated with the Renal Apical Na⁺ Channel," *Proc. Natl. Acad. Sci. USA* 84: 4836–4840.

Ricinus communis *Agglutinin*
1. Novick, D. (1987). "The Human Interferon-γ-receptor: Purification, Characterization, and Preparation of Antibodies," *J. Biol. Chem.* 262: 8483–8487.

Wheat Germ Agglutinin
1. Rönnstrand, L. (1987). "Purification of the Receptor for Platelet-derived Growth Factor from Porcine Uterus," *J. Biol. Chem.* 262: 2929–2932.
2. Tollefsen, S., et al. (1987). "Separation of the High-affinity Insulin-like Growth Factor I Receptor from Low-affinity Binding Sites by Affinity Chromatography," *J. Biol. Chem.* 262: 16461–16469.

Concanavalin A
1. Lin, S., and Fain, J. (1984). "Purification of (Ca²⁺-Mg²⁺)-ATPase from Rat Liver Plasma Membranes," *J. Biol. Chem.* 259: 3016–3020.
2. Wimalasena, J., et al. (1985). "The Porcine LH/hCG Receptor," *J. Biol. Chem.* 260: 10689–10697.

Ulex europaeus *Agglutinin*
1. Duong, L., et al. (1989). "Purification and Characterization of the Rat Pancreatic Cholecystokinin Receptor," *J. Biol. Chem.* 264: 17990–17996.

5.3.2 Ligand Chromatography

5.3.2.1 Introduction

Ligand affinity chromatography is one of the most impressive methods of protein biochemistry. A good affinity column reaches purification factors of 1,000 and more, with yields between 10 and 50%. A specific ligand of the sought-after protein with a reactive group is coupled to a matrix (agarose or polyacrylamide) via a spacer (Figure 5.7). If the binding properties of the ligands are preserved, the derivatized matrix selectively binds to the sought-after protein. The matrix is extensively washed on a column and the sought-after protein is eluded with the same or a different ligand of the protein (i.e., specifically).

Problems typically stem from the derivatized matrix not binding the sought-after protein. You usually find it again quantitatively in the eluate, but now and then the protein in the column loses its activity. Then the experimenter believes that his protein had bound and fiddles around with the elution conditions for months and without effect. Finally, it happens that the matrix binds the protein, but cannot be eluded specifically—either because the binding to the ligand is too strong (e.g., if the protein has several binding sites for the ligand and the affinity matrix is highly substituted) or because the protein binds unspecifically. After all, the fact that the sought-after protein binds to the matrix does not mean that it binds via its ligand binding site. Many ligands carry a charge and transform the matrix into an ion exchanger, and hydrophobic

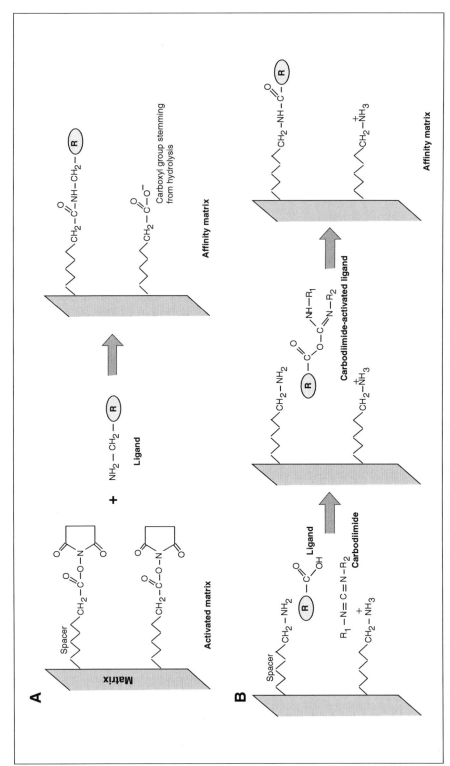

Figure 5.7. Coupling of ligands to matrices. (A) Via its primary amino group, a ligand couples to a matrix activated with N-hydroxysuccinimide ester. (B) Carbodiimide coupling of the carboxyl group of a ligand to matrix-bound primary amino groups. In the process, the carbodiimide changes into the corresponding urea derivative.

spacers or ligands turn the affinity chromatography into a hydrophobic chromatography. You have affinity chromatography if the sought-after protein eludes with low concentrations of a ligand.

Affinity chromatography can fail for reasons other than inappropriate ligand. You need to watch out for the length and type (hydrophile/hydrophobic) of the spacers, the chemical nature and pore size of the matrix, and the buffer (ion types, pH, detergent, ion strength) in which the sought-after protein is offered to the matrix. Moreover, ligands rarely couple as well and as irreversibly as the vendor brochures claim.

The development of an affinity chromatography takes about a year—except in lucky cases—but it is often the only way to display rare proteins in a pure form. The purification factors and yields of conventional procedures are generally too low to reach an enrichment of over 1,000, even by combining several purification steps. Consider this: with a medium yield of 50% per purification step, only 6% of the initial amount is left after only four steps. If you add further steps, the sought-after protein quickly drops under the detection threshold.

5.3.2.2 The Role of the Ligand

The first and most difficult task is finding a suitable ligand. The ligand should bind specifically and with high affinity ($K_D < 50$ nM; better, $K_D < 10$ nM) to the sought-after protein and it should be possible to couple it to a matrix without losing affinity or specificity. Coupling of ligands with a primary amino group is the easiest, whereas the coupling of ligands with a carboxylic group is a little more difficult. Sometimes sulfhydryl, hydroxyl, aromatic amino, or bromide alkyl group are used for coupling (see "Sources" under Section 5.3.2.4). Bio-Rad offers matrices (e.g., Affi-Prep Hz) that couple protein ligands via their sugar residues (i.e., they leave the protein part untouched).

With ligands of lower MW and only one derivable primary amino group, you can attempt to first derivatize the primary amino group with a large side chain (instead of the matrix) and determine the K_i of this compound. It is a good sign if the K_i of the derivative is not substantially higher than that of the source compound. There are proteins for which no ligand exists. In this case, and if the detection assay permits, you can clone peptide ligands from epitope libraries (Scott and Smith 1990).

5.3.2.3 The Role of the Matrix

The second unknown in affinity chromatography is the matrix. Here, many commercial products are available (Table 5.6), and you would only make or develop your own affinity matrices after you have tried these products and found them all unsatisfactory.

The pores of the matrix must be big enough to let the sought-after protein diffund into the inside of the gel globules. If the pores are too small, the ligand in the inside is lost for the chromatography.

A long spacer is often needed with ligands of small MW, so that the ligand can settle into its binding pocket without problems. To avoid hydrophobic or ion exchanger effects, the spacer should be uncharged and hydrophile.

The density of the matrix in activated groups determines the maximally attainable substitution with ligands. A high substitution degree often, but not always, yields the best results (Parcej and Dolly 1989). If enough ligand is available, it is worth derivatizing small amounts of different matrices (different matrix material, spacer, density in activated groups, coupling chemistry) and checking for adsorption via the batch method.

The stability of the covalent binding between ligand and matrix depends on the nature of the activated groups of the matrix. Acid amide and ether compounds are stable. The former originate with matrices whose alkylcarboxy residues were activated to N-hydroxysuccinimide esters. According to Wilchek and Miron (1987), the activated groups of commercial products such as Affigel 10 consist only partially of N-hydroxysuccinimide esters from alkylcarboxy residues. They report that up to 80% of the activated groups are unstable β-alanine esters

Table 5.6. Activated matrices for affinity chromatography.

Name	Manufacturer	Matrix	Reactive Group	Spacer Length	Spacer Type	Good For
Affigel 10	Biorad	Cross-linked agarose (Bio-Gel A-5m)	N-hydroxysuccinimide ester (15 μMol/ml gel)	10 C	Hydrophile	Primary amino group
Affigel 15	Biorad	Cross-linked agarose (Bio-Gel A-5m)	N-hydroxysuccinimide ester (15 μMol/ml gel)	15 C	Hydrophile, positive charge	Primary amino group
Activated CH-Sepharose	Pharmacia	Agarose (Sepharose 4B)	N-hydroxysuccinimide ester (6–10 μMol/ml gel)	6 C	Hydrophobe	Primary amino group
Epoxy-activated Sepharose	Pharmacia	Agarose (Sepharose 6B)	Oxirane groups (15–20 μequiv/ml gel)	11 C	Hydrophile	Primary amino groups, hydroxyl groups, thiol groups
CNBr-activated Sepharose 4B	Pharmacia	Agarose (Sepharose 4B)	Imidocarbonate groups	None	—	Primary amino groups
Act-Ultrogel ACA 22	IBF-biochemics	Acrylamide/ agarose	Aldehyde groups	—	—	Primary amino groups
Spectra/Ge I MAS	Spectrum	Not known	Aldehyde groups	5 C	Not known	Primary amino groups
Spectra/Ge I MAS B	Spectrum	Not known	Bromoacetyl groups	10 C	Not known	Primary amino groups
Reacti-Gel	Pierce	Agarose	Imidazolyl carbamate	None	—	Primary amino groups

whose coupling products are alkali-labile and hydrolyze over time (the ligand "leaks"). During coupling with epoxides, which is supposed to lead to ethers, unstable esters form as well. The Spectrum company claims to have developed an activated matrix (Spectra/gel MAS) that enters into stable covalent bindings with the ligands.

Sources
1. Parcej, D., and Dolly, J. (1989). "Dendrotoxin Acceptor from Bovine Synaptic Plasma Membranes," *Biochem. J.* 257: 899–903.
2. Wilchek, M., and Miron, T. (1987). Limitations of N-hydroxysuccinimide Esters in Affinity Chromatography and Protein Immobilization," *Biochemistry* 26: 2155–2161.

5.3.2.4 Special Case: The Purification of His-tagged Proteins

Even the big guys have to work protein-biochemically once in a while. In four of five cases, this means purifying proteins tagged with His. Such His tags are 6 to 10 histidine residues long and can sit on the C or N terminal end of the protein. Proteins with His tags are largely produced in *E. coli*, but other expression systems are also possible (e.g., insect cells). His-tag proteins are purified via affinity chromatography over Ni-chelate columns, because His tags bind specifically and reversibly to these columns. Agarose often serves as a matrix. There are two purification methods: native and denatured.

The native method is used when the His-tag protein is in solution or can easily be brought into solution. For purification, the protein is put into slightly alkaline buffer (e.g., 50 mM Tris-Cl, pH 8.0) and after thorough centrifugation equilibrated with the Ni-chelate column material (1 to 24 h at 4° C). The binding capacity of most column materials lies at 10 to 15 mg of His-tag protein/ml of gel. Once the protein is adsorbed, it is washed. (This step is important, because not only does His-tag protein stick to the Ni-chelate columns but to a lot of dirt as well) (e.g., metal proteases or proteins containing histidine). For washing, every researcher has his own detergents and methods. I can recommend the following: 20 column volumes each of Tris pH 8.0, 0.5 M NaCl, followed by 20 column volumes of 50 mM Tris pH 8.0, 0.5 M NaCl, and 5 mM imidazole. The latter removes the proteins bound to the column (which contain histidine but do not have a His tag). It does not do so always, however, and not always completely. To calm your conscience, you can wash a third time with 50 mM Tris pH 8.0, 0.5 M NaCl, and 10 mM imidazole. To elude, the imidazole concentration is raised to 0.5 to 1 M. At this concentration, imidazole with a His tag competes for the Ni-chelate binding sites (after all, histidine contains an imidazole ring).

Problems: The His-tag protein often does not bind to the column. This can be due to the fact that you only believe that the His-tag protein is a His-tag protein. However, it could also be that the His-tag protein pulled in its tail and kept it tucked away inside. It is said that it sometimes helps to add 2 mM EDTA to the preparation and to dialyze it against the buffer. Why this is supposed to help, however, is not clear to me. Another possibility is the denaturing method. The native conformation of the protein is destroyed and the His tag becomes accessible. This method is also used when the protein is located within inclusion bodies of *E. coli*.

With the denaturing method, the sample is dissolved in buffer containing urea or guanidine chloride. Typical compositions of such a lysis buffer would be 8 M urea, 1 M NaCl, 10% glycerin pH 8.0 or 6 M guanidine chloride, 100 mM NaH_2PO_4, and 10 mM Tris-Cl pH 8.0. In the case of membrane proteins, you could still add 0.5% NP-40 or TRITON-X-100.

The broth is sonicated extensively and centrifuged extensively. Next, you equilibrate the Ni-chelate column with lysis buffer and then pour the lysate on top. Washing is the decisive factor for the purity of the product. You usually wash with lysis buffer, and when you want to be especially thorough you also wash with lysis buffer containing 5 to 10 mM imidazole or with a somewhat more acidic lysis buffer. Kees et al. (2001) recommend washing the column again with 10 mM Tris-Cl pH 8.0 and 60% isopropanol in 10 mM Tris-Cl pH 8.0. This, they say, efficiently removes the endotoxins from *E. coli* and detergents. As with other columns, for the Ni-chelate columns it is also true that the main thing about washing is variation, not duration! The denatured protein is eluded with a pH step (e.g., with lysis buffer pH 4.95). It is wise to elude an unknown protein first with a pH gradient to determine the precise elution pH. For subsequent purifications, you use a pH step. You can also insert a smaller pH step before as a washing step.

Tightly binding His-tag proteins elude with 1% trifluoride acetic acid. You can use the columns repeatedly without having to load them again with nickel ions each time. You should only avoid washing the columns with EDTA, but that is probably obvious.

Problem: Now and then, the His-tag protein appears in the eluate as an oligomer that cannot even be dissolved in SDS. Here, Ni ions that leak from the column form SDS-stable chelates with the proteins. EDTA helps.

Recommendation: If the Ni-chelate column still yields unsatisfactory results, try different metal chelate columns. There are metal pentadentate chelator agaroses for copper, nickel, tin, and cobalt ions. The same protein can exhibit surprisingly different behaviors on different columns, as far as purification factor and yield are concerned.

Source
1. Kees, L., et al. (2001). "Purification of His-tagged Proteins by Immobilized Chelate Affinity Chromatography: The Benefits from the Use of Organic Solvent," *Protein Expr. Purif.* 18: 95–99.

5.3.2.5 How Do You Proceed?

Protocols for coupling ligand and matrix can be found in the literature, and the previous chapters point out potential problems. As a rule, coupling the ligand to an activated matrix is done within a few hours. After the coupling, you should thoroughly wash the derivatized matrix on a fritte—alternating between high and low ion strengths and high and low pHs (change from pH 4 to pH 8)—in order to get rid of nonbound ligands, provided the ligand can stand the change in pH. After the washing, measure how much ligand stuck to the gel. For this, you can use the light or UV adsorption of the derivatized matrix in 87% glycerine (for absorbing ligands), add some radioactive ligand during the coupling (if available), and perform a protein determination (protein ligand) or a phosphate determination (for ligands containing phosphorus).

Derivatized matrices often leak small amounts of ligand. This is due to the hydrolysis of the covalent matrix/ligand binding (see Section 5.3.2.3) or because the matrix was not washed satisfactorily and slowly releases adsorbed ligand. This leaking effect is often irrelevant, even for the functionality of the matrix, which can often be used for months. However, with highly loaded matrices whose highly affine ligand hangs from one single unstable binding the leak effect can be disruptive, for example, when the released ligand hinders the binding of the sought-after protein to the affinity matrix or its detection during the pass-through. The latter leads to the mistaken conclusion that the protein had bound to the column.

Chromatography: Load the sample onto the column in a buffer that permits an optimal binding between derivatized matrix and the sought-after protein, according to pH, ion strength, and so on. Allow them enough time to bind. Use prior binding assays to estimate the conditions.

After the adsorption of the protein, you should wash the column with several column volumes of buffer. You can increase the washing power of the buffer with detergents (e.g., 0.05% TRITON-X-100) and/or high salt concentrations (e.g., 0.2 M KCl). Of course, the high ion strength or the detergent must not separate the sought-after protein from the matrix. The sought-after protein is eluded either with the same ligand or (better) with a chemically different ligand. You can also add an agent that selectively lowers the binding affinity between sought-after protein and ligand (i.e. the binding is dependent on Ca^{2+}, and thus elution with EDTA). If this is not practical (e.g., because there is no such agent and the ligand is too valuable), you have to resort to nonspecific elution methods such as changing the ion strength or the pH.

The elution ligand often disturbs the detection of the sought-after protein in the eluate. Diluting the sample decreases the ligand concentration, but also the measuring signal (often until it is unrecognizable). You neutralize the bothersome ligand with the following methods.

- Changing of the detection assay (e.g., detection with antibodies instead of a binding assay with radioactively selected ligand).
- Dialysis of the eluate (takes a long time and the sought-after protein could be adsorbed or digested by proteases).
- Centrifugation of the eluate through gel filtration columns (incomplete separation of the ligand and danger of adsorption).
- Precipitation with PEG (requires a protein carrier for diluted protein solutions, can only be done with part of the eluate, and some ligands adsorb to the precipitate).
- Ion exchanger chromatography or lectin chromatography with glycoproteins of the eluate. Although both methods allow extensive washing, they are laborious and involve heavy losses.

The protein concentration in the eluate of an affinity chromatography is generally low. Therefore, the adsorption of the purified protein to column materials, pipette tips, or test tubes becomes noticeable (reduce with high ion strength or 0.05% TRITON-X-100 or siliconize). For lack of distraction, the remaining proteases will primarily go for the sought-after protein.

> If thou followest these precepts and rules, Sancho, thy days will be long, thy fame eternal, thy reward abundant, thy felicity unutterable.

Sources

Overview Articles
1. Cuatrecasas, P. (1970). "Protein Purification by Affinity Chromatography," *J. Biol. Chem.* 245: 3059–3065.
2. Robinson, J., et al. (1980). "Affinity Chromatography in Nonionic Detergent Solutions," *Proc. Natl. Acad. Sci. USA* 77: 5847–5851.
3. Scott, J., and Smith, G. (1990). "Searching for Peptide Ligands with an Epitope Library," *Science* 249: 386–390.

Coupling of Low-molecular Ligands via Their Primary Amino Group
1. Pfeiffer, F., et al. (1982). "Purification by Affinity Chromatography of the Glycine Receptor of Rat Spinal Cord," *J. Biol. Chem.* 257: 9389–9393.
2. Raeber, A., et al. (1989). "Purification and Isolation of Choline Acetyltransferase from the Electric Organ of *Torpedo marmorata* by Affinity Chromatography," *Eur. J. Biochem.* 186: 487–492.
3. Sigel, E., et al. (1983). "A γ-aminobutyric Acid/Benzodiazepine Receptor Complex of Bovine Cerebral Cortex," *J. Biol. Chem.* 258: 6965–6971.

Coupling of Low-molecular Ligands via Their Carboxyl Group
1. Hampson, D., and Wenthold, R. (1988). "A Kainic Acid Receptor from Frog Brain Purified Using Domoic Acid Chromatography," *J. Biol. Chem.* 263: 2500–2505.
2. Senogles, S. (1986). "Affinity Chromatography of the Anterior Pituitary D_2-dopamine Receptor," *Biochemistry* 25: 749–753.

Coupling of Low-molecular Ligands via Their Hydroxyl Group
1. Abood, L., et al. (1983). "Isolation of a Nicotine Binding Site from Rat Brain by Affinity Chromatography," *Proc. Natl. Sci. USA* 80: 3536–3539.

Coupling of Low-molecular Ligands via Their Aromatic Amino Group
1. Haga, K., and Haga, T. (1983). "Affinity Chromatography of the Muscarinic Acetylcholine Receptor," *J. Biol. Chem.* 258: 13575–13579.

Coupling of Low-molecular Ligands via Their Bromide Alkyl Group
1. Bidlack, J., et al. (1981). "Purification of the Opiate Receptor from Rat Brain," *Proc. Natl. Acad. Sci. USA* 78: 636–639.

Coupling of Peptides via Primary Amino Groups
1. Sheikh, S., et al. (1991). "Solubilization and Affinity Purification of the Y2 Receptor for Neuropeptide Y and Peptide YY from Rabbit Kidney," *J. Biol. Chem.* 266: 23959–23966.

Coupling of Proteins via Primary Amino Groups
1. Puma, P., et al. (1983). "Purification of the Receptor for Nerve Growth Factor from A 875 Melanoma Cells by Affinity Chromatography," *J. Biol. Chem.* 258: 3370–3375.
2. Rehm, H., and Lazdunski, M. (1988). "Purification and Subunit Structure of a Putative K^+ Channel Protein Identified by Its Binding Properties for Dendrotoxin I," *Proc. Natl. Acad Sci. USA* 85: 4919–4923.

5.4 The Purity Test

> It is true that, in order to see if it was strong and fit to stand a cut, he drew his sword and gave it a couple of slashes, the first of which undid in an instant what had taken him a week to do.

The usual purity test is the SDS gel electrophoresis with subsequent silver stain. If you want to do more, add an analytic IEF or perform a two-dimensional gel electrophoresis. If several bands appear on the SDS gel, this could be due to pollution or proteolysis products of the sought-after protein. The sought-after protein could also be an oligomer protein complex. Thus, you have to determine which of the bands belong to the sought-after protein. For this, you examine the distribution of band intensity and biological activity. You run a sucrose gradient with the purified protein sample. For each fraction of the gradient, you run a part on the SDS gel, stain the gel for protein, and measure the intensity of the bands (e.g., with a scanner). At the same time, you determine the biological activity in the respective fractions. If a band belongs to the sought-after protein, its intensity distribution across the gradients must agree with the distribution of the biological activity. Of course, this is only a necessary condition, not a sufficient one.

A gel filtration or an ion exchanger column eluded with salt or pH gradients does the same job as a sucrose gradient. However, sucrose gradients have the advantage that you can run different gradients (sucrose, glycerine) in a centrifuge under different conditions (salt, pH).

For binding proteins, you identify the ligand/binding subunit with cross-linking assays (see Section 2.4). If it is likely that the bands in the SDS gel stem from the sought-after protein, the question remains whether you are dealing with different subunits or whether the bands are related and originated from (for example) proteolysis. Here, the partial proteolytic digestion is the decisive factor. The two proteins in question are partially digested with a protease (e.g., *Staphylococcus aureus* V8, chymotrypsin, papain, subtilisin) under identical conditions, and the proteolysis products are separated on an SDS gel. If the two proteins are different, the band patterns in the gel must likewise be different. If one protein is a proteolysis product of the other, the patterns should be similar.

Partial proteolytic digestion provides quick information. However, because the sensitivity of a protein as opposed to proteases strongly depends on its conformation, this information is only dependable if both samples are completely denatured during the digestion and contain sufficient SDS (> 0.05%) (Walker and Anderson 1985).

For partial digestion, the following is also true: Without control there is a lot of shouting and little wool. Support the result of your partial digestion well! Run several digestions with proteases of different specificity. Use different SDS/gel systems to separate breakup products of different size ranges (linear gel, gradient gel). Finally, a positive control and a negative one (e.g., the comparison of BSA with BSA and BSA with ovalbumin) show you that the method works under the selected conditions.

Sources
1. Cleveland, D., et al. (1977). "Peptide Mapping by Limited Proteolysis in Sodium Dodecyl Sulfate and Analysis by Gel Electrophoresis," *J. Biol. Chem.* 252: 1102–1106.
2. Walker, A., and Anderson, C. (1985). "Partial Proteolytic Protein Maps: Cleveland Revisited," *Anal. Biochem.* 146: 108–110.

5.5 Profiting

Once the protein purification is finally done, the experimenter documents it in print. Figure 5.8 indicates research possibilities.

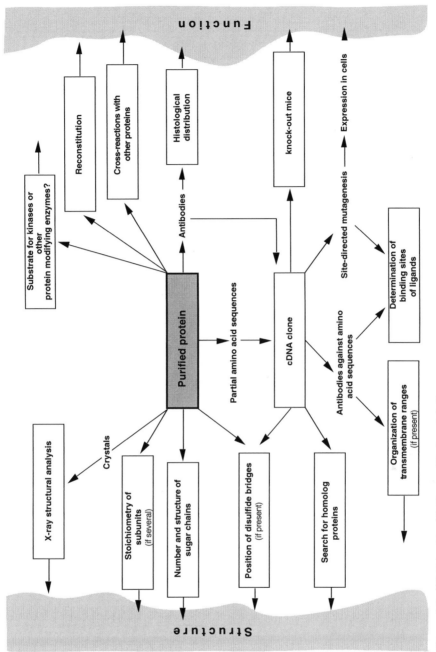

Figure 5.8. What to do with the purified protein?

- Isolating the cDNA clone: Recommendable if there is still enough time before the doctoral defense, if sequencing the protein is unproblematic, and if the protein is not too rare. The cDNA clone opens a host of further research possibilities.
- Antibodies (monoclonal and polyclonal): Producing antibodies against the purified protein is not enough for a paper, unless the antibodies have interesting properties (e.g., they influence the function of the antigen).
- Whether a protein is glycosylated can generally be answered quickly. However, for a paper the researcher must still address additional questions. At which position and how are the sugar chains linked to the protein? Of which type are the chains? What is their sequence? Does the glycosylation influence the function of the protein? If the methods have to be introduced first, the project can turn into a treadmill for the doctoral candidate that provides little gain. Often, he only weeds the field for the successor (see Chapter 9).
- Is the protein a substrate for kinases or other protein-modifying enzyme? A few initial experiments show whether the protein is suitable as a kinase substrate. Careful: most proteins with a serine or threonine can be phosphorylized if you only incubate them long enough with Mg ATP and any kinase. Not every phosphorylation has something to do with the biological reality. If it looks reasonable (the kinases phosphorylize fast and stoichiometrically), you determine the stoichiometry of the phosphorylation, which kinases phosphorylize which amino acids under which conditions, the kinetics, and so on. Then you can still determine the position of the phosphorylized amino acid in the amino acid sequence. Material for a solid paper! However, with this paper you can only spark interest if you can prove that the phosphorylation of the protein influences its function.

Sources

1. *Methods Enzymol.* (1991). vol. 200, "Protein Phosphorylation Part A: Protein Kinases—Assays, Purification, Antibodies, Functional Analysis, Cloning, and Expression."
2. *Methods Enzymol.* (1991). vol. 201, "Protein Phosphorylation Part B: Analysis of Protein Phosphorylation, Protein Kinase Inhibitors, and Protein Phosphatases."

Chapter 6 Antibodies

"There's a remedy for everything except death," said Don Quixote.

It is always a good idea to produce antibodies against a purified protein. With antibodies you can do immunoaffinity columns, immunological detection assays, histologic investigations on slices, and screening of expression banks. Antibodies against peptide sequences of a transmembrane protein inform us about its position in the membrane. Antibodies against single subunits of oligomer proteins provide information about their composition and stoichiometry. The best thing about antibodies is that the antibody producer does not have to do these experiments herself. Often it is enough to produce the desired antibodies and to pass it on to other workgroups.

Antibodies are glycoproteins. They consist of four polypeptide chains: two light and two heavy ones. The chains are connected with each other via disulfide bridges (Figure 6.1). The MW of the heavy chains is about 55 kd, that of the light chains is about 25 kd, and intact antibodies weigh in at 160 kd. Light and heavy chains have an N-terminal variable and a C-terminal constant region. The variable region binds the antigen.

Serum contains antibodies of the classes IgG, IgM, IgA, IgE, and IgD. The classes differ in their heavy chains. The IgG antibodies have the highest concentration in the serum (8 to 16 mg/ml). They are synthesized by B lymphocytes and have two heavy chains of the γ type (Figure 6.1). The subclasses of the IgG antibodies (IgG$_1$, IgG$_{2a}$, IgG$_{2b}$, IgG$_3$, and so on) differ in the C-terminal end of the heavy chains. IgG antibodies are stable proteins. They withstand low concentrations of SDS (0.05%) and several freeze/thaw cycles.

IgM antibodies are pentameres of the basic structure shown in Figure 6.1, with an MW of about 900 kd. IgM antibodies are known as unstable and sensitive, and they hardly even survive any purification. They are a curse to every experimenter.

This chapter describes the production of anti-serums, the purification of antibodies, and the most important immunological screening technologies. For monoclonal antibodies, I refer you to *Methods Enzymol.* I (1986) 121, "Hybridoma Technology and Monoclonal Antibodies." I also touch only briefly upon the newer in-vitro immunization methods in the following (with friendly consultation from Dr. Rose-John, Mainz).

For polyclonal as well as monoclonal antibodies, the experimenter needs a functioning organism. This is often disadvantageous. For example, human antibodies against human antigens are only difficult to produce because of the immunological tolerance, although they would be of medical interest. Toxic antigens cause the death of the animal rather than an immunological reaction, unless the experimenter inactivates the toxin before. However, this means he loses the most interesting property or the most interesting epitope. In-vitro immunization with phages offers a way out of this dilemma (Figure 6.2).

The methodical advantages of in-vitro immunization:

- Speed: If antigen and phage libraries exist, an experienced worker can produce useful sc or Fab fragments within only four weeks. However, success often heavily depends on the quality of the phage library. This, in turn, depends on how the library was won, how the animal was prepared from which the genes were taken, and on the number of different phages. Even good phage banks with 10^{10} to 10^{12} different phages are no guarantee of a hit. Finally, if you have to set up in-vitro immunization from scratch and you are not experienced in the method, you can count on two years of hard work.
- Low price: In comparison with monoclonal antibodies, in-vitro immunization does not need expensive media, cell culture materials, and so on.

Sources
1. Griffiths, A., et al. (1993). "Human Anti-self Antibodies with High Specificity from Phage Display Libraries," *EMBO J.* 12: 725–734.

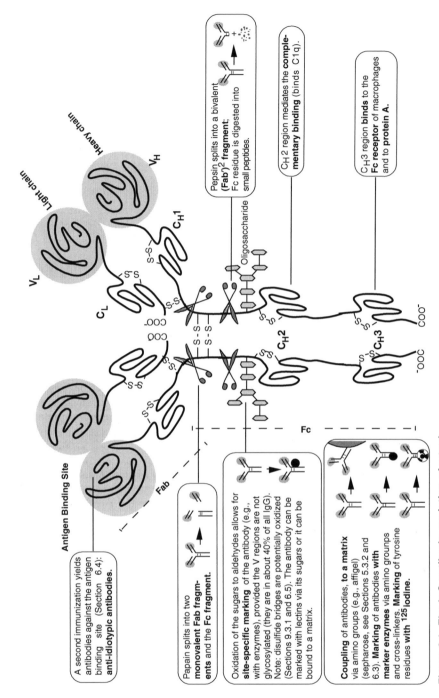

Antigen Binding Site

Light chain

Heavy chain

V_L

V_H

C_L

C_H1

C_H2

C_H3

COO⁻

Oligosaccharide

Fab

Fc

A second immunization yields antibodies against the antigen binding site (Section 6.4): **anti-idiotypic antibodies.**

Pepsin splits into a bivalent **(Fab')2 fragment;** Fc residue is digested into small peptides.

C_H2 region mediates the **complementary binding** (binds C1q).

C_H3 region **binds** to the **Fc receptor** of macrophages and to **protein A.**

Papain splits into two **monovalent Fab fragments** and the **Fc fragment.**

Oxidation of the sugars to aldehydes allows for **site-specific marking** of the antibody (e.g., with enzymes), provided the V regions are not glycosylated (they are in about 40% of all IgG). Note: disulfide bridges are potentially oxidized (Sections 9.3.1 and 6.5). The antibody can be marked with lectins via its sugars or it can be bound to a matrix.

Coupling of antibodies, **to a matrix** via amino groups (e.g., affigel) (sepharose, see Sections 5.3.2 and 6.3). **Marking** of antibodies **with marker enzymes** via amino grounps and cross-linkers. **Marking** of tyrosine residues with **125 iodine.**

Figure 6.1. The IgG antibody: variation of a tool.

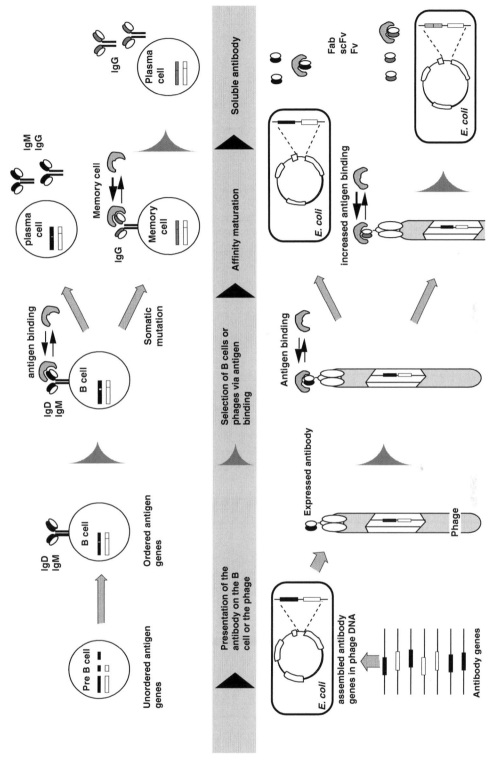

Figure 6.2. Imitating the strategy of the immune system (after Marks et al. 1992).

2. Marks, J., et al. (1992). "Molecular Evolution of Proteins on Filamentous Phage," *J. Biol. Chem.* 267: 16007–16010.
3. Nissim, A., et al. (1994). "Antibody Fragments from a Single Pot Phage Display Library as Immunochemical Reagents," *EMBO J.* 13: 692–698.

6.1 Production of Polyclonal Antibodies

6.1.1 Antigen

A substance is antigenic if it triggers antibodies against itself in an organism. Soluble proteins are quite antigenic, and aggregated proteins are very antigenic. Denatured proteins are less antigenic than native proteins, but antibodies against denatured proteins are better suited for the development of immunoblots. The antigenicity of long-chain sugar is weak to medium. That of nucleic acids is weak.

Substances with MW < 5 kd are rarely antigen. To trigger the formation of antibodies, peptides or even smaller molecules thus must be coupled to a carrier or linked into high-molecular adducts. BSA or the glycoprotein KLH (keyhole limpet hemocyanin) serve as carriers.

The amino acid sequence of a protein is often known, but the protein itself is still unpurified. With antibodies against peptides from the protein sequence, this problem can be solved. For the antibodies to specifically bind the corresponding protein, the peptides should be at least eight amino acids long. In addition, the affinity of antipeptide antibodies generally increases with the length of the peptide. For antibodies against native proteins, the peptide sequence must be hydrophile because hydrophile peptide sequences are probably located on the surface of the native protein. As rule of thumb, the N-and C-terminal ends are located on the surface of the protein and are thus accessible for immunization. The antigenicity of peptide sequences is difficult to predict, but the "hydrophilicity scale" from Parker et al. (1986) provides useful pointers. Peptides that are up to 20 amino acids long can be produced quickly and in mg amounts with peptide synthesizers.

Sources

Conversion of Small Molecules into Antigens
1. Parker, J., et al. (1986). "New Hydrophilicity Scale Derived from HPLC Peptide Retention Data: Correlation of Predicted Surface Residues with Antigenicity and X-ray Derived Accessible Sites," *Biochemistry* 25: 5425–5432.
2. Posnett, D., et al. (1988). "A Novel Method for Producing Anti-peptide Antibodies," *J. Biol. Chem.* 263: 1719–1725.
3. Seguela, P., et al. (1984). "Antibodies Against γ-aminobutyric Acid: Specificity Studies and Immuno-cytochemical Results," *Proc. Natl. Acad. Sci. USA* 81: 3888–3892.
4. Wang, J. (1988). "Antibodies for Phosphotyrosine: Analytical and Preparative Tool for Tyrosyl-phosphorylated Proteins," *Anal. Biochem.* 172: 1–7.

Immune system: The variable domains of the heavy and light chain form the antigen binding site. The multitude of antigen binding sites is due to the combination of different gene elements. The VH domain (H from heavy chain) is formed via the combination of a V gene (from variable) with a D gene (from diversity) and a J gene (from joining). The VH domain and its products are black. The VL domain (from light chain) is formed via the combination of a Vκ-or Vλ gene with a Jκ or Jλ gene. The VL domain and its products are white. The gene combinations lead to millions of B cell clones. Each clone has a certain combination of VH and VL genes and presents the corresponding antibody as an antigen receptor on the cell surface. If antigen binds, the B cell multiplies (first selection). It converts into long-lived memory cells as well as into short-lived plasma cells that secrete the antibodies. The affinity of the antibodies is increased by a second selection of the memory cells.

Phage system: You get combined V genes (VH and VL) via PCR from lymphocyte mRNA (from immunized or nonimmunized organisms). VH and VL genes are combined at random and are cloned in phages, which creates millions of different phages. The VHNL pairs encode either for scFv antibodies (VH and VL product artificially linked via a polypeptide) or for Fab fragments. The phages carry the antibody fragments on their surface. Thus, phages with antigen-binding antibody fragment can be purified with immobilized antigen (first selection). Nonsupressor bacteria infected with purified phages synthesize soluble antibody fragments. The isolated antibody genes can be mutated, and the corresponding mutant phages subjected to a second selection. This yields antibody fragments of higher affinity.

If the antiserum needs to be specific, the antigen used for the immunization must not contain any antigenic pollutants. With protein samples containing strongly antigenic pollutants, the immune response against the pollutant can be stronger than the response against the main protein. Depending on the animal, 1 to 100 µg of antigen is needed for a successful immunization.

The injection of untreated antigen rarely causes a good immune response (especially with soluble antigens). In addition, some antigens are toxic. The antigen is thus processed, by emulgation with adjuvant and/or aggregation (e.g., treated with formalin, following Hirokawa 1978).

SDS gel electrophoresis is often the last step in antigen purification. Antigens in gels can be processed in the following ways (I prefer C).

- The antigen is identified in the gel with a staining method without fixation (e.g., sodium acetate; see Section 1.42), the acrylamide piece with the antigen is cut out, and chopped-up gel pieces are injected into the animal. It is chopped up either by repeated pressing through two syringes, which are connected via a short tube, or by freeze-drying (followed by grinding in the mortar and rehydration). Both methods are laborious.
- The antigen is blotted on nitrocellulose, and the antigen band is stained via a gentle method (e.g., copper iodide; see Section 1.6.1) and then cut out. The experimenter can implant the membrane piece hypodermically. Many people also dissolve the nitrocellulose in DMSO, emulsify with adjuvant, and inject (Chile et al. 1987). If you leave DMSO out at RT for a longer period of time, dimethylsulfate forms under the influence of light. Dimethylsulfate is a potent cell toxin and causes cancer. Characteristics: the DMSO does not harden at 4° C anymore (H. Maidhof, Mainz). In the end, an ultrasonic treatment (six times for 30 second with microtip) transforms the nitrocellulose into a powdery suspension that can be injected with or without adjuvant (Diano et al. 1987). These methods often lead to the formation of antibodies against nitrocellulose that interfere during the development of immunoblots. It helps to adsorb the serum to blocked nitrocellulose membranes.
- The experimenter identifies the antigen in the gel with a staining method without fixation (e.g., sodium acetate; see Section 1.4.2). The acrylamide piece with the antigen is cut out and electrically eluded (see Section 5.2.2.2). An emulsion of the eluate with adjuvant is injected.

Sources
1. Chiles, T., et al. (1987). "Production of Monoclonal Antibodies to a Low-abundance Hepatic Membrane Protein Using Nitrocellulose Immobilized Protein as Antigen," *Anal. Biochem.* 163: 136–142.
2. Diano, M., et al. (1987). "A method for the Production of Highly Specific Antibodies," *Anal. Biochem.* 166: 224–229.
3. Hirokawa, N. (1978). "Characterization of Various Nervous Tissues of the Chick Embryos Through Responses to Chronic Application and Immunocytochemistry of β-bungarotoxin," *J. Camp. Neurol.* 180: 449–466.

6.1.2 Adjuvant

An adjuvant is vital for the success of an immunization. It works as a depot for the antigen and increases the immunological reaction of the animal. There are many adjuvants. Among experimenters, the most popular one is Freund's adjuvant (FA). Pierce offers two adjuvants, Imject Alum and AdjuPrime, which supposedly are better tolerated by the animal to be immunized.

FA consists of nondegradable mineral oil, which creates the depot effect. The complete FA also includes dead mycobacterial tuberculosis, which triggers the immune response. FA and complete FA must not be given intravenously. For the initial injection, the antigen in PBS is mixed 1:1 with complete FA, and a stable oil emulsion is produced. The quality of this emulsion determines the depot effect and with it the success of the immunization. With some patience and luck, the two-syringe method or ultrasonication yields good emulsions. The emulsion is good if a drop of the emulsion remains a drop on a water surface and does not dissolve.

6.1.3 Injection and Serum Harvesting

With life a great many things come right.

The standard animal for polyclonal antibodies is the rabbit. In the course of an immunization protocol, a rabbit delivers up to 500 ml of serum, hamsters or guinea pigs 20 or 30 ml, and a mouse 0.5 ml. The immune response is different from rabbit to rabbit, which is why the careful experimenter immunizes at least two animals. For an immunization, rabbits require 10 to 100 μg of antigen per injection. Hamsters or guinea pigs get by with 2 to 10 μg.

For the antigen injection, the experimenter has a choice among intramuscular (im), intradermal (id), subcutaneous (sc), intravenous (iv), or intraperitoneal (ip) injection. In bigger animals, with some anatomical knowledge, injection into the lymph nodes is also possible. The id injection requires skill, but it is said to give a better immune response. In addition, it provides for a longer depot effect. Chopped-up gel pieces cannot be injected intradermally and mouse skin is too thin for id injections.

The sc injection is easy, and per injection site you can inject 50 μl for the mouse and 800 μl for the rabbit. With sc as well as id injections for the rabbit, you place several (up to 10) injections on both sides of the spine. Intradermally, you can inject about 100 μl into the rabbit per injection site. The beginner best learns this skill from an experienced colleague. In many laboratories, these techniques are passed on almost like a biological trait from one doctoral candidate generation to the next doctoral candidate generation.

About one week before the first injection, the experimenter takes blood from the animal and prepares the preimmuneserum (important for later control experiments). Ten days after the first injection, he draws blood again and determines the antibody titer. After another 10 days, he boosts (i.e., antigen in FA gets injected). Ten to fourteen days after the boost, blood is taken again and the titer is determined. Two weeks later, the experimenter can boost again, and 10 to 14 days later another blood drawing follows. It is a good idea to document the immunization protocols in a special notebook, with time points, antigen preparations, blood drawing, and so on.

Drawing blood from the ear vein of the rabbit is an uncomfortable business. In inexperienced hands, and under unhappy circumstances, it can lead to the death of the antibody donator (the neck is an especially sensitive part of the rabbit). The experimenter shaves a site around the ear vein, clamps the other ear vein, disinfects the shaved site with 70% alcohol, and cuts the vein with a sterile scalpel at a 45-degree angle. He catches the exiting blood (20 to 30 ml) in a 50-ml glass tube.

The rabbit is often scared, or the blood flow in the ear stops for other reasons. The experimenter becomes impatient, and the rabbit even more anxious. A vicious cycle develops that causes the experimenter to break out in a cold sweat and drives the rabbit into a panic. Drawing blood goes smoothly if the rabbit is warm (wrapped in a soft towel) and sits on a rough surface. A mechanical rabbit holder is bad form. Rubbing of the ear promotes the blood flow and rubbing some Xylol on the ear's central artery also expands the arteries.

The mouse is put under red light for 10 to 20 minute before the blood drawing to promote the blood circulation. Then, the animal is inserted into a mouse block (Bio-Tec, Basel) and a maximum of 100 μl of blood is taken through a cut in the tail. Antiseptic ointment is applied to the cut after the blood drawing.

The sausage technique is not for everybody, and many mice cannot get used to it and barely give blood. Ella Klundt has published (Klundt 2001) an alternative for small quantities of mice blood (40 to 60 µl). With this vein puncture, either the right or the left collateral vein of the mouse's tail is punctured. The method is mouse friendly but requires some practice.

Fourty µl of blood are not enough for you? You want more? In this case you have to puncture the blood sinus behind the mouse's eye (Klundt 2001). For this somewhat revolting method, a glass capillary is directed to a certain point behind the eye bulbus and pushed into the blood sinus. The capillary sucks out the blood and you direct it into an Eppendorf container. If you let the mouse bleed out completely, you can gain approximately 2 ml of blood per animal. If you draw less blood, so that the mouse survives, you can puncture repeatedly in intervals of a few days (alternating eyes).

The mouse is anesthetized before the puncture, but the method still requires skill and mental stability. Get trained by an experienced puncturist. Nobody there who can show you the technique? Then get at least informed about the anatomy of the mouse eye or (better) move on to the next method.

There is also a Norwegian method for drawing mouse blood (Hem et al. 1998). Here, the vena saphena magna in the hind leg is punctured. For this, the mouse is stuck head-first into a Falcon tube, the hind leg shaved, the vein punctured, and the blood sucked off through a capillary. The method delivers approximately 100 µl of blood and you can repeatedly draw blood at the same site in the course of one day. With suitable holding tubes, you can also bring bigger animals to donate blood (from the rat up to the ferret).

Recommendation: Mouse blood coagulates unbelievably quickly. Suck the blood drops into the capillary immediately upon exit of the vein.

Sources
1. Hem, A., et al. (1998). "Saphenous Vein Puncture for Blood Sampling of the Mouse, Rat, Hamster, Gerbil, Guinea Pig, Ferret, and Mink," *Laboratory Animals* 32: 364–368. The same method can be found on the Web (includes many pictures) at *www.uib.no/vivariet/mou_blood/Blood_coll_mice_.html*.
2. Klundt, E. (2001). "Tipps für Blutsauger," *Laborjournal* 11/2001, S61.

For harvesting serum, you leave the blood at RT for 60 minutes. Then you carefully shake the blood cake off the tube wall or loosen it with a glass stick and incubate for another 2 to 3 h at RT or overnight at 4° C. The serum is decanted and centrifuged for 10 minutes at 5,000 rpm to remove the remaining blood cells. At −80° C, serum is stable for years.

6.1.4 Purification of Antibodies

After an immunization, only a small share of the serum proteins are antibodies and only a small portion of the antibodies bind antigen. Hence, it is worthwhile extracting the antibodies from the serum. The popular methods are ammonium sulfate precipitation and chromatography over HA, protein A sepharose, or an antigen column. The antigen column distinguishes antigen binding and nonantigen-binding antibodies.

HA chromatography processes large amounts of serum in one step and concentrates the antibodies. It is not necessary to dialyze the serum before performing the chromatography over a column buffer. The yield is good, and the antibodies relatively clean. HA chromatography does not completely separate monoclonal antibodies from tissue culture supernatants with fetal calf serum from albumin.

Source
1. Bukovsky, J., and Kennet, R. (1987). "Simple and Rapid Purification of Monoclonal Antibodies from Cell Culture Supernatants and Ascites Fluids by Hydroxyl Apatite Chromatography on Analytical and Preparative Scales," *Hybridoma* 6: 219–228.

The ammonium sulfate precipitation of serum (with 45% saturation for rabbit serum) is a quick-and-dirty method, with an emphasis on "dirty." The precipitation takes a few hours and afterward you still have to dialyze.

Source

1. Dunbar, B., and Schwoebel, E. (1990). "Preparation of Polyclonal Antibodies," *Methods Enzymol.* 182: 663–670.

Protein A (MG 42 kd) from *S. aureus* binds antibody reversibly via its Fc domain (1 M protein A binds 2 M IgG). High salt concentrations (2 to 3 M NaCl) and alkaline pH values (pH 8 to 9) strengthen the binding between protein A and the antibody. Nevertheless, for the purification of antibodies from rabbit, mouse (except IgG$_1$), people, horse, and guinea pig physiological salt concentrations are enough. Thus, the chromatography of serum over protein A sepharose in one step delivers (most) IgG antibodies. Many companies sell premade protein A columns. Protein A columns are eluded with a buffer pH of 2 to 3. The stable protein does not mind a treatment with 4 M urea or pH 2.5.

This easy and efficient purification method has the disadvantage that not all IgG subclasses bind to the column with high affinity and thus get lost (e.g., mouse IgG$_1$). There are species differences: protein A binds rabbit antibody well and rat antibody badly.

Source

1. Ey, P., et al. (1978). "Isolation of Pure IgG$_1$, IgG2a, and IgG2b-immunoglobulins from Mouse Serum Using Protein A-sepharose," *Immunochemistry* 15: 429–436.

Protein G (MW 35 kd) binds antibody and albumin. An artificial variant of protein G binds only antibody. The advantage of protein G is its complementary specificity to protein A: the antibodies that do not bind to protein A, such as many monoclonal antibodies, often bind to protein G.

An antigen column allows one to isolate antigen-specific antibodies (affinity-purified antibodies). The affinity purification of antibodies is required for some ELISAS and for the production of enzyme-conjugated antibodies. Affinity-purified antibodies give good immune precipitations, and the chromatography over the antigen column finally transforms low-titer serums into reasonable reagents.

Affinity purification of antibodies is no wizardry as long as the antigen can be coupled to a commercial matrix with standard methods, and as long as it does not couple at exactly the site targeted by the antibodies. It takes approximately two days.

Problems: For an antigen column, you need large amounts of antigen. Antibodies with low affinity against the antigen do not bind to the column. Those with very high affinity cannot be eluded from the column in their native state.

The antigen is covalently coupled to a matrix (e.g., proteins or peptides in Affigel 10) via a primary amino, carboxyl, or sulfhydryl group. Coupling methods are described in Section 5.3. The experimenter pumps the diluted serum or the hybridoma supernatant over the column, washes out the unbound protein, and successively eludes with acidic pH, basic pH, and 3 M MgCl$_2$.

After a few runs, the antibody binding capacity of the column often already decreases. Reasons: proteolysis or denaturing of the antigen or blockage through highly affine antibodies. However, many antigens are more stable in the column than in solution. Washing with 4 M urea or 1.5 M KSCN helps against blockage.

Affinity purification: Dilute serum 1:10 with 10 mM Na-HEPES pH 8.0, centrifuge at 20,000 rpm for 30 minutes (Sorvall, SS34), and slowly load the supernatant onto the column. Afterward, wash extensively with 10 mM Na-HEPES pH 8.0 and 10 mM Na-HEPES pH 8.0, 300 mM KCl (at least 10 column volumes each). Some also wash with 200 mM NaSCN pH 5.8. Acid-sensitive antibodies are eluded with 100 mM glycine-HCl pH 2.5 (about 5 column volumes). Afterward, wash with 10 mM Na-HEPES pH 8.0 until the pH in the eluate is 8.0. Then, elude with 100 mM triethylamine pH 11.5 for the alkalinity-sensitive antibodies. Again, wash the column with Na-HEPES pH 8.0 and then elude with 3 M MgCl$_2$. Start with a strong buffer in the fraction tube when you elude with pH extremes (e.g., 1/10 fraction volume 1 M Tris-Cl pH 8.1). In the end, adjust the column to pH 8.0 with 10 mM Na-HEPES, check the pH of the eluates (neutralize when required), and remove the MgCl$_2$ (e.g., via dialysis).

6.2 Immunoprecipitation

Immunoprecipitation (IP) isolates a certain antigen from the multitude of antigens in a solution. The IP (possibly with subsequent gel electrophoresis) answers among others the following questions. Does an antibody recognize a binding or enzyme activity? Under the influence of a parameter, does the cell change the MW of the antigen, its phosphorylation state, or the construction of its sugar chains? IP also provides information about the specificity of antibodies or the distribution of the subunits of oligomer protein families over their members.

The advantage of IP compared with an immunoaffinity column (see Section 6.3) lies in the fact that you do not have to pour a column and a lot of preparations can be run at the same time. However, the IP is an analytic tool. For isolating larger amounts of antigen, you use immunoaffinity columns.

In the precipitate, you determine the antigen either via a function (e.g., ligand binding, enzyme activity) or via the MW in the SDS gel electrophoresis. With a protein stain of the gel, the antigen often disappears among the large amounts of antibodies in the precipitate. Then the experimenter radioactively marks the antigen before the IP and identifies it afterward via an autoradiogram. In cell cultures, she can mark proteins with radioactive amino acids (e.g., ^{35}S methionine). Antigen solutions or cell surface antigens can be iodized.

Some calculating saves you work and disappointments during the IP of rare proteins (i.e., proteins in low concentration such as transmitter or hormone receptors) with antibodies of low affinity ($K_D > 50$ nM, as with many antipeptide antibodies). With the assumed concentrations of antibody, antigen, and K_D, mass-action law and retention equations (Figure 2.4) allow us to calculate the expected percentage of antigen antibody complex (second-degree equation). Sometimes it turns out that the antibody can bind only a few percent of the antigen under the given conditions.

Source
1. Firestone, G., and Winguth, S. (1990). "Immunoprecipitation of Proteins," *Methods Enzymol.* 182: 688–700.

6.2.1 Immunoprecipitation with Immobilized Protein A

Because of their bivalent nature, polyclonal antibodies precipitate at an optimal ratio of antibody and antigen. The precipitate forms a net that carries away a multitude of proteins other than antigen and antibody. Furthermore, the precipitation requires large quantities of antigen and antibody. Thus, the method is hardly used anymore, even in form of the Ouchterlony technique. Instead, antigen is commonly added in excess to the antigen solution to prevent precipitate from forming. Afterward, the antigen/antibody complexes and the free antibodies are precipitated with protein A sepharose or fixed *S. aureus* cells (Pansorbin).

Monoclonal antibodies often do not bind well to protein A. To precipitate the antigen/antibody complexes anyway, you perform one more incubation with anti-mouse IgG (Promega, Jackson Lab), protein G sepharose (Sigma), or anti-mouse IgG sepharose before adding the immobilized protein A.

The precipitate contains antigen, antibody, the fixed *S. aureus* cells or protein A sepharose, and nonspecifically adsorbed protein (with cell lysates, largely actin). You can reduce the portion of nonspecifically adsorbed protein by using affinity-purified antibodies instead of serum and precipitating with higher ion strength (0.15 to 0.4 M NaCl), or by processing them in the presence of detergents (1 to 2% TRITON-X-100 or NP-40, or 0.1% deoxycholate). It also helps to load the precipitation medium onto a sucrose cushion and to centrifuge the loaded protein A sepharose and so on through the sucrose. If the background is still too high, the precipitate is dissolved and the precipitation repeated (Platt et al. 1986; Doolittle et al. 1991).

An IP without control is like landing a plane without light. The following control experiments should be performed for an IP.

- IP with preimmunoserum or with a monoclonal antibody that is not directed against the antigen.
- IP without the anti-antigen antibody (e.g., only with protein A sepharose or Pansorbin).
- IP with cell lysates that do not contain the antigen.
- If the antibodies are directed against a peptide sequence, it must be possible to prevent the IP via the suitable peptide.

Sources
1. Doolittle, M., et al. (1991). "A Two-cycle Immunoprecipitation Procedure for Reducing Nonspecific Protein Contamination," *Anal. Biochem.* 195: 364–368.
2. Platt. E., et al. (1986). "Highly Sensitive Immunoadsorption Procedure for Detection of Low-abundance Proteins," *Anal. Biochem.* 156: 126–135.
3. Sakamoto, J., and Campbell, K. (1991). "A Monoclonal Antibody to the β-subunit of the Skeletal Muscle Dihydropyridine Receptor Immunoprecipitates the Brain ω-conotoxin GVIA Receptor," *J. Biol. Chem.* 266: 18914–18919.

6.2.2 Immunoprecipitation with Immobilized Antibody

If you do not want to have any free antibodies in the precipitate (e.g., because you have to detect the antigen by SDS gel electrophoresis and you do not want to or have not been able to mark the antigen radioactively), couple the (affinity-purified) antibodies covalently to a matrix (e.g., Affigel 10). From the precipitate, only antigen and nonspecifically adsorbed proteins go in solution with Lämmli sample buffer (without reductant). The antibodies remain on the matrix. You can now run the precipitate extract on SDS gels and stain these for protein, without antibodies smearing all over. You can also blot the gel and develop the blot with precipitating antibody. With a lot of antibodies in the precipitate extract, you get an ugly blot because the denatured antibodies will appear on the blot as a wide band. Denatured antibodies bind the second antibody (with which the blot is developed; see Section 1.6.3), often almost as well as the native antibodies (which sit on the antigen).

The antibodies couple to the usual activated matrices (see Section 5.3), not only via the Fc domain but at random (i.e., also via the antigen binding site). Part of the coupled antibody thus binds the antigen not very well or not at all. Therefore, Schneider et al. (1982) cross-link the antibodies to protein A sepharose. Due to this trick, the antibodies couple covalently to the matrix in the right orientation, via the Fc domain.

The IP with monoclonal antibodies presupposes a certain affinity ($K_D < 100$ nM) of the antigen/antibody binding. With polyclonal antibodies, the multivalent antigen binding ensures that the antigen sticks well to the matrix, even with low affinity of the single antibodies. However, it often leads to higher nonspecific binding.

You can reduce the nonspecific binding with the techniques described in Section 6.2.1. A simple washing is a common measure, less because of its effectiveness than because this is the first thing that occurs to a beginner. For washing, the precipitate is shaken up with buffer, centrifuged again, and the supernatant sucked off. You can do this several times and the precipitate will slowly become cleaner and cleaner, but you will not necessarily become happier. Affigel 10 or sepharose pearls do not form stable pellets and you therefore often also suck off some of the pearls with the supernatant. The yield of adsorbed protein as well as quantitative comparisons between precipitations thus become a gamble. Furthermore, because of the soft voluminous pellet the supernatant cannot be sucked off completely, which decreases the washing effect and dilutes the eluate.

Brymora et al. (2001) help themselves here with a little washing machine. It consists of a small column that fits into a microcentrifuge tube (Figure 6.3). The precipitation medium is incubated in the sealable column and then centrifuged. Now the gel on the fritte can be washed comfortably and efficiently. Afterward, you can elude the bound protein from the column (e.g., with SDS sample buffer). You can also use hot sample buffer, because the column can be heated in the bain-marie up to 85° C. According to Brymora et al., this technique delivers up to three times higher yields than the method of centrifuging and sucking off the super-

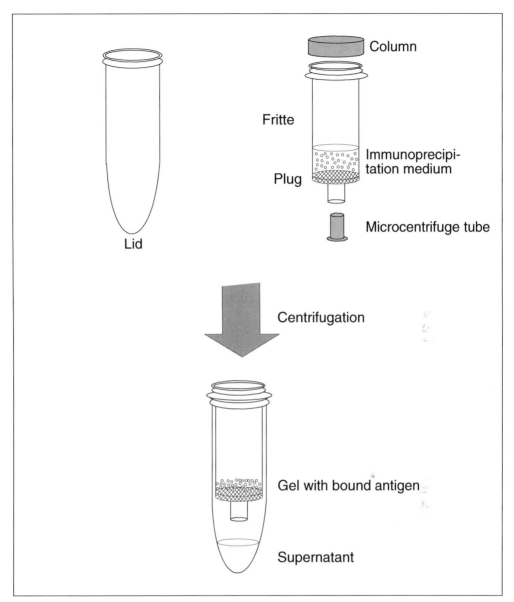

Figure 6.3. Washing machine for immunoprecipitation (after Brymora et al.).

natant. Suitable supplies are available from (for example) Amersham Pharmacia (Probe Quant G-50).

Sources
1. Brymora, A., et al. (2001). "Enhanced Protein Recovery and Reproducibility from Pull-down Assays and Immunoprecipitations Using Spin Columns," *Anal. Biochem.* 295: 119–122.
2. Grassel, S., et al. (1989). "Immunoprecipitation of Labeled Antigens with Eupergit C1Z," *Anal. Biochem.* 180: 72–78.
3. Peltz, G., et al. (1987). "Monoclonal Antibody Immunoprecipitation of Cell Membrane Glycoproteins," *Anal. Biochem.* 167: 239–244.
4. Schneider, C., et al. (1982). "A One-step Purification of Membrane Proteins Using a High-efficiency Immunomatrix," *J. Biol. Chem.* 257: 10766–10769.

6.3 Immunoaffinity Chromatography

Immunoaffinity chromatography quickly and with certainty yields a clean antigen if specific monoclonal or polyclonal antibodies against the antigen are available. The success of immunoaffinity chromatography depends on the quality of the antibody, but it is a good-natured method, not overly sensitive, and usually works at the first attempt. Antibodies (at least IgG) can easily be coupled to common matrices, usually without losing the binding activity. Immunoaffinity chromatography also makes extensive washing possible and delivers antigen free of antibody. However, the experimenter generally has to elude the antigen with nonspecific and rough means. The attainable purification factors lie between 1,000 and 5,000. The yield depends on the elution method, but is typically above 50%. Coupling the antibodies to the matrix and doing the chromatography typically takes about two days.

How do you proceed? You couple the antibodies covalently to a matrix, pour the antibody matrix into a column, and wash out uncoupled antibody (see Section 5.3). Then you load the column with antigen. To bind to the immunoaffinity column, the antigen needs at least 2 to 3 hours (ideally, overnight).

You elude with glycine HCl buffer pH 2.5 or 3.5 M $MgCl_2$. Cruder methods such as elution with pH 1.8 or with 3 M thiocyanate, or even with 1% SDS, also get you there. However, the column cannot be used anymore afterward. Even with "milder" elution agents (pH 2.5), the biological activity of the antigen often perishes, and the column also dies after a few runs. Immunoaffinity columns with polyclonal antibodies are difficult to elude because of the strong multivalent antigen binding. Mild elution agents suffice for columns with monoclonal antibodies or polyclonal antibodies against a peptide sequence of the antigen. In these cases, it is also possible to elude with ligands (e.g., the peptide). If you elude with pH 1.8 to 2.5 and would like to be gentle with your antigen, you neutralize the eluate by inserting a strong buffer into the fraction tube.

Sources
1. Leah, J., et al. (1988). "Purification of Ornithine Aminotransferase by Immunoadsorption," *Anal. Biochem.* 170: 495–501.
2. McGillis, J., et al. (1987). "Immunoaffinity Purification of Membrane Constituents of the IM-9 Lymphoblast Receptor for Substance P," *Anal. Biochem.* 164: 502–513.

6.4 Antibodies Against Unpurified Proteins

How do you get antibodies against an unpurified protein? There are three possibilities.

• You produce monoclonal antibodies against the unpurified or partially purified protein sample until you find an antibody that binds the sought-after protein. Success depends on the purity of the protein sample, the antigenicity of the sought-after protein, the lab's experience in the production of monoclonal antibodies, and how elaborate the detection assay for the sought-after protein is. The search can take between half a year and several years. Once in a while this leads to interesting results on the side (e.g., antibodies against unknown proteins). You also learn a useful and popular technique and thus increase your market value.

• If the sequence of the protein is known (e.g., from its expression cloning), a natural next step would be to produce antibody against suitable peptides (see Sections 6.1.1 and 6.3). This project is almost guaranteed to yield results. If they are sufficiently specific and affine, the antipeptide antibodies enable more than just the purification of the corresponding protein. Often they provide information about structure and folds of the protein and, with membrane proteins, intra- or extracellular localization of the peptide sequence. However, antibodies against an amino acid sequence do not necessarily have to also recognize the corresponding native protein. The sequence in the native protein may be inaccessible to the

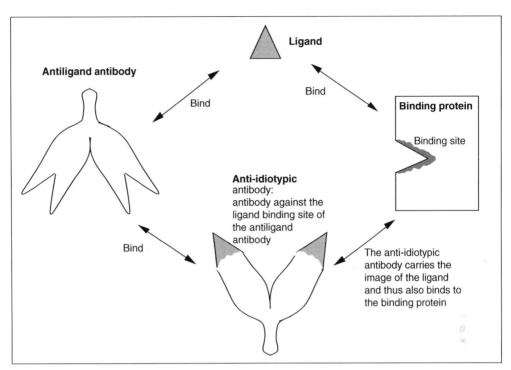

Figure 6.4. Anti-idiotypic antibodies (after Marx 1985).

antibody, or sugar chains might shield the sequence from the antibody. Finally, the idea of working with antipeptide antibodies is not exactly original, and published sequences are readily available.

- If you have an antigen ligand, you can produce anti-idiotypic antibodies (Figure 6.4). First, you produce antiligand antibodies. Antiligand antibodies bind the ligand (i.e., they have a binding site for the ligand). In some antibodies, this binding site resembles the ligand binding site of the sought-after protein. Thus, you produce monoclonal or polyclonal antibodies against the antiligand antibodies (monoclonal or affinity-purified polyclonal). Many of these antibodies bind to the ligand binding site of the antiligand antibodies. Some of these anti-idiotypic antibodies also bind to the ligand binding site of the sought-after protein, which makes itself known (for example) through inhibition of the ligand binding itself. These antibodies are the important ones. If their titer against the sought-after protein is high, they can be used for immunoaffinity chromatography.

If you want to use polyclonal anti-idiotypic antibodies for an immunoaffinity chromatography, you have to affinity-purify them before in an antiligand antibody column. Polyclonal anti-idiotypic antibodies do not bind the sought-after protein multivalently, but only via the binding site of the ligand. Hence, the protein of the immunoaffinity column eludes even under mild conditions (e.g., with an excess of ligand).

All theoretical elegance notwithstanding, it is laborious and risky to take the route of anti-idiotypic antibodies. You have to process the ligands for the immunization, set up one or several detection assays, and carry out at least two immunizations whose success depends on luck (Table 6.1). A beginner needs years to produce useful anti-idiotypic antibodies. Typically, antibodies have only minimal practical value by the time they become available. During the years of their production, the tactical situation in the area has often changed drastically.

Sources
1. Haasemann, M., et al. (1991). "Anti-idiotypic Antibodies Bearing the Internal Image of a Bradykinin Epitope," *J. Immunol.* 147: 3882–3892.
2. Kleyman, T. R., et al. (1991). "Characterization and Cellular Localization of the Epithelial Na$^+$ Channel," *J. Biol. Chem.* 266: 3907–3915.

Table 6.1. Ligand/receptor pairs.

Class	Ligand	Receptor/Binding Site	References
Molecules of low MW	Adenosine	Adenosine	Schick and Kennedy 1989, p. 40
	Amiloride	Na^+ channel	Kleyman et al. 1991
	Alprenolol	β-adrenerg	Schick and Kennedy 1989, p. 40
	Bis-Q	Acetylcholine	—
	Spiperone	D2 Dopamine	—
	Haloperidol	D2 Dopamine	—
Peptides	Angiotensin II	Angiotensin	—
	Bradykinin	Bradykinin	Haasemann et al. 1991
Proteins	Substance P	Substance P	Schick and Kennedy 1989, p. 40
	Choleratoxin	Ganglioside GM1	—
	IgE	Fcξ	—

Table 6.2. Steps for ELISA.

Adsorption of the Antigen	Adsorption of the Antibody
1. Adsorption of antigen A and accompanying protein to solid material takes hours with PVC microtiter plates but minutes with nitrocellulose.	1. Adsorption of affinity-purified or monoclonal anti-antigen antibody to solid material (e.g., PVC microtiter plates) takes hours.
2. Block of unsaturated protein binding sites on the solid material (e.g., with BSA takes 30 minutes).	2. Block of free protein binding sites on the solid material.
3. Incubation with anti-antigen antibody for 2 h.	3. Incubation with antigen solution A for 2 h.
4. Washing off of unbound anti-antigen antibodies (3 × 10 minutes).	4. Washing off of unbound proteins.
5. Incubation with enzyme-conjugated anti-IgG antibody for 2 h. Peroxidase is usually used as enzyme.	5. Addition of an enzyme-conjugated anti-antigen antibody. With monoclonal antibodies, the first and second antibodies must not be identical.
6. Washing off of unbound enzyme-conjugated anti-IgG antibodies (3 × 10 minutes).	6. Washing off of unbound enzyme-conjugated anti-antigen antibody.

The amount of bound enzyme-conjugated anti-IgG antibody is determined via the enzyme reaction.

3. Kussie, P., et al. (1989). "Production and Characterization of Monoclonal Idiotypes and Anti-idiotypes for Small Ligands," *Methods Enzymol.* 178: 49–63.
4. Schick, M., and Kennedy, R. (1989). "Production and Characterization of Anti-idiotypic Antibody Reagents," *Methods Enzymol.* 178: 36–48.

6.5 Immunological Detection Techniques

Immunological measuring methods stand out in terms of their simplicity, versatility, and sensitivity. With dot blots or ELISAs you can easily measure 200 to 400 samples per day, and with a good dot blot 500 pg of antigen in 100 μg protein provides a measurable signal. The assays are generally used to determine the concentration of antigens or the titer of serums. Often-used procedures are outlined in Tables 6.2 and 6.3. Generally, the following are true: the more specific anti-antigen and enzyme conjugated an anti-IgG antibody the more specific the assay, the more stringent the washing processes and the less layers the more specific and weaker the signal, and the higher the concentration of the antibodies the stronger the signal and background stain.

Sources
1. Frutos, M., et al. (1996). "Analytical Immunology," *Methods Enzymol.* 270: 82–101.
2. Parker, C. (1990). "Immunoassays," *Methods Enzymol.* 182: 700–718.

Table 6.3. Different ELISAs.

Measured Quantity	Layers in Well	Advantages	Disadvantages
Antigen concentration	—Enzyme-conjugated anti-antigen antibody 2* (constant) —Antigen (variable) —Anti-antigen antibody 1* (constant)	—	The assay requires two monoclonal or one affinity-purified antibody. One of the antibodies must be conjugated with enzyme.
Anti-antigen antibody (screen for hybridoma supernatants; determine serum titer)	—Enzyme-conjugated anti-IgG antibody (constant) —Anti-antigen antibody (variable) —Antigen (constant)	Enzyme-conjugated anti-IgG antibody is available in retail.	—
Antigen concentration	—Enzyme-conjugated anti-antigen antibody (constant) + antigen (variable) —antigen (constant)	Little work; specific.	The anti-antigen antibody must be enzyme-conjugated. Polyclonal antibodies must be affinity-purified first.
Antigen concentration	—Enzyme-conjugated anti-IgG antibody (constant) —Enzyme-conjugated anti-antigen antibody (constant) + antigen (variable) —Antigen (constant)	Enzyme-conjugated anti-IgG antibody is available in retail. Assay also works with serum.	—
Antigen concentration	—Enzyme-conjugated anti-species 2 antibody (constant) —Anti-antigen antibody of species 2 (constant) —Antigen (variable) —Anti-antigen antibody of species 1 (constant)	High sensitivity.	The assay requires one monoclonal or one affinity-purified anti-antigen antibody of species 1. Due to the four layers, it is time consuming, laborious, and error prone.

*Anti-antigen antibodies 1 and 2: monoclonal antibodies against different epitopes of the antigen, or one of the anti-antigen antibodies is an affinity-purified polyclonal antibody.

Dot blots adsorb the antigen to nitrocellulose membranes. The dot blot distinguishes itself by quick antigen loading, and you can use serum as an anti-antigen antibody solution. The dot blot is also compatible with the 96-well microtiter plate technique. In comparison to ELISA, it requires two additional devices and steps.

It is wise to load the antigen with a filtration device (Schleicher and Schüll). Afterward, you fix the proteins (e.g., with ethanol/acetic acid solutions, TCA, or heating up) and block the protein binding sites of the membrane (BSA, milk powder, and so on, see Section 1.6.2). This is followed by incubations with anti-antigen and peroxidase conjugated anti-IgG antibody. You wash between the individual steps (Table 6.2). Afterward, you punch the dots on the nitrocellulose membrane into the depressions (wells) of microtiter plates. With a suitable substrate solution, a peroxidase-catalyzed color reaction develops in the wells. The color solution is transferred into new microtiter plates (to get rid of the confetti) and finally measured in the ELISA reader. A calibration curve with defined amounts of antigen quantifies the results (Becker et al. 1989).

The adsorption of antigen to nitrocellulose is not covalent. Thus, the adsorbed antigens can be washed off again (e.g., with higher concentrations of detergents such as Nonidet P 40 and Tween 20) (Lui et al. 1996). How the antigen is loaded is important for it to stick to the membrane and for preserving of the binding capability to the antibody. A good loading solution for protein antigens is 0.1 M NaOH in 20 to 40% methanol (Wiedenmann et al. 1988) or 0.5% deoxycholate, 20% methanol in Tris buffer pH 7.4 (Becker et al. 1989). Under these conditions, nitrocellulose binds up to 1,000 times more protein than polyvinyl chloride (300 ng/cm^2) or polystyrene (400 ng/cm^2).

Even more popular than dot blots are microtiter plate assays, so-called ELISAs (enzyme-linked immunosorbent assay), in which antibody or antigen is loaded into the depression of polyvinyl chloride or polystyrene plates (Kemeny 1994). The depressions are then further coated with antibody, antigen, and enzyme-conjugated antibody in a defined sequence. The antigen is detected via an enzymatic color reaction (Table 6.2). Many companies (Nunc, Flow, Costar, Falcon) offer a palette of products such as 8- or 12-channel pipettes, automatic washing devices, ELISA readers, and so on that make life easier for the friends of ELISA.

As with the dot blot, there is the danger with ELISA that stringent washing steps and incubation steps also loosen part of the adsorbed antigen or antibody again. There are also antigens (e.g., small peptides) that do not adsorb to polystyrene plates or that change their binding properties during adsorption. Here, plates that covalently couple antigen or antibody help (Covalink from Nunc).

Washing the ELISA plates well with antigen or antibody between incubation steps provides for good signals. The washing protocols are as numerous as the immunolabs. P. Häring from Hoffmann-La Roche of Basel recommends that between incubation steps you wash the plates twice with tap water, once with PBS and 0.05% Tween 20, twice again with tap water, and finally with PBS.

For some ELISAs, homemade affinity-purified antibodies have to be conjugated with an enzyme (peroxidase, alkaline phosphatase). You can solve this problem with cross-linkers (Jeanson et al. 1988), or more elegantly through the formation of aldehyde groups in the sugar residues of the enzyme (or antibody) (Tijssen and Kurstak 1984) (Figure 6.1). Aldehyde groups form through oxidation with periodate (Figure 9.1). Wolf and Hage (1995) provide a useful overview of the reaction conditions; response time, pH, temperature, and periodate concentration. By choosing suitable reaction conditions, you can vary (between 1 and 8) the number of aldehyde groups per antibody.

With dot blots and ELISAs, you can easily fall victim to artifacts and false positive signals, especially if you use serum dilutions as anti-antigen antibody solution. Controls:
- Protein sample without antigen.
- Assay without anti-antigen antibody.
- Assay with preimmunoserum or with a monoclonal antibody that does not recognize the antigen.
- The careful experimenter double checks an unexpected positive signal of an unknown sample with SDS gel electrophoresis and subsequent blot.

Sources

Dot Blots
1. Becker, C-M. et al. (1989). "Sensitive Immunoassay Shows Selective Association of Peripheral and Integral Membrane Proteins of the Inhibitory Glycine Receptor Complex," *J. Neurochem.* 53: 124–131.
2. Lui, M., et al. (1996). "Methodical Analysis of Protein-nitrocellulose Interactions to Design a Refined Digestion Protocol," *Anal. Biochem.* 241: 156–166.
3. Smith, C., et al. (1989). "Sodium Dodecyl Sulfate Enhancement of Quantitative Immunoenzyme Dot-blot Assays on Nitrocellulose," *Anal. Biochem.* 177: 212–219.
4. Varghese, S., and Christakos, S. (1987). "A Quantitative Immunobinding Assay for Vitamin D Dependent Calcium Binding Protein (Calbindin-D28K) Using Nitrocellulose Filters," *Anal. Biochem.* 165: 183–189.
5. Wiedenmann, B., et al. (1988). "Fractionation of Synaptophysin-containing Vesicles from Rat Brain and Cultured PC 12 Pheochromocytoma Cells," *FEBS lett.* 240: 71–77.

ELISA with Adsorbed Antigen
1. Kemeny, D. M. (1994). *ELISA—Anwendung des Enzyme Linked Immunosorbent Assay im Biologisch-medizinischen Labor.* Stuttgart: Gustav Fischer Verlag.
2. Kingan, T. (1989). "A Competitive Enzyme-linked Immunosorbent Assay: Applications in the Assay of Peptides, Steroids, and Cyclic Nucleotides," *Anal. Biochem.* 183: 283–289.
3. Yoshioka, H., et al. (1987). "An Assay of Collagenase Activity Using Enzyme-linked Immunosorbent Assay for Mammalian Collagenase," *Anal. Biochem.* 166: 172–177.

ELISA with Adsorbed Antibody
1. Goers, J., et al. (1987). "An Enzyme-linked Immunoassay for Lipoprotein Lipase," *Anal. Biochem.* 166: 27–35.
2. Kemeny, D. M. (1994). *ELISA—Anwendung des Enzyme Linked Immunosorbent Assay im Biologisch-medizinischen Labor.* Stuttgart: Gustav Fischer Verlag.

3. Kwan, S., et al. (1987). "An Enzyme Immunoassay for the Quantitation of Dihydropteridine Reductase," *Anal. Biochem.* 164: 391–396.
4. Zhiri, A., et al. (1987). "A New Enzyme Immunoassay of Microsomal Rat Liver Epoxide Hydrolase," *Anal. Biochem.* 163: 298–302.

Conjugation of Antibodies with Marker Enzymes
1. Jeanson, A., et al. (1988). "Preparation of Reproducible Alkaline Phosphatase-antibody Conjugates for Enzyme Immunoassay Using a Heterobifunctional Linking Agent," *Anal. Biochem.* 172: 392–396.
2. Tijssen, P., and Kurstak, E. (1984). "Highly Efficient and Simple Methods for the Preparation of Peroxidase and Active Peroxidase Antibody Conjugates for Enzyme Immunoassays," *Anal. Biochem.* 136: 451–457.
3. Wolfe, C., and Hage, D. (1995). "Studies on the Rate and Control of Antibody Oxidation by Periodate," *Anal. Biochem.* 231: 123–130.

Chapter 7 Proteomics

Hearing this, Sancho with tears in his eyes entreated him to give up an enterprise compared with which the one of the windmills, and the awful one of the fulling mills, and, in fact, all the feats he had attempted in the whole course of his life, were cakes and fancy bread.

7.1 Introduction

As a rule, the word is the servant of the circumstances; it does not create them. And this is good. Still, there are exceptions. One concerns protein research—an area in which exceptions abound. From the end of the 1970s to end of the 1990s, protein research had somewhat of a backward flair. Investigating proteins only with protein-chemical methods was done only by fossilized researchers who had not made the jump into the modern age; namely, molecular biology. How they had to suffer!

It could happen that they were working on a receptor purification, only to read after one year of intensive efforts that the cDNA had been expression cloned. With cDNA, you could simply do a lot more than with purified protein. You got the entire sequence. You could change the sequence, express the protein, produce antibodies against the entire protein or parts of it, measure the function, and so on.

In 1994, Marc Wilkins threw the term *proteome* into the scientific community. And the word became flesh. Today, proteome and proteomics are as hip as PCR was at the end of the 1980s. The word *proteome* is defined as the quantitative totality of proteins of a cell, tissue, or organism (i.e., the knowledge of all expressed proteins and their respective concentrations under certain external conditions).

However, of course, in reality the word had not changed the circumstances. What happened was rather that the current of science has slowly changed its direction and nobody noticed until the word *proteome* put it in the spotlight. One of the rocks that deflected the current consists of the new MALDI mass spectrometers. They determine the MW of proteins and peptides quickly and with high accuracy, and they lend a flair of high tech to protein research. Less spectacular but of similar reach was the introduction of immobilized pH gradients into IEF. Finally, with the end of the Human Genome Project molecular biology had lost its grand vision. A new overarching goal was needed.

People are mindful again of the fact that the goal of biology is to understand the life signs of cells, and the life signs of a cell are based on proteins. Nucleic acids only provide the architectural plan. "There's more to paella than the recipe" (Anderson and Anderson 1998). The new overarching goal consists of understanding the functional network of the cell—the teamwork between proteins, RNA and DNA. How do concentrations and modifications of a protein depend on the other proteins? How do concentration changes of protein X affect the concentrations of the other proteins? To understand the cell as a molecular machine, to capture it in a set of equations, that is the new Holy Grail. There will be a lot of heartache and many a Percival.

Also in its financial aspects, protein research becomes more attractive. There is probably nothing that expresses the state of a cell or an organism with more sensitivity than the quantitative spectrum of its proteins. An illness suppresses the expression of certain proteins and increases or initiates the expression of other proteins. The protein spectrum is thus suitable as an illness or health indicator. Furthermore, the effects of pharmaceuticals can be traced via the protein spectrum (Figure 7.1) and their side effects can be estimated. Of course, you could

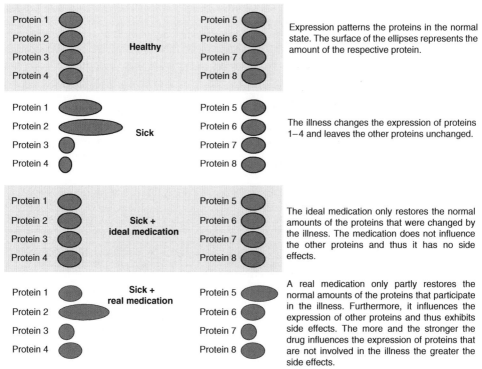

Figure 7.1. The value of quantitative protein spectra for the diagnosis of illnesses and the effect of drugs.

also use the mRNA spectrum for this purpose. After all, it is not a problem to measure the different mRNA in a cell extract. For this there are (for example) the biochips from Affymetrix. However, the protein spectrum is better suited because many processes take place only among proteins. Furthermore, changes in protein spectra allow easier inferences about the mechanisms.

For example, the administration of halothane (a widely used inhaled narcotic) causes trifluoroacetylation of certain liver proteins. The acetylated proteins in turn trigger a hyperimmunological reaction in some people. After administration of halothane, the acetylated proteins appear as new spots in the protein spectrum. In the RNA spectrum, on the other hand, you see at most a change in the distribution of different antibody mRNAs. But careful! The amount of RNA correlates only weakly with the amount of encoded protein. After Anderson and Anderson (1998), the correlation coefficient between mRNA and the respective protein lies on average at 0.48 (i.e., in the middle between perfect correlation of 1.0 and no correlation of 0). In the case of halothane, the protein spectrum points to the direct cause and the mRNA spectrum shows the consequential effect. These spectra together give you the entire mechanism.

Another example: etomoxir irreversibly inhibits carnitine palmitoyltransferase I. This enzyme provides the transport of palmitoyl acid residues into the mitochondrial matrix, where the fatty acid residues are then reduced via β-oxidation. With inhibited transferase, lipids collect in the liver. The lipids in turn stimulate the production of the protein ADRP. This is because ADRP lines lipid drops. In the protein spectrum, the formation of ADRP after etomoxir administration can be traced more reliably than with the mRNA spectrum.

Once the mechanism of a drug is found, drugs with similar mechanisms can easily be identified. They show similar effects in the protein spectrum. In the opposite direction, you can infer that two substances that have similar effects in the protein spectrum use similar mechanisms.

Thus, it is beyond doubt that protein spectra are interesting for medicine and the pharmaceutical industry. But how do you measure them? A relevant measuring method should be able to do the following.

- At least the majority of the proteins should be captured.
- The amounts of the proteins should be quantitatively measured over a maximally large range.
- Because only the comparison of (at least) two spectra (e.g., with and without drug) is meaningful, the method must be reproducible. Ideally, this should be the case between different laboratories, but at the very least in the hands of one and the same experimenter.

It is difficult to even approach satisfying these demands. The main problems are as follows.

- The protein concentrations in most natural samples are spread out over six orders of magnitude. Albumin, for example, occurs in the serum in a concentration of 30 to 40 g/l. Somatotropin, in contrast, occurs only in a concentration of 0.3 to 5 µg/l. The linear range of the staining methods, on the other hand, stretches over only two and at most three orders of magnitude.
- For proteins, in contrast to DNA or RNA, there is no possibility of amplification. If you want to register rare proteins, you have no choice but to load a lot. This strategy corresponds to a head-on attack in the military and in either case it only rarely leads to success (Hart 1991).
- Almost no method captures all proteins: very acidic and very basic ones, very big and very small ones.
- Proteases digest proteins to smaller proteins and thus increase variety and confusion.
- It is often difficult to reproduce how the sample was taken (see Section 7.2). Protein and mRNA spectra have this in common.

Two methods are used to capture proteomes: 2D electrophoresis (Section 7.3) and the SELDI protein chip system (Section 7.5). These methods are complemented by the good old microsequencing (Section 7.6) and different mass spectroscopic methods (Sections 7.4 and 7.6.5). After all, via partial sequences and the exact MW the proteins of a spectrum can be unambiguously identified in databases.

2D electrophoresis, mass spectrometry, and protein chips—all are fine and dandy, and all are fraught with wonderful problems. Many experimenters fail during the first step: the banal, widely underestimated sample taking. Hence, an entire chapter is dedicated to it.

Sources
1. Anderson, L., and Anderson, N. (1998). "Proteome and Proteomics: New Technologies, New Concepts, and New Words," *Electrophoresis* 19: 1853–1861.
2. Hart, L. (1991). "*Strategy*," Meridian Books.
3. Lottspeich, F. (1999). "Proteomanalyse: Ein Weg zur Funktionsanalyse von Proteinen," *Angew. Chem.* 111: 2630–2647.

7.2 Sample Taking

The hope of proteome research lies in the comparison of proteomes: ill against healthy, medicated against control, and cells with protein X against the same cells without protein X. However, comparisons are difficult. They easily lead to the grocery dilemma: How do I avoid comparing apples to oranges?

Assume you want to measure the effect of a drug on the protein spectrum of the liver. With mice, this is relatively unproblematic. You take mice that are genetically uniform and brought up in the same way and you inject one with PBS and the other with PBS plus drug. Then you extract in each case a large piece of tissue from the same area in the liver and you dissolve it in sample buffer (see Section 7.3). Now you run two 2D gels, one for the control mouse and one for the medicated one, in each case with identical amounts of protein. The differences in the 2D gels should be due to the drug.

But how do you perform such an experiment with a patient? People are not genetically uniform. Furthermore, they often eat different things at different intervals, live differently, and work differently. All that has an effect on their liver proteomes. The way out from the nonuniformity of the patients is to use every patient as his own control. Thus: first a puncture, then the drug, then the second puncture. This yields two 2D gels, and their differences are due to the drug's effect. This assumes that the timing of punctures and drug administration are well chosen and the patient does not catch a cold between the punctures. Furthermore, the first puncture must not have any after-effects. Finally, both punctures must yield comparable tissue (Figure 7.2).

It becomes even more difficult if you want to compare liver tumors with healthy liver cells. There you cannot puncture a control, because the patient is already admitted as a cancer patient. However, if you compare the proteome of his liver puncture to an average of healthy people, the differences do not have to represent tumor-specific proteins but may be based on the banal causes mentioned previously (e.g., genetic polymorphisms or nutritional predilections of the patient). And there she sits, the clinical researcher, and scratches her head over her 2D gels: "Two dozen new spots in comparison to the healthy. Wonderful! But which ones stem from the tumor?" It would help a lot to be able to compare the tumor's proteome to healthy cells of the same liver. But oh! Many tumors do not form big, uniform cell heaps but infiltrate healthy tissue with many little cell heaps or individually wander around between hepatocytes. The protein spectrum you get from a puncture then strongly depends on the location of the puncture (Figure 7.2). Sometimes you get many cancer cells, and sometimes a few. Sometimes you get many endothelial cells, and sometimes a few. Sometimes you get a lot of stem cells, and sometimes a few. The proteomes of the hepatocytes themselves also presumably depend on position and age. Add to this that the proteome of a tumor changes over time. After all, tumors constantly form new cell lines with new proteomes. The results are thus not reproducible from the get-go—the bugbear of every experimenter. You can torture the patient with punctures over months and the result will not necessarily become clearer. You can feel sorry for doctors sometimes.

Microdissection via the laser adhesive technique may be a solution to this plight (Banks et al. 1999). The tissue sample (e.g., the punctate) is cut into slices with the microtome and the slices are stained with hematoxylin and eosin. You lay the stained cut on a glass plate and push it under an inverted microscope. Now you identify the tumor and lay a little tube on the interesting area. The bottom of the little tube is sealed with a UVA polymer film. The film thus touches the tissue. Now the cancerous parts of the tissue are glued to the film. This is done with a laser beam directed at the desired areas. Once the adhesion is complete, you take the little tube off, with film and the tissue that is stuck to it. The rest of the cut remains on the glass plate. Now the little tube is put on an Eppendorf cup like a lid. In the cup there is sample buffer. If you turn the cup around, the sample buffer loosens the protein from the film (Figure 7.3).

According to Banks et al., neither the stain nor the adhesion influence yield antigenicity and sequenceability of the proteins. This is astonishing. I would have at least expected that rinsing with water, 70% ethanol, 100% ethanol during the staining procedure would wash a lot of the protein away. And is it really true that the histological stains do not modify the proteins? Astounding.

In any case, the method has one problem: per slice you get minute amounts of protein. This means you have to stain, inspect, glue, and extract many, many slices. Rosamonde Banks needed 13 h (without a break!) to prepare enough protein for a 2D gel. Many doctorate students or TAs would probably not have a similar stamina. The team leaders who want to use the laser adhesive technique as a standard method will have to invest a substantial share of their research money into help-wanted ads in lab journals.

Source

1. Banks, R., et al. (1999). "The Potential Use of Laser Capture Microdissection to Selectively Obtain Distinct Populations of Cells for Proteomic Analysis: Preliminary Findings," *Electrophoresis* 20: 689–700.

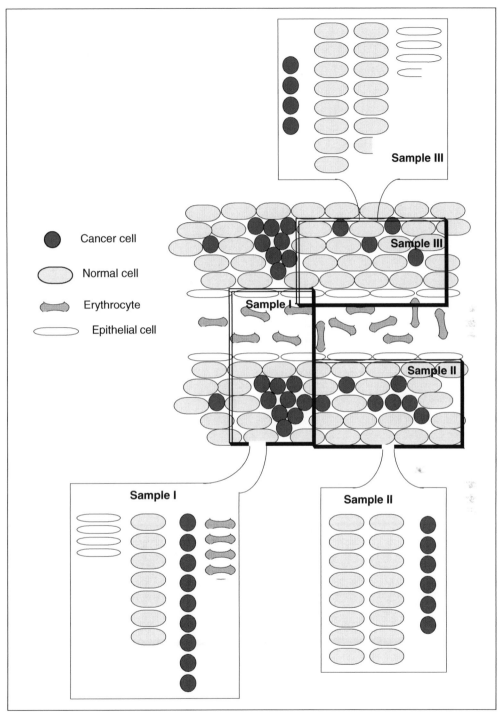

Figure 7.2. The problem of sample taking. Because of the heterogeneous composition of a tissue, the protein spectrum depends qualitatively and quantitatively on where and how the sample is taken.

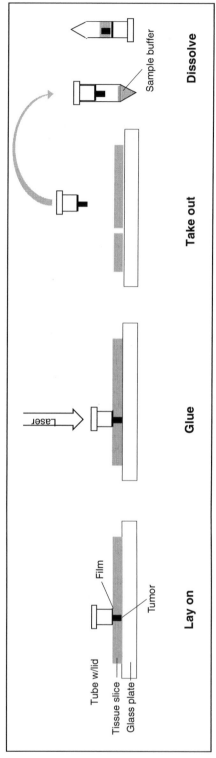

Figure 7.3. Microdissection with the laser adhesive technique.

7.3 2D Gel Electrophoresis

Two-dimensional gel electrophoresis (2D phoresis) separates proteins according to isoelectric point and MW (i.e., in two dimensions). This thus allows the analysis of complicated protein mixtures. O'Farrel and Klose invented this technique independently of each other in 1975. However, for a long time 2D phoresis led a wallflower existence. It did deliver impressive pictures of *E. coli* lysates, but that was it. This was due to the fact that 2D phoresis was not able to reproduce well—not only from laboratory to laboratory but in the hands of the same experimenter. The protein position in the IEF gel wobbled like a one-year-old who is just learning to walk. I once ran two 2D gels from the same sample (cell extract of PC12 cells kept at −80° C) with one week between. Both gels looked impressive (many little dots and spots, and a lot of smudges), but I was able to discern a resemblance only after looking at them for a while. You felt like an art expert: an interpreter of irreproducible smears. With old-style 2D phoresis, nobody has achieved a sensational result as far as I know. It promised big possibilities, but it did not keep this promise.

This sad state improved with the introduction of immobilized IEF (see Section. 5.2.3.3). The pH gradient fixed in the gel increased the reproducibility as well as the resolution of 2D phoresis by an order of magnitude (Corbett et al. 1994). Furthermore, the fixed pH gradient made it possible to dry the gels on a plastic sheet, and store and reswell them again when required. That faciliated the usage and retail enormously, and freed the experimenter from the inane but nevertheless attention-requiring task of gradient pouring. IEF gels with immobilized pH gradients are by now standard for 2D phoresis.

The result of 2D phoresis crucially depends on sample, sample treatment, and sample resolution. How do you extract (for example) the proteins from a liver sample? Dissolve it directly in sample buffer? Grind first in liquid nitrogen and then mix with sample buffer? Lyophilize first and then dissolve in sample buffer? No agreement has been reached yet. The next question is also unclear. What is the best sample buffer? Because the first step of 2D phoresis is IEF, the proteins have to keep their own charge. That means you cannot dissolve with SDS. Nevertheless, the sample buffer has to dissolve as many proteins as possible and split them into subunits. Furthermore, it must prevent aggregation. Finally, the sample buffer should denature all proteases. Which sample buffer can do that? None! But some come close to this ideal (however, not very close).

All sample buffers contain urea (3 to 8 M), a detergent (NP-40 or CHAPS), and a reducing agent (mercaptoethanol or DTT). The sample buffer from the Expasy home page (*www.expasy. ch/ch2d/protocols*) is well liked: 8 M urea, 4% (w/v) CHAPS, 40 mM Tris, and 65 mM DTTS, including a trace of bromophenol blue. Now and then, sample buffers with thiourea are also used: 8 M urea, 2 M thiourea, 2% (w/v) CHAPS, 1% (w/v) DTT, and 0.8% pharmalyte of the corresponding pH range.

Keep the salt concentration of the sample low! This is not possible? Then dilute the sample in a large volume of sample buffer. Avoid elaborate manipulations such as dialyzing, column runs, and so on. Manipulate a lot; you lose a lot (of protein). You have the sample dissolved in sample buffer? Now it is a matter of focusing. For this you select a dry IEF stripe of suitable pH range. Commonly used strips are 18- to 24-cm long and available in retail for broad (3 to 10) and narrow (e.g., 5.5 to 6.5) pH ranges. The strip is rehydrated overnight in a rehydration cassette. For the rehydration solution, the Expasy home page recommends 25 ml 8-M urea, 2% (w/v) CHAPS, 10 mM DTT, and 2% (v/v) resolyte of the relevant pH range together with a trace of bromophenol blue. The completely soaked (rehydrated) strips are transferred into the IEF chamber and covered with paraffin oil to prevent water from evaporating during the focusing. You pipette the samples into the test chamber at the cathodic or anodic end. On analytic gels, you should not load more than 50 µg, but you can shovel up to 15 mg onto preparative ones. The location of the application, anodic or cathodic, apparently matters. Some samples appear to focus better if you apply them close to the anode, and others when applied close to the cathode.

Presumably, this has to do with precipitation effects (i.e., part of the protein does not withstand the extreme pH values close to the electrodes). I find the method from Sanchez et al.

(1997) most applealing. Jean-Charles Sanchez does not apply either at the anodic or at the cathodic end. Instead, he reswells the IEF gel strip right in the sample solution. This way, he avoids precipitation and gets by with shorter focusing times. Furthermore, he does not need any special sample chambers for applying larger amounts of protein. A sample at 50 μg is treated the same way as one of 15 mg.

You focus in devices that are commonly available in retail (Pharmacia, Bio-Rad). However, the conditions are as countless as the focusers. Generally, you start slowly (e.g., you linearly increase the voltage over 3 h from 300 to 3,500 V, followed by 3 h at 3,500 V and, in the end, 5,000 V overnight). Others focus at 150 V for the first 30 minutes, then at 300 V for 1 h, and, in the end, 3,500 V. As far as the duration of the focusing is concerned, represented in volt hours (Vh), the literature offers all values between 40,000 and 400,000 Vh. I think that 100,000 Vh are enough.

After focusing, the proteins in the IEF gel have to be saturated with SDS. For this, you incubate the focused gel strip for 10 to 12 minute in 50 mM Tris-Cl pH 6.8, 6 M urea, 30% (v/v) glycerine, 2% (w/v) SDS, and 2% (w/v) DTT. Afterward, the free SH groups are blocked with 2.5% (w/v) iodoacetamide in 50 mM Tris-Cl pH 6.8, 6 M urea, 30% (v/v) glycerine, and 2% (w/v) SDS for five minutes. The proteins are now ready for the separation in the second dimension, the SDS gel. The IEF gel strip equilibrated with SDS is laid on an SDS plate gel. Again, this raises the question: On what type of an SDS gel? With which SDS gels do you capture the most proteins? Nine to 16% gradient gels are stylish at the moment (capture proteins of 200–8 kd) and 12% gels (capture proteins of 150–14 kd).

For the sake of better reproducibility, you should not polymerize the SDS gel in the presence of SDS. SDS forms micelles, which contain acrylamide monomers. The micelles disrupt the homogeneous polymerization and provide for unpolymerized acrylamide in the gel. Unpolymerized acrylamide can block N-terminals or cross-link proteins. How does the SDS get into the gel later? From the running and sample buffer! SDS runs faster than the SDS protein complexes. Thus, the latter always move in an SDS-containing environment.

The Expasy home page mentioned previously recommends using piperazine diacrylyl instead of bisacrylamids as a cross-linker. Gels cross-linked with piperazine diacrylyl apparently separate the proteins better and block less N-terminals. Furthermore, gels cross-linked with piperazine diacrylyl can be silver stained better. The addition of 5 mM Na-thiosulfates into the running buffer is also done for better staining.

Many experimenters do not use a stacker gel. The IEF gel (in Tris-Cl pH 6.8) together with the agarose used for fixing is sufficient as a stacker, they think. Others (e.g., Corbett et al. 1994) equilibrate the completely focused IEF strips in solutions similar to those mentioned previously (however, with 50 mM Tris-Cl pH 8.8) and then lay the strip on a 4-cm-long stacker (in 125 mM Tris-Cl pH 6.8, 0.1% (w/v) SDS). What should you choose? The method that makes least work (i.e., that without additional stacker gel). I suspect that Corbett et al. use stacker gel for historical reasons.

The SDS separation gel finally separates the focused proteins according to size. Depending on the sample, the experimenter gets up to 3,000 spots.

Problems with IEF: Cysteine oxidation and carbamylation (urea!) create artifact spots. The urea-containing buffers should not be left lying around at ambient temperature for a long time, but should be frozen in aliquots at −80° C or made fresh. In IEF, small temperature changes have big effects on the position of some proteins, presumably because the dissociation constant of an ionizable group strongly depends on the temperature. This reduces reproducibility. Who can guarantee that, say, the thermostat in Uppsala sets the temperature to a tenth of a degree like the thermostat in Rome? Particularly, if you are possibly dealing with different models of thermostats?

Some proteins (e.g., membrane proteins) aggregate under the conditions of IEF. Then you get the same smeary result as native gel electrophoresis (see Section 1.3.2). Here it sometimes helps to dissolve the protein sample in SDS. However, because SDS protein complexes do not focus they do not have a useful isoelectric point. You have to push out the SDS again after diluting with NP-40 or CHAPS (Garrels 1979; Corbett et al. 1994). Many membrane proteins cannot be impressed with this trick and form aggregates again after NP-40/urea is added. They remain only in solution as SDS protein complexes or as complexes with other charged detergents. Furthermore, SDS forms mixed micelles with NP-40 that migrate to the + pole of the IEF gel due to their negative charge. This leads to an unequal distribution of detergents in the IEF gel. Maybe this is the reason for the bad resolution of IEF gels in the presence of SDS/NP-40 mixed micelles. Dockham et al. (1986) suggest dissolving membrane proteins in an alkaline buffer containing lysine, TRITON-X-100, and urea.

Hartinger et al. (1996) developed a special 2D phoresis for membrane proteins. In the first dimension the proteins are separated in acidic buffer in the presence of the cationic detergent benzyldimethyl-n-hexadecylammonium chloride. The second dimension is an SDS gel.

Even synaptophysin, a small integral membrane protein of synaptic vesicles (which I know as a disagreeable gel smudger), focuses nicely on Hartinger gels. However, this method is also not the best you could imagine. Both dimensions separate according to MW. The resolution is not exactly great and you receive no information about the isoelectric point of your protein. The reproducibility of the gels is also questionable. Their usage cannot be any more inconvenient. For example, Hartinger et al. also pour urea into the sample buffer and the gel of the first dimension, and this sample buffer has to be made fresh every time. Still more annoying: because even the two-fold sample buffer solidifies at RT you have to keep it at 60° C. The sample must not be boiled in sample buffer nor frozen. It has to go on the gel directly after dissolving. Finally, you have to wash out the cationic detergent before you lay the gel of the first dimension onto the SDS gel. With all of its disadvantages, the Hartinger method seems to be suited for the analysis of the membrane composition of defined vesicles.

Problems with the SDS gel: It is true, even this well-worn method can still create problems. In particular, this is true where reproducibility is concerned. It is clear when the current must be turned on for the SDS gel: as soon as you put the IEF gel on. But when do you turn it off? When the marker, bromophenol blue, almost reaches the anode? What does "almost" mean? Do you turn it off when the marker disappears in the anode? Or 10, 20, 30 minutes after the marker has disappeared? Absolutely even runs happen only rarely. The marker band runs a little bit crooked, and on one corner it disappears sooner than on the other. What should you put down as the run time then? The run time is important, because it determines the position of the proteins in the 2D gel, even if only in one direction. SDS gel electrophoresis simply is not a balanced method like IEF, where focusing for one hour more or less does not play a role. The use of Rf values does not substantially improve the situation, especially with gradient gels. Researchers who let their 2D gels run for different times in the second dimension will have difficulty comparing these gels.

General problems: The reproducibility of 2D gels has been improved with the IEF with fixed gradients, but it still leaves things to be desired and probably always will. The method simply has too many steps and hence too many sources of error. Errors include weighing out the acrylamide, errors polymerizing, errors setting the temperature, errors determining the run time of the SDS electrophoresis, and so on.

Another disadvantage is that it captures only part of the proteins in the sample. Very big and very small ones get lost, as well as very basic and very acidic ones (Figure 7.4).

Even with small genomes (e.g., yeast), 2D phoresis captures two-thirds of all expressed proteins in the best case. With mammalian cells, the ratio is probably lower.

Rare proteins of a proteome are difficult to capture. You can certainly do thick applications and then use sensitive stains (e.g., Sypro orange), but unfortunately you will find out that this does not help much. After all, a protein does not lie in sharply contoured spots in a 2D gel but exhibits a Gauss distribution (Figure 7.5). The thicker you apply and the more sensitively you stain the more space the spots will take up. They expand and the spots of the common proteins cover the rare ones with their thick Gauss tails. Add to this that the linear range of a staining method covers at most three orders of magnitude. Everything is simply deep blue or deep orange or deep black; that is, as dark as your prospect of finding your rare protein under the fat blob.

Finally, it takes a long time—especially the IEF. From the preparation of the sample to the finished 2D gel, you can well count on two days. This tugs on your nerves and more than offsets the advantage of 2D phoresis (low price). How can you do it better? I believe that the solution lies in the combination of a high-resolution method with the mass spectrometer. You would gain a lot if you could run the IEF gels directly with the mass spectrometer. Here,

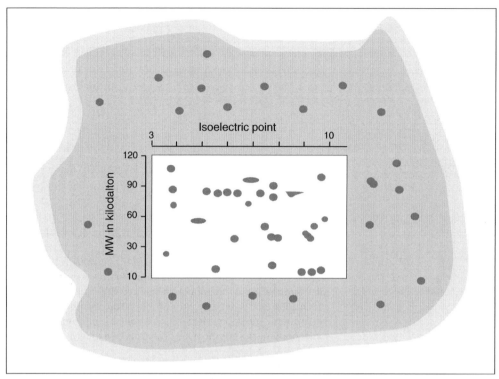

Figure 7.4. Lost land all around. Even 2D gel electrophoresis only shows a part of the proteome.

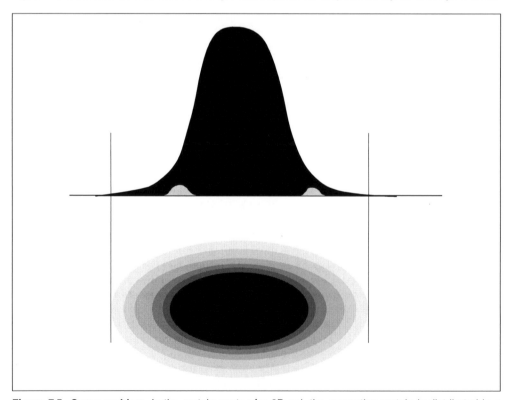

Figure 7.5. Cover problem. In the protein spots of a 2D gel, the respective protein is distributed in a Gaussian curve. Hence, the Gauss tails of proteins of higher concentration cover the spots of proteins of lower concentration.

promising work is in progress. Thus, it is possible to focus in the IEF gel and then blot on PVDF membranes. You then scan the blot, in the end, with an IR laser mass spectrometer (see Section 7.4.l). The method of Loo et al. (1999) is even more promising. These authors have developed a "virtual" 2D phoresis (see Section. 7.4.2). Both methods still struggle, however, with the quantification of the proteins.

Sources

1. Corbett, J., et al. (1994). "Positional Reproducibility of Protein Spots in Two-dimensional Polyacrylamide Gel Electrophoresis Using Immobilised pH Gradient Isoelectric Focusing in the First Dimension: An Interlaboratory Comparison," *Electrophoresis* 15: 1205–1211.
2. Dockham, P., et al. (1986). "An Isoelectric Focusing Procedure for Erythrocyte Membrane Proteins and Its Use for Two-dimensional Electrophoresis," *Anal. Biochem.* 153: 102–115.
3. Garrels, J. (1979). "Two-dimensional Gel Electrophoresis and Computer Analysis of Proteins Synthesized by Clonal Cell Lines," *J. Biol. Chem.* 254: 7961–7977.
4. Hartinger, J., et al. (1996). "16-BAC/SDS-PAGE: A Two-dimensional Gel Electrophoresis System Suitable for the Separation of Integral Membrane Proteins," *Anal. Biochem.* 240: 126–133.
5. Sanchez, J., et al. (1997). "Improved and Simplified In-gel Sample Application Using Reswelling of Dry Immobilized pH Gradients," *Electrophoresis* 18: 324–327.

7.4 Mass Spectroscopy of Peptides and Proteins

7.4.1 Mass Spectrometers

Mass spectrometers entered the laboratories during the 1960s. Chemists used them for mass determination and structural investigations of volatile molecules and molecule fragments. This was done as follows. Electrons (70 eV) from a heating wire are shot at the steam of a sample. The electrons break up bindings and ionize the sample molecules. The resulting fragmentions are deflected into a magnetic field. The deflection depends on mass and charge of the ions (i.e., with known charge the mass spectrometer measures the MW). The measurement is so exact that (for example) positive ions such as CO (MW 27.9949), H_2CN (MW 28.0187), C_2H_4 (MW 28.0313), and N_2 (MW 28.0061) can be distinguished.

Traditional mass spectrometers fail with peptides, proteins, and DNA. Molecules of that size (and charged ones on top of that) are not volatile. Furthermore, under the high-energetic electron bombardment they would disintegrate into countless components. With a trick it is still possible to let high-molecular ions jump into the vacuum. The trick is called matrix-assisted laser-desorption ionization (MALDI) (Figure 7.6), developed by Franz Hillenkamp at the end of the 1980s. The proteins are first incorporated in crystals of UV-adsorbent molecules. In the process, the acidic UV-adsorbent molecules transfer protons to the proteins and give them a positive charge. The protein-doped crystals are then pushed into the high vacuum of the mass spectrometer and irradiated with a UV laser pulse. This explosively releases the UV-adsorbent molecules and with them the built-in protein ions. Molecules with such properties—co-crystal formation, proton transfer, UV adsorbtion—are called matrix (Table 7.1).

The proteins (+ protons) enter the gas phase in the nude (i.e., without hydrate water and counterions such as Na^+ or Cl^-). It is largely single polypeptide chains that appear in the gas. Quarternary proteins already disintegrate into their subunits in the acidic, denaturing matrix solution.

Special proteins create special problems. It is difficult to incorporate membrane proteins into the crystals. In addition, it is unclear whether the membrane proteins remain associated with lipids and detergents and protein/lipid/detergent complexes jump into the gas (if anything jumps at all). It seems like the membrane protein bacteriorhodopsin was successfully analyzed with MALDI-TOF. With glycoproteins, the sugar residues sometimes shift position or are cut off by the acidic matrix or the photon current. Finally, a part of the matrix molecules disintegrates under the laser, reacts with the proteins, and thereby increases their MW. This becomes noticeable through so-called adduct peaks in the spectrum.

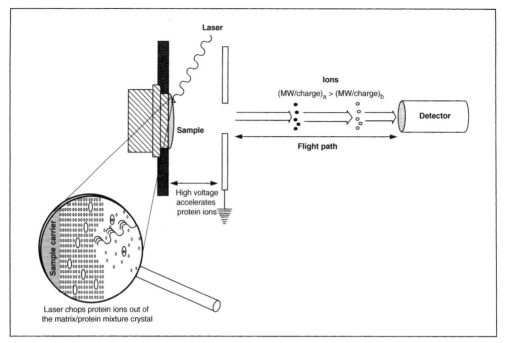

Laser

Ions

(MW/charge)ₐ > (MW/charge)_b

Detector

Sample

Flight path

High voltage
accelerates
protein ions

Sample carrier

Laser chops protein ions out of
the matrix/protein mixture crystal

Figure 7.6. Matrix-assisted laser-desorption ionization mass spectrometer (schematic).

When MALDI works, and it usually works, the result is a gas of positively charged protein ions. An electric field accelerates these ions toward a slit. The same field accelerates all ions, which thereby receive a speed proportional to one divided by the root from mass divided by charge. Two proteins with identical charge but of different mass fly with different speed, and a bivalently positively charged protein flies faster than the same protein with only one charge. All ions fly through the slit into a field-free vacuum tube—the time-of-flight analyzer (or TOF). All ions fly the same distance; namely, the length of the TOF. They then hit the detector. However, because they have different mass-to-charge ratios, and hence different speeds, they reach the detector at different times. These flight times are measured.

The matrix can load a protein differently with protons. Although most proteins fly through the vacuum as univalent positive ions, bivalent positive and (less often) trivalent positively charged protein ions are also found. Larger proteins thus often result in several peaks.

A useful MALDI-TOF determines the mass of a protein with an accuracy of 0.1 per thousand, and expensive devices even with 0.001 per thousand. And this within minutes. The devices work best with proteins from 30 to 40 kd, but they also provide useful data for larger proteins such as immune globulins. The record lies at just under 1,000 kd. Of course, protein mixtures can also be analyzed with the MALDI-TOF.

By the height of the peaks in the MALDI-TOF spectrum you can in principle (with suitable internal standards) measure the protein concentration (Nelson et al. 1994). However, this seems to be a difficult art.

A matrix is UV- or IR-absorbent crystallizable material that incorporates biopolymers in its crystals. In the high vacuum of the MALDI-TOF mass spectrometer, the crystals expell matrix molecules and incorporated biopolymers under the influence of a laser beam (UV or IR). In the process, proteins and peptides receive a positive charge via proton transfer.

For producing protein-doped crystals following Beavis and Chait (1996), the matrix is dissolved to saturation in a suitable solvent and then mixed with protein/peptide solution and dried on the sample carrier of the mass spectrometer. Mixtures of water with organic solvents such as acetonitrile, methanol, or propanol in the ratio given in the table serve as a solvent for the matrix.

Table 7.1. Properties of MALDI matrices (after Beavis and Chait 1996).

	Matrix	Solvent for Matrix	Crystal Formation with Incorporation of:		Signal Intensity and Average Charges per Protein/Peptide	Remarks
			Peptides	Proteins		
Matrices for UV Laser	Gentisic acid	Water 9:1 Organic solvents	Works most of the time	Fails sometimes	Acceptable (+)	Matrix produces stable protein ions. Photochemical adduct peaks occur: {(protein) +136}.
	Sinapinic acid	2:1	Fails sometimes	Works most of the time	Acceptable (+)	Suitable for protein mixtures. With small peptides, often weak signals. Photochemical adduct peaks occur: MW (protein) +206.
	Indoleacrylic acid	2:1	Works most of the time	Works most of the time	Good (++)	Suitable for complicated protein mixtures. Photochemical adduct peaks occur: MW (protein) +185.
	α-cyano-4-hydroxy cinnamic acid	2:1	Works most of the time	Works most of the time	Very good (+++)	Hardly any photochemical adduct peaks. Adsorption of Cu to some peptides leads to peptide/Cu peaks. Matrix produces multiply charged labile protein ions.
Matrices for IR laser	Succinic acid	40 mM succinic acid in water			Good (++)	The irradiation with the IR laser seems to create dimers and trimers.

It does not necessarily have to be a UV laser that catapults the proteins into the gas phase. Infrared (IR) lasers are also suited for MALDI. Infrared lasers transmit larger amounts of energy than UV lasers. In particular, they transmit enough energy to loosen proteins from PVDF blot membranes. Hence, with an IR laser you can analyze proteins directly from the blot. UV lasers, on the other hand, do not get a signal out of PVDF-adsorbed proteins (Sutton et al. 1997).

Furthermore, IR laser matrices are hydrophile and get by without organic solvents. Hence, the protein spots or bands on the blot membrane are preserved, and they do not flow into each other as can happen with hydrophobic UV matrices. The most widely used IR matrix is succinic acid.

For the IR-MALDI off the blot, residues of blot buffer are disruptive. Thus, wash thoroughly! Furthermore, the PVDF membrane must lie in the right orientation on the MALDI sample carrier. Because of the high binding capacity of the PVDF membranes, the protein binds only to the surface of the PVDF membranes, namely, on the blot side, where the gel came into contact with the membrane. Remember: buttered side up.

The high-energetic transmission of the IR lasers has disadvantages as well as advantages. Sensitive proteins can fragment, and others aggregate covalently—into dimers and trimers—and either leads to artifact peaks.

The IR-MALDI is not suitable for the analysis of peptide digestion by proteins. The digestion buffer probably does not go down well with the matrix. Thus, it would be ideal if you were able to measure with IR lasers as well as UV lasers in one device.

In addition to UV/IR-MALDI, there is another possibility for bringing protein ions into the gas phase: electrospray ionization (ESI). Figure 7.7 illustrates how the method works. The proteins are not gasified via incorporation into an evaporable matrix but by spraying the protein solution as finest droplets. Weak acids serve as ionization helpers, and organic solvents as spraying helpers. Acetonitrile/water 50:50 with 0.1% acetic acid is a typical carrier solution. Salts and detergents disrupt ESI and have to be removed. For this, Troxler et al. (1999) use small return-phase columns (C_8, elution with TFA, acetonitrile).

With ESI mass spectrometry, the ion mass is determined with quadrupoles (Figure 7.7). The accuracy of the mass determination with the quadrupole is comparable to that of TOF. However, most quadrupoles measure only up to mass-to-charge ratios of 3,000 to 5,000. If you want to measure big proteins, you have to give them a high charge. The advantage of quadrupoles lies in their combination. Several quadrupoles, one after the other, allow you to fragment individual protein ions via collision with rare gases and to determine the MW of the fragments (Figure 7.7).

ESI is said to be less sensitive than MALDI. However, the sensitivity depends not only on the ionization method but on the protein, the carrier solution (ESI), the matrix, and the crystallization method (MALDI). Generally, 10 to 100 fM protein suffice for a measurement with ESI.

Fragmenting, digesting, and identifying countless protein spots on countless gels seem to be the lot of the proteomist for the foreseeable future. This can become boring. It is good that ESI mass spectrometers offer another possibility for play. In comparison to MALDI, spraying is a mild ionization method. If anyone succeeds in ionizing at physiological pH, it should be possible to investigate intact protein complexes. At least for the Ca^{2+} complexes of proteins this seems to be possible. Troxler et al. (1999) use the volatile buffer ammonium acetate pH 7.0 (with or without 100 μM Ca^{2+}) for this purpose instead of the usual acidic solution (pH 3.0). Under these conditions, they can distinguish the Ca^{2+} complexes from native α-parvalbumin and different recombinant α-parvalbumins with the ESI mass spectrometer. But apparently you need to be lucky. With some proteins this works; with others it does not. For example, with calmodulin (also a Ca^{2+} binding protein) the method is a lot worse (Troxler, personal communication). α-parvalbumin can presumably only be gasified as an ion, because α-parvalbumin is very acidic (Troxler et al. use gigantic amounts, such as 800 fM). See the overview article of Joseph Loo (1997). This, apparently, is the bible for believers in the ESI complex.

Investigations of protein complexes are generally to be treated with caution. What is true in gas may be completely different in solution. Furthermore, 10 mM ammonium acetate pH

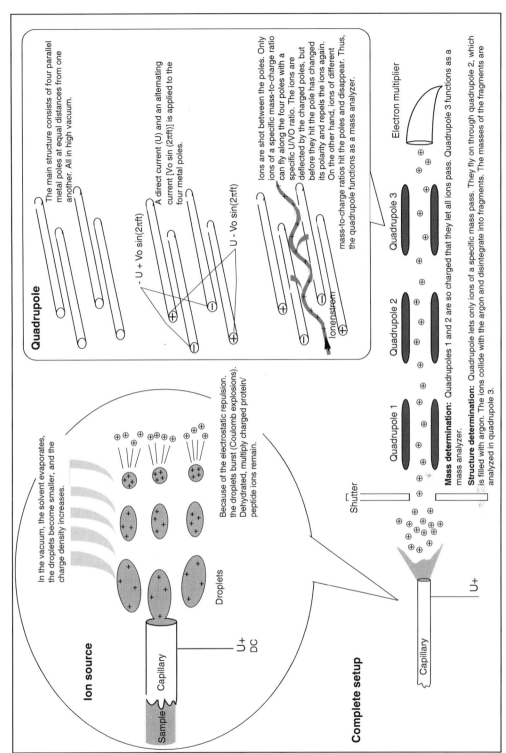

Figure 7.7. Electrospray ionization mass spectrometer (ESI).

7.0 is not exactly the most physiological of all buffers. Many proteins would not do too well under these conditions. But that is the curse of the life as a researcher: you are never quite sure, not even with all controls. You always have to find support with additional methods. Try to refute objections.

Sources

1. Chait, B., and Kent, S. (1992). "Weighing Naked Proteins: Practical, High-accuracy Mass Measurement of Peptides and Proteins," *Science* 257: 1885–1894.
2. Loo, J. (1997). "Studying Non-covalent Protein Complexes by ESI," *Mass Spectrometry Reviews* 16: 1–23.
3. Nelson, R. W., et al. (1994). "Quantitative Determination of Proteins by Matrix-assisted Laser-desorption Ionization Time-of-flight Mass Spectrometry," *Anal. Chem.* 66: 1408–1415.
4. Sutton, C., et al. (1997). "The Analysis of Myocardial Proteins by Infrared and Ultraviolet Laser Desorption Mass Spectrometry," *Electrophoresis* 18: 424–431.
5. Troxler, H., et al. (1999). "Electrospray Ionization Mass Spectrometry: Analysis of the Ca^{2+}-binding of Human Recombinant α-parvalbumin and Nine Mutant Proteins," *Anal. Biochem.* 268: 64–71.

7.4.2 Sample Preparation for MALDI

Avoid anything that could derivatize your protein and change its MW (e.g., high concentrations of formic acid, urea, TFA, and compounds reacting with free amino groups). Many proteins are also partially oxidized during purification and storage and/or dimidated, which makes the preparation heterogeneous and widens the peaks of the MALDI-TOF spectrum.

The quality of the protein-doped crystals is experimentally crucial and decisive for the quality of their spectrum. Only proteins incorporated in matrix crystals jump into the gas phase. Not every matrix is suited for every protein. Which matrix delivers the best results is not predictable. You have to try things! Table 7.1 shows suitable matrices.

The protein concentration of the crystallizing solution should lie between 1 and 10 µM. Low concentrations of salt, buffers, or lipids do not matter for the crystallization. Larger amount of nonvolatile substances such as glycerine, polyethylene glycol, 2-mercaptoethanol, and DMSO, on the other hand, inhibit the crystallization. Ionic detergents are deadly. In the presence of SDS, no proteins are incorporated into the crystals. You must remove SDS thoroughly (e.g., via protein blots, ion pair extraction [Henderson et al. 1979], or HPLC). With zwitterionic detergents you should also be careful. No crystals form with (for example) Zwittergent 3–16. Azid disrupts the ion formation during the laser bombardment, and lipid disrupts crystallization. Finally, the pH of the protein/matrix solution must be lower than 4, because above pH 4 a substantial part of the matrix molecules occurs in ionized form. These do not crystallize at all or in a different way. How do you produce the crystals?

Bartlet-Jones et al. (1994) dry the protein/peptide solution onto the sample carrier of the MALDI-TOF. They then pipette a droplet of matrix solution on top and let it dry again. Beavis and Chait (1996) advertize the dried-drop method: protein/peptide solution is mixed with saturated watery/organic matrix solution so that the protein/peptide concentration of the mixture lies between 1 and 10 µM. Apply a droplet of it onto the sample carrier of the MALDI-TOF and let it air dry at room temperature. Important: always start with a fresh matrix solution. The protein/peptide must be completely dissolved and neither protein/peptide nor matrix may precipitate during mixing. The matrix occasionally falls out of solution if the protein/peptide solution does not contain organic solvents. Do not heat the protein/matrix droplets. This changes crystallization as well as protein/peptide incorporation, largely following Murphy's law. Protein concentrations higher than 10 µM are pointless. Rather, they decrease the signal. The optimal mixing ratio between sample and matrix solution lies somewhere between 1:1 and 1:10. See Beavis and Chait (1996) for further recommendations.

According to Vorm et al. (1994) and Vorm and Mann (1994), fast drying yields smaller and more evenly distributed protein-doped crystals. Hence, the protein/matrix droplets are often dried in a vacuum. Hewlett-Packard offers a a device with which you can visually track the crystallization in the vacuum.

If the protein/peptide solution contains high concentrations of nonvolatile substances, it is advisable to wash the crystals briefly (10 seconds) in cold water. The nonvolatile substances

Table 7.2. You have a choice! Three sample preparations for MALDI.

Dried Drops	Fast Evaporation	Dissolved Nitrocellulose
Dissolve 40 mg/ml α-cyano-4-hydroxy cinnamic acid (αC) in 50% acetonitrile, 0.1% TFA.	Dissolve 40 mg/ml α-cyano-4-hydroxy cinnamic acid (αC) in acetone.	Dissolve 40 mg/ml α-cyano-4-hydroxy cinnamic acid (αC) in acetone.
Add internal calibration standards (0.25 to 0.5 mM).	Dissolve 20 mg/ml nitrocellulose in acetone.	Dissolve 20 mg/ml nitrocellulose in acetone.
Mix the matrix (αC solution) with the peptide sample (optimize ratio between 1:1 and 1:10).	Mix αC solution, nitrocellulose solution, and 2-propanol in ratio 2:1:1.	Mix αC solution, nitrocellulose solution and 2-propanol in ratio 2:1:1.
Pipette 1 μl of the mixture onto the MALDI sample carrier.	Add internal calibration standards (0.25 to 0.5 mM).	Add internal calibration standards (0.25 to 0.5 mM).
Let it dry.	Pipette 5 μl of the mixture onto the MALDI sample carrier.	Drop 2 μl of the mixture onto 2 μl of peptide sample.
Run spectra.	Let it dry.	Pipette 1 μl of the mixture onto the MALDI sample carrier.
—	Pipette 0.5 μl 5% formic acid onto the MALDI sample carrier.	Let it dry
—	Pipette 0.5 μl peptide sample in 5% formic acid onto the MALDI sample carrier.	Wash MALDI sample carriers one after the other with 5% formic acid and MilliQ water.
—	Let it dry.	Run spectra.
—	Wash MALDI sample carriers one after the other with 5% formic acid and MilliQ water.	—
—	Run spectra.	—

accumulate on the crystal surface and are partially removed by the washing, which improves the signal (Beavis and Chait 1996). Careful: the crystals separate easily from the sample carrier. Quickly suck off the water or shake it off. Sample preparation may seem complicated when you read it, but in practice crystallization turns out to be astoundingly easy.

Two basic methods have asserted themselves over the last years. In their details, they differ from laboratory to laboratory. The basic methods are the dried-drop method and the fast evaporation method. Shortly before this book went to print, yet another method caught my eye. Its inventors claim to have combined the advantages of both methods and removed the disadvantages (i.e., generated a hybrid). Try it out. Table 7.2 stems from Landry et al. (2000) and juxtaposes the three methods.

The sample often sits in gels and is difficult to get out. First, it creates work. In addition, so many things can happen, such as contamination, protease digestion, and adsorption. For ESI, you can see that there is no way around extracting from the gel, but with MALDI it should be possible to do something directly from the gel. And, indeed, it can be done. Loo et al. (1999) desorb proteins directly from the IEF gel via the following technique. Wash the IEF gel for 10 minutes in 1:1 acetonitrile, 0.2% TFA, to remove the urea and the detergents. Then soak the washed gel for 5 to 10 minutes in matrix (saturated sinapinic acid in 1:1 acetonitrile, 0.2% TFA) and dry at RT. Lay the dry gel on the sample carrier of the MALDI and scan it with the UV laser.

Loo calls this "virtual" 2D gel analysis. The first dimension means the IEF. The second, however, is the MALDI-TOF instead of an SDS gel. True, MALDI-TOF measurements are a lot more exact than SDS gel electrophoresis (for Loo et al., 0.1 to 0.2% in comparison to about 10%), but they are virtual. You only have a computer printout in hand.

Of course, virtual 2D phoresis also has a problem. MALDI peaks are difficult to quantify. For the virtual spots, the virtual 2D phoresis can only provide virtual protein amounts.

Sources
1. Bartlet-Jones, M., et al. (1994). "Peptide Ladder Sequencing by Mass Spectrometry Using a Novel, Volatile Degradation Reagent," *Rapid Communications, in Mass Spectrometry* 8: 737–742.
2. Beavis, R., and Chait, B. (1996). "Matrix-assisted Laser-desorption Ionization Mass Spectrometry of Proteins," *Methods Enzymol.* 270: 519–551.
3. Henderson, L., et al. (1979). "A Micromethod for Complete Removal of Dodecyl Sulfate from Proteins by Ion-pair Extraction," *Anal. Biochem.* 93: 153–157.
4. Landry, F., et al. (2000). "A Method for Application of Samples to Matrix-assisted Laser Desorption Ionization Time-of-flight Targets That Enhances Peptide Detection," *Anal. Biochem.* 279: 1–8.
5. Loo, J., et al. (1999). "High Sensitivity Mass Spectrometric Methods for Obtaining Intact Molecular Weights from Gel-separated Proteins," *Electrophoresis* 20: 743–748.
6. Vorm, O., and Mann, M. (1994). "Improved Mass Accuracy in Matrix-assisted Laser Desorption/Ionization Time-of-flight Mass Spectrometry of Peptides," *J. Am. Soc. Mass Spectrom.* 5: 955–958.
7. Vorm, O., et al. (1994). "Improved Resolution and Very High Sensitivity in MALDI TOF of Matrix Surfaces Made by Fast Evaporation," *Anal. Biol. Chem.* 66: 3281–3287.

7.4.3 The Possibilities of MALDI and ESI

The MALDI-TOF mass spectrometer determines the MW of the protein. Because it also analyzes protein mixtures (separates protein mixtures substantially better than SDS gel electrophoresis), it can perform many functions of SDS gel electrophoresis. For example, the MALDI-TOF provides convincing data about the purity of proteins. As far as the sensitivity is concerned, you need 10 to 100 fM less protein for the MALDI-TOF spectrum than for the silver staining of an SDS gel. However, the MALDI-TOF mass spectrometer offers more than the SDS gel electrophoresis.

- Do you know the amino acid sequence (e.g., from cDNA cloning) and do you have purified protein? Then you can determine the MW of the protein via MALDI-TOF and compare to the MW calculated from the cDNA sequence. If the MW are identical, the primary structure of your protein is correct. Deviations are evidence of modifications (e.g., phosphorylations, glycosylations, point mutations). With smaller proteins, the acetylation of the N-terminal amino acid becomes apparent in the MALDI-TOF. You determine the position of a modification by cutting the protein into peptides, measuring the MW of the individual peptides, and comparing to the MW predicted from the sequence.
- You determine the cutting sites of proteases via the MW of the fragments.
- You can trace the effect of glycosidases through the decrease in MW of a glycosylated protein. At the same time, you measure the number of separated sugar groups. Because of the heterogeneity of the glycoproteins, MALDI-TOF peaks are usually wide (analogously to SDS gel electrophoresis).
- Kinases increase the MW of their substrate protein by 80 Dalton, or a multiple thereof. Phosphatases reduce the MW of phosphorylized proteins by the corresponding amount. By testing for MW decrease, you can find out whether your protein is phosphorylized. If the size of the protein does not permit you to irrefutably prove an addition or subtraction of 80 Dalton you can tryptically digest aliquots before and after the enzyme treatment and measure the MWs or the individual peptides. MALDI-TOF could become even more valuable for quick determinations of the phosphorylation state (ratio of isoforms with none, one, two, and so on phosphate groups) of a protein in the cell metabolism. Up to now, the cell extracts were analyzed with 2D gel electrophoresis or immunoaffinity chromatography with subsequent isoelectric focusing. To mark the phosphoproteins, you had to dump gigantic amounts of ^{32}P into cell cultures.
- These can limit the epitope of an antibody. Digest the epitope-carrying protein (e.g., with trypsin) and then fish with your antibody in the peptide soup. You determine the MW of the immunoprecipitated peptide with the MALDI-TOF. From the sequence of the epitope-carrying protein, you calculate the MW of all possible tryptic fragmentation products and then fit the immunoprecipitated peptide into the sequence. Via digestion with proteases of different specificity, the epitope can be limited further.
- You can characterize an unknown protein. For this, you digest it with specific proteases and determine the MWs of the resulting peptides via MALDI-TOF. Tryptic (and other) digestions

can be measured in one go. The MW of three peptides readily characterizes the protein almost as unambiguously as the amino acid sequence. To identify the protein, compare the MW of its fragments to the corresponding data in a database. For example, the data collection of the SERC Daresbury laboratory allows for comparison with fragments of more than 50,000 proteins (Pappin et al. 1993). This way, you can quickly and reliably determine the protein spots in 2D gels.

- With the MALDI-TOF, you can even determine bacteria. For this, you smear intact bacteria of a primary culture onto the sample plate, let the bacteria cocrystallize with matrix, and irradiate with a UV laser. The desorbing bacteria molecules result in a mass spectrum, which unambiguously defines the bacteria strain. The spectrum is probably formed by the mucopeptides, glycoproteins, and lipoproteins of the bacterial cell wall and supporting layer—just those molecules that traditional determination methods are also based on. The mass spectrum is entered into a database, and an algorithm searches it for comparable spectra. The bacteria strain is thus identified. Micromass, which sells the system, claims that you can distinguish among genetic transformants, antibiotic-sensitive/resistant strains, and strains with different plasmid profiles. An assay needs only one small colony. One thing is certain: it is fast. From the sample preparation to the finished result it takes only a few minutes.

- The MALDI-technique is not limited to proteins/peptides. With a suitable matrix you can also bring DNA and RNA into the gas phase. What for? For example, you can sequence oligos by means of oligonucleotide ladders (Limbach et al. 1995). The smallest MW difference between DNA bases—between adenine and thymine—amounts to at least 9 Dalton. For RNA bases, the smallest difference—between cytosine and uracil—lies at 1 Dalton.

Oligosaccharides can also be incorporated into matrix crystals and gasified. Actually, you should even be able to sequence the oligosaccharides with the MALDI-TOF (ladder sequencing; see Section 7.6.5 and Stahl et al. 1994). In principle, MALDI brings any polymer into the gas phase, even polyethylene glycol. However, you have to supply such neutral polymers with a positive charge. Otherwise, the electric field cannot accelerate them into the TOF. These polymers are therefore complexed with metal ions (whose MW, of course, needs to be taken into account).

The big hope of ESI is to be able to analyze noncovalent protein complexes (Loo 1997). This would allow quick and reliable examinations of the stoichiometry of the protein complexes. No more elaborate and dubious cross-linking assays (see Chapter 8), but two ESI runs—one under native conditions—and one under denaturing conditions—and there is your answer.

With good instruction, you can learn within two weeks how to work with the MALDI-TOF mass spectrometer and how to crystallize. With the ESI it is even faster, because you do not have to crystallize. With most devices you get helpful analysis software. In summary: a mass spectrometer is not cheap, but it is worth the expense.

Sources

1. Beavis, R., and Chait, B. (1996). "Matrix-assisted Laser Desorption Ionization Mass Spectrometry of Proteins," *Methods Enzymol.* 270: 519–551.
2. Chait, B., and Kent, S. (1992). "Weighing Naked Proteins: Practical, High-accuracy Mass Measurement of Peptides and Proteins," *Science* 257: 1885–1894.
3. Limbach, P. A., et al. (1995). "Characterization of Oligonucleotides and Nucleic Acids by Mass Spectrometry," *Curr. Opin. Biotechnol.* 6: 96–102.
4. Loo, J. (1997). "Studying Non-covalent Protein Complexes by ESI," *Mass Spectrometry Reviews* 16: 1–23.
5. Pappin, D., et al. (1993). "Rapid Identification of Proteins by Peptide-mass Fingerprinting," *Current Biology* 3: 327–332.
6. Stahl, B., et al. (1994). "The Oligosaccharides of the Fe(III)-Zn(III) Purple Acid Phosphatase of the Red Kidney Bean," *Eur. J. Biochem.* 220: 321–330.
7. Stults, J. (1995). "Matrix-assisted Laser-desorption Ionization Mass Spectrometry (MALDI-MS)," *Current Opinion in Structural Biology* 5: 691–698. The review contains a host of references for special topics, such as differentiation of sulfate and phosphate groups, MALDI with infrared lasers and succinic acid matrix, subfemtomolar detection thresholds with a thin layer matrix, highly sensitive protein detection in the attomolar range, and high-resolution mass spectrometry.

7.5 Protein Chips

7.5.1 Protein Chips with SELDI

The almost endless possibilities of mass spectrometry have been increased by yet another hopeful variant: SELDI (surface-enhanced laser desorption ionization). SELDI combines protein chips with a UV-MALDI-TOF. The chips are solid aluminum strips coated with cationic or anionic ion exchangers, with hydrophile or hydrophobic molecules. Chips with activated surfaces are also available. These bind proteins covalently via their amino groups and enable you to coat chips with antibodies or receptors as you need them. Every chip has eight coated holes with a diameter of 1 mm. You apply the sample into the holes. Part of the proteins/peptides is adsorbed. The remainder is washed off. The adsorbed proteins/peptides are transferred into matrix and can then be analyzed in the mass spectrometer (Figure 7.8). What can SELDI do better than MALDI?

Because of the washing step, disruptive ions or detergents are absent. More still: the proteins are adsorbed on the surface linearly and in one layer. Hence, SELDI produces uniformly distributed and uniformly oriented crystals. No unordered crystal salad anymore, as with the other MALDI techniques. The uniform crystal coating produces an excellent and fairly reproducible signal: the peak heights vary between successive runs of the identical sample only between 10 and 30%.

Another advantage is that the SELDI protein/peptide spectrum differs from that of a 2D gel. For example, 2D gels do not dissolve below 6 kd or do not dissolve very well. SELDI, on the other hand, separates peptides/proteins especially well in the range between 1 and 25 kd. Furthermore, with a suitable chip coating you can also capture hydrophobic proteins. Finally, you can select. With one chip you measure (for example) only hydrophobic proteins/ peptides, and with the other only acidic ones. The measuring happens in a blink of an eye. Pipette 1 µl of sample into each chip hole (by hand), wash, apply matrix, dry, measure. None of this endless long focusing and no handling of brittle gels. After 10 minutes, the thing is done.

With chips not coated covalently with proteins, you can wash matrix and sample residues off after measuring and then reuse the chip. However, for important and clean experiments you should use new chips. Which new possibilities does SELDI offer?

You can investigate protein-protein interactions (see also Sections 2.5 and 2.2.5.2). Example: you have purified or expressed a protein X and you would like to know its function. Neither the sequence nor the purification provides any clues. It would be cool to know which proteins bind to protein X. Maybe you know their function, which would give you valuable pointers. Thus, you couple your protein X to a SELDI chip, if possible under different conditions. Then you wash off the unbound protein X and block unused groups. Now you incubate the chip holes with cell extract or wherever else you suspect binding partners to occur. Wash it off. Co-crystallize with matrix. Off with the chip into the MALDI-TOF. After a few seconds you know whether a protein binds to protein X. Now this raises the question: Is this protein the physiological binding partner? To decide this, you need controls. In a control chip hole, you couple a protein that is similar to protein X in isoelectric point and MW. This hole should bind no proteins, or at least no other proteins. On the other hand, cell extracts or serums known not to contain binding partners for protein X should leave no trace in the mass spectrum. Furthermore, the protein should bind to X somewhat stoichiometrically.

Is everything correct? Then you identify the binding partner in databases and order a bottle of champagne. However, if you fail you pay dearly—in the word sense. Failure does not mean that no partner exists. In the cell, every Jack will find his Jill. It is more likely that your conditions were not right, either with the coupling of protein X to the chip or with the incubation of the coated chip with the cell extract. Thus, you have to try out one condition after the other and purchase one chip after the other. Each chip costs $70.

Another possible application of SELDI comes with the included analysis software. With it you can determine the quantitative differences between two spectra (e.g., ill/healthy or drug/without-drug). This makes sense with SELDI, because the spectra are fairly reproducible

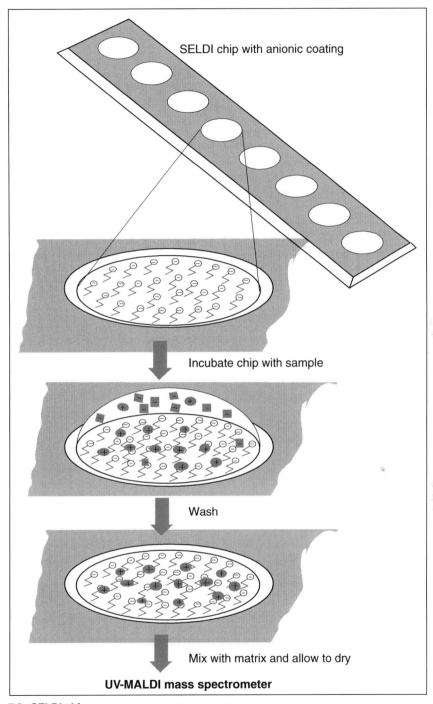

SELDI chip with anionic coating

Incubate chip with sample

Wash

Mix with matrix and allow to dry

UV-MALDI mass spectrometer

Figure 7.8. SELDI chip.

here. Example: you search for an early tumor marker in the blood. Early tumor markers occur in the blood often in very low concentrations, because there are still only few cancer cells (early markers!) and because the latter let only part of the marker into the blood (if at all!) and the marker is digested by proteases. Thus, you already have your first problem: the gigantic amounts of albumin and immunoglobulins cover up your rare proteins. The way out: you take a chip (e.g., with hydrophobic or cationic surface), adsorb your serum dilution, and then wash off albumin and immunoglobulins. For this, you fill one hole of your chip with buffer, the remaining seven with serum dilution, and then try six different washing protocols (e.g., with different salt concentrations). With the lowest salt concentration that still removes albumin and immunoglobulins you now run your experiment (i.e., you compare multiple healthy serums with multiple tumor serums). Of course, you can also adsorb with this salt concentration (possibly enriching your marker) and then compare. If you see differences between the spectra, you need to identify the responsible proteins and make sure that these were truly tumor markers and not markers for such things as dietary differences (see Section 7.2).

Of course, it can (and will) happen that you do not see any differences, maybe because the tumor does not give off marker into the blood or because you washed your marker off together with the albumin. However, you have not invested a lot of time. Just try the next type of chip. If a marker exists at all, you will find a condition under which the marker sticks and the albumin washes off.

This optimism is rooted in the fact that I have never worked with SELDI. My knowledge is based on the literature and several conversations with company people (SELDI is sold by Ciphergen). Company people are always full of enthusiasm for their product. Thus, I am like a man who has read a lot about the making of shoes and once even looked over the shoemaker's shoulder. Such people usually believe that making shoes is the easiest thing in the world—a belief quickly revised as soon as they have to do the work themselves. At least, I am quite certain that SELDI is a hopeful method—albeit costly. You have to invest one hundred fifty thousand dollars. But so what: If the experiments go awry, you still have the pretty aluminum chips. You can pierce them, put them on a silver chain, and give them to your significant other as a birthday present. I even believe that there are gold-plated chips.

7.5.2 Fortune Cookies

The SELDI chip is very nice, but it does not compare to the DNA chips that everyone is talking about right now. There are chips with 100,000 and more spots that indicate the presence and the relative amount of (almost) every RNA in the cell extract. The protein biochemist would also like to have such a toy, ideally a chip on which he would only have to pour some cell extract and half an hour later it would tell him number, type, modification, and concentration of every protein in the broth. Daring people even demand chips that determine the concentration of metabolites such as glucose and lactate as well as oligo- and polysaccharides.

In a cell there are a lot more protein species than RNA species (i.e., a protein chip would have to have many more spots than a DNA chip). Is that doable? Theoretically, sure. You would just have to covalently stick a corresponding number of monoclonal antibodies to the chips—a "minor" task with a few hundred thousand highly specific, highly affine monoclonal antibody species. Real life has even more trouble in store. Although it may be possible to bind the countless antigens to the countless antibodies on the chip, how would you determine the binding?

One possibility would be to make two antibodies for every antigen—one bound to the chip and another fluorescence-marked detection antibody that binds to another epitope of the antigen. Once you have produced some hundred thousand monoclonal antibodies you should be in good practice, so a few hundred thousand more should not matter much.

Another possibility would be to construct the protein chip as a sensor chip and detect the binding of the antigens via the plasmon resonance (Section 2.2.5.2). Can this method be miniaturized such that it measures the mass increase of the tiniest spots? Can it register hundreds of thousands of spots fast enough? I do not know, I know only that as yet there is no real protein chip and this does not surprise me at all. Maybe instead of insisting on coating chips with antibodies we should look for molecules that can do both: bind and indicate binding.

7.5.3 Aptamers

The search for molecules for the production of protein biochips led us to aptamers. Aptamers are DNA or RNA oligonucleotides with a length of 15 to 60 nucleotides that bind specifically (for example) to protein. The binding affinities lie between K_D 1 pM to K_D 1 μM. In vitro, aptamers can be produced easily, inexpensively, and in large amounts. If you incorporate modified nucleotides, they become resistant against nucleases. Because DNA and RNA do not have souls as complicated as proteins, it is also not a problem to incorporate reporter molecules at any location.

Aptamers can apparently be produced against anything—against ions such as Zn^{2+} as well as against nucleotides (e.g., ATP), oligopeptides, proteins (e.g., thrombin), and glycoproteins (e.g., CD4). Indeed, aptamers are already used as an antibody substitute in ELISAs and Western blots.

However, if you look closely the shiny aptamers have stains. With oligonucleotides, you cannot produce such gigantic numbers of 3D structures as with proteins. Furthermore, the inevitable negative charge of their phosphate groups limits affinity as well as binding variety. Because of the electrostatic rejection of negatively charged nucleotide spine and negatively charged amino acids or oligosaccharide chains, it will be difficult to produce aptamers with high affinity against relevant protein epitopes. It is not surprising that most proteins against which aptamers could be developed have binding sites for polyanions (e.g., thrombin- or heparin-binding growth factors). Aptamers largely form only against epitopes that also bind other polyanions (such as heparin). If one can mark a half-dozen different epitopes of a certain protein with antibodies, aptamers often bind only to one, and this generally with low enthusiasm. The affinities of aptamers are for small molecules (amino acids, dopamine) in the μM range, for nucleic acid-binding proteins in the nM range, and for heparin-binding proteins in the subnanomolar range. For proteins that bind neither nucleic acids nor heparin, aptamers have only low affinity. The K_Ds are between 0, 1 μM, and 1 μM.

Low epitope variety, low affinity, and a specificity that is not always encouraging—these are the main problems with the application of aptamers. Does the application of more flexible RNA molecules and special nucleotides help solve these difficulties? This question will still burn through some doctoral candidates.

The disadvantages of the aptamers are opposed by a big advantage: you can use them as a binder as well as a binding indicator. Hamaguchi et al. (2001) came up with the following. At the 5'-end of the aptamer they attached nucleotides complementary to the nucleotides on the 3'-end. Furthermore, a fluorophor (e.g., fluoresceine) was attached to the 5'-end and a quencher to the 3'-end. In the absence of the ligand, the aptamer (Hamaguchi et al. call it a signal aptamer) forms a loop. There, fluorophor and quencher lie close to each other and no fluorescence can be stimulated. If the signal aptamer binds to its ligands, the loop is opened. Fluorophor and quencher thus gain distance between each other. The bound signal aptamer fluoresces (Figure 7.9).

This elegant method seems to work at least for a thrombin signal aptamer. If Hamaguchi et al. add thrombin to the thrombin signal aptamer, the fluorescence increases continuously from 0 to 40 nM thrombin. With saturating concentrations of thrombin, they achieve a 2.5-fold amplification of the fluorescence signal. The binding of thrombin and signal aptamer was specific. At least, related plasma serine proteases such as factors IX and Xa had no effect on the fluorescence of the signal aptamer. Unfortunately, this was not true for the nonspecific binding. Single-string binding proteins and lactate dehydrogenase increased the fluorescence by at least 15-fold of the original value. The binding of these proteins was of low affinity, but with the plentiful occurrence of (for example) lactate dehydrogenase this was only weak consolation. The authors suggest adding chopped-up DNA to the assay to suppress the unspecific binding.

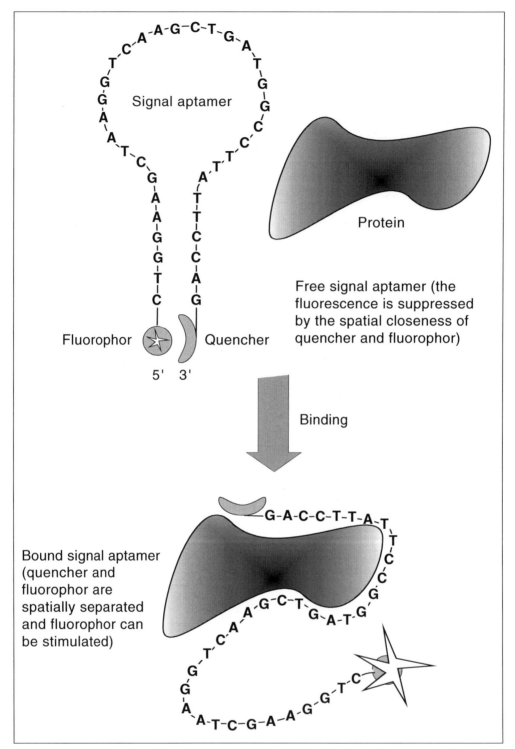

Figure 7.9. Signal aptamers bind and indicate binding.

Overall, the current literature about protein biochips has a tendency to make you sad: high-flying wishes, great plans, and a heap of petty problems that cannot be impressed by wishes. The crux seems to be in the production of aptamers that get at least as close to covering a similar variety of epitopes as antibodies do. Would that succeed with RNA and a number of unusual nucleotides or with RNA-peptide hybrids? I have my doubts. Would nature not have already used something similar? Are there organisms with aptamer immune systems?

Sources
1. Hamaguchi, N., et al. (2001). "Aptamer Beacons for the Direct Detection of Proteins," *Anal. Biochem.* 294: 126–131.
2. Morris, K., et al. (1998). "High Affinity Ligands from In Vitro Selection: Complex Targets," *Proc. Natl. Acad. Sci. USA* 95: 2902–2907.

7.6 Microsequencing

> "Your worship would make a better preacher than knight-errant," said Sancho. "Knights-errant knew and ought to know everything, Sancho," said Don Quixote.

Doctoral candidates and post-docs, the errant knights of science, also have to know something about everything—or at least they have to know how to give the impression that they know something of everything. Everything includes microsequencing. Don't get me wrong: you don't necessarily have to be able to microsequence, but you should be able to put in your two cents during relevant discussions in the coffee shop. This chapter provides you with the necessary basis. What is microsequencing good for?

- For smaller proteins, you can construct the entire sequence from the sequences of proteolytic fragments. Nobody does this anymore nowadays, you say? Wrong. If you cannot clone the protein, or if you are not allowed to clone it (as is the case with certain toxins), you fall back on this method even today.
- Microsequencing delivers the information needed for the synthesis of oligonucleotides.
- Microsequencing provides clues about post-translational changes of the protein such as phosphorylation, sulfation, glycosylation, and the position of the disulfide bridges. Partial sequences can sometimes throw light on the relationships to other proteins.

For microsequencing of a protein, you must perform several preparation steps. The protein has to be purified and prepared, and for the most important methods the protein must have a free N-terminus (or you have to cleave the protein into peptides).

7.6.1 Preparing the Protein

The end point of many protein purifications is SDS gel electrophoresis. The protein to be sequenced thus occurs as a band in the gel. Hunkapiller et al. (1983) gently stain the band (e.g., with Na-acetate), cut out the gel piece, and elude the protein in a dialysis chamber. The eluate is applied to a filter and then sequenced. The method is reliable but complicated and has a low yield. Prussak et al. (1989) do without a dialysis chamber and elude the protein via passive diffusion in the presence of 0.01% SDS.

Many experimenters blot the sought-after protein from the SDS gel onto a membrane that can be put directly into the sequencing machine after the blot buffer's glycine is removed. From the IEF gel, you can also blot (for example) with protein mixtures whose components do not differ in size. Nevertheless, before blotting the IEF gel the ampholines must be washed out with perchloric acid (Hsieh et al. 1988). Some protein gets lost, and sensitive bindings (e.g., Asp-Pro) are partially hydrolyzed due to the acidic pH.

The protein transfer from the gel to the blot should be as efficient as possible in order for the protein on the blot to be easily identifiable. The blot membrane should be stable against the reagents used for sequencing. Nitrocellulose and nylon membranes do not withstand the solvents of the Edman chemistry. The remaining choice for the experimenter is between coated glass fiber and PVDF membranes.

Glass fiber membranes are coated either with polybrene (Vandekerckhove et al. 1985) or with positively charged Silan compounds (Xu and Shively 1988). PVDF membranes can be used without coating (Matsudaira 1987), but coating them with a polybrene coating makes them superior to glass fiber membranes (Xu and Shively 1988). Furthermore, on PVDF membranes the experimenter can identify the sought-after protein with Coomassie without influencing the subsequent microsequencing (Xu and Shively 1988). Hence, PVDF membranes are generally used. With both membranes, a certain percentage of protein gets lost (10 to 30%) during the blotting.

Sources
1. Hsieh, J., et al. (1988). "Electroblotting onto Glass-fiber Filter from an Analytical Isoelectrofocusing Gel: A Preparative Method for Isolating Proteins for N-terminal Sequencing," *Anal. Biochem.* 170: 1–8.
2. Hunkapiller, M., et al. (1983). "Isolation of Microgramm Quantities of Proteins from Polyacrylamide Gels for Amino Acid Sequence Analysis," *Methods Enzymol.* 91: 227–236.
3. Matsudaira, P. (1987). "Sequence from Picomole Quantities of Proteins Electroblotted onto Polyvinylidene Difluoride Membranes," *J. Biol. Chem.* 262: 10035–10038.
4. Prussak, C., et al. (1989). "Peptide Production from Proteins Separated by Sodium Dodecyl Sulfate Polyacrylamide Gel Electrophoresis," *Anal. Biochem.* 178: 233–238.
5. Vandekerckhove, J., et al. (1985). "Protein-blotting on Polybrene-coated Glass Fiber Sheets," *Eur. J. Biochem.* 152: 9–19.
6. Xu, Q., and Shively, J. (1988). "Microsequence Analysis of Peptides and Proteins: Improved Electroblotting of Proteins onto Membranes and Derivatized Glass-fiber Sheets," *Anal. Biochem.* 170: 19–30.

7.6.2 Blocked N-Termini

For about 50% of all proteins, the N-terminus is blocked by N-acetyl amino acids glycosylated amino acids, pyrrolidone groups, or others. This blockage prevents Edman degradation. Edman degradation is still today the basis for most sequencing technologies (see Sections 7.6.4 and 7.6.6).

N-terminal blockage can occur in vivo or during the purification. The careful experimenter thus uses only PA solvents that do not contain any aldehydes (e.g., PA ethanol or acetic acid). Also, she exposes her protein only for a short time to oxidative conditions (e.g., fixing with acetic acid in open trays). In Section 7.3 I recommended using piperazine diacrylyl instead of bisacrylamide as a cross-linker if you have to run an SDS gel with the protein. Also, you should not polymerize the gel in the presence of SDS. After all, SDS forms micelles containing acrylamide monomers and unpolymerized acrylamide can block N-termini or cross-link proteins. Finally, 0.002% thioglycolic acid in the upper buffer helps against oxidative changes during SDS gel electrophoresis.

Naturally, the N-terminus does not need to wait for SDS electrophoresis. It can be readily plugged into one of the purification columns. Here, the buffer seems to play a role. When I isolate (for example) voltage-dependent K^+ channel protein, I use glycerine-containing buffers. The channel is easily sequenced (the first attempt was successful). My competitor did not have glycerine in the buffers and a year later he complained to me that he had also isolated the protein, but the N-terminus would always be blocked. How could this be? And he looked at me as if I were a magician or something worse. I shrugged and mumbled, "I don't know. In my hands it works." I really did not know the reason. I had not added glycerine in order to protect the N-terminus against oxidation or acetylation, but because glycerine-containing buffers pour as smoothly as honey.

Furthermore, I imagined, glycerine would stabilize the protein structure. But to be honest: even if I had known about the glycerine's protective effect at the time, I would have given the identical response.

My recommendation: protect the N-terminal ends during purification. But what if the N-terminus is already blocked? You can try sequencing C-terminally (see Section 7.6.5) or with the mass spectrometer (see Section 7.6.6). If you cannot do this or do not want to, you have no choice but to give up and cleave the protein into peptides, and separate and sequence them. Which of the three methods is to be preferred depends on the local climate. If you can find someone in the institute who knows a lot about sequencing by mass spectrometer, enlist the services of this person. For one or a few sequences it is not worth studying the peptide ladder technique for yourself (see Section 7.6.6), especially if you have to purchase a mass spectrometer first.

Cleaving peptides, on the other hand, does not require big equipment purchases (every lab has an HPLC nowadays), and it works almost always. However, the cleaving and separating is laborious. As a beginner, you should plan on three months.

Unblocking N-termini rarely works. Wellner et al. (1990) free N-termini that are blocked with N-acetyl serine or N-acetyl threonine using anhydrous trifluoroacetic acid. The yield lies between 3 and 40%, but usually near the lower value. A related method is alcoholytic deacetylation (Georghe et al. 1997). Here, the protein is treated with trifluoroacetic acid/methanol (1:1) at 47°C for two days. Up to 50% of the proteins are supposedly deacetylated, with little cleaving of peptide bindings (< 10% for peptides and < 30% for proteins). The authors offer some examples as proof, and their data also look trustworthy. However, from experience I know that the effectiveness of such methods differs from protein to protein. Such methods are typically not suited for the protein with which you happen to be working. In other words, the deacetylation is substantially lower than expected and the protein cleaving substantially higher. Try out it, and do not hold me responsible for the consequences. A treatment with pyrrolidon carboxylate peptidase may help for proteins blocked with a pyrrolidon carboxylate group, according to Doolittle and Armentrout (1968).

Sources

1. Doolittle, R., and Armentrout, R. (1968). "Pyrrolidonyl Peptidase: An Enzyme for Selective Removal of Pyrrolidone Carboxylic Acid Residues from Polypeptides," *Biochemistry* 7: 516–521.
2. Georghe, M., et al. (1997). "Optimized Alcoholic Deacetylation of N-acetyl-blocked Polypeptides for Subsequent Edman Degradation," *Anal. Biochem.* 254: 119–125.
3. Wellner, D., et al. (1990). "Sequencing of Peptides and Proteins with Blocked N-terminal Amino Acids: N-acetylserine or N-acetylthreonine," *Proc. Natl. Acad. Sci. USA* 87: 1947–1949.

7.6.3 Cleaving the Protein into Peptides

You have been unlucky and your N-terminus is blocked? You need the sequence of a protein and are not able to clone or are not allowed to? Then you have no choice but to cleave the protein into peptides. I know, this causes an inner struggle. There you have paid attention to the proteases, added inhibitors, shivered in the cold room, and worked through the night until you had rings under your eyes—only to get it done faster you are supposed to add proteases and intentionally destroy the precious product? But there's nothing you can do: if you want to get a sequence, you have to cleave, cleave, cleave.

However, cleaving with selective proteases is not the only method of obtaining sequenceable peptides. You can also try it with bromine cyanide or diluted acids. These agents likewise cleave only at certain amino acid residues. How to proceed?

The experimenter separates heterooligomers before cleaving them into subunits. Then she denatures the purified subunit and reduces possible disulfide bridges. A carboxymethylation with iodoacetic acid protects against the reoxidation of the cysteines (Lind and Eaker 1982). This reaction requires a sure instinct, in that other amino acids such as methionine, lysine, and histidine also react at high concentrations of iodine acetate or at the wrong pH. If you just need a few partial sequences, you can skip the carboxymethylation.

Source

1. Lind, P., and Eaker, D. (1982). "Amino-acid sequence of the α-subunit of Taipoxin, an Extremely Potent Presynaptic Neurotoxin from the Australian Snake Taipan," *Eur. J. Biochem.* 124: 441–447.

7.6.3.1 Protease Digestion

Of the three selective protein cleaving methods, digestion with proteases is most commonly used and is thus discussed here in most detail. You can digest a protein with a protease in several ways to receive sequenceable peptide.

One method is to incubate the denatured protein in solution with minimal amounts of selective protease until the protein is completely cleaved into peptides that are resistant to the protease. You separate these on the HPLC and generally get several sequenceable peptides (Leube et al. l987; Prussak et al. 1989).

You do not have to digest them completely. You can also stop with bigger cleaving products. You separate the digestion product either by HPLC or (if you have larger fragments) by SDS gel electrophoresis. In the latter case, the protein to be sequenced is digested in the pocket of an SDS gel, and the cleaving products are electrophoresized, blotted onto a suitable membrane, and sequenced (Kennedy et al. 1988). This incomplete protease digestion saves time and reduces losses.

The most popular method is to completely digest the proteins after the blot. For this, you identify your protein on the (unblocked!) blot by means of protein stain, cut out the blot piece, and add a selective protease. This creates peptides that are separated by HPLC. You can sequence these peptides following Edman (see Section 7.6.4) or by means of a peptide ladder and MALDI-TOF (see Section 7.6.6).

There is one problem: if you digest proteins directly on PVDF membranes or nitrocellulose membranes, the unblocked membrane binds the protease and inactivates it. Furthermore, proteases work better with protein in solution. You avoid the problem by releasing the protein from the blot piece and preventing the adsorption of protease and peptides. According to Fernandez et al. (1994), you can use a solution of l% hydrogenated TRITON-X-l00 in 10% of acetonitrile, 100 mM Tris pH 8.0. Lui et al. (1996) recommend 1% Zwittergent 3–16 in 100 mM NH_4HCO_3. However, Zwittergent 3–16 does not get along with MALDI (see Section 7.4.2).

Trypsin cleaves proteins at the carboxy terminal side of lysine and arginine residues. Arg-Pro or Lys-Pro sites are trypsin resistant. Also, trypsin only slowly attacks peptide bindings between a basic amino acid (Lys, Arg) and an acidic one (Glu, Asp). The pH optimum of trypsin lies between 8 and 9, and the optimum relation of enzyme to substrate is 1:50 to 100. Ca^{2+} ions inhibit the self-digestion of trypsin. The trypsin must not contain any chymotrypsin activity (i.e., TPCK must be treated and highly purified). Lysine residues can be protected from trypsin by derivatization (e.g., with citraconic acid), and in reverse, treating the cysteine groups with iodoethylene trifluoroacetamide introduces new trypsin cleaving sites.

The V8 protease of *Staphylococcus aureus* (MW l2 kd) is active at a pH between 3.5 and 9.5 and develops maximal activity at pH 4.0 and 7.8. At pH 4.0, the protease partially precipitates. In phosphate buffer, V8 protease cleaves peptide bindings on the carboxyl terminal side of aspartate or glutamate residues. In 50 mM ammonium bicarbonate buffer pH 7.8 or ammonium acetate buffer pH 4.0, on the other hand, the enzyme cleaves only behind glutamate residues. Below 40° C, the protease does not exhibit any self-digestion. Its watery solution can be frozen and thawed without activity loss. Divalent cations or EDTA have no effect on the enzyme activity. The enzyme also still works in 0.5% SDS. Diisopropyl fluorophosphate inhibits V8 protease.

For some time, the endoproteases Lys-C, Glu-C, and Arg-C have also been used. Lys-C cleaves specifically at the carboxyl terminal side of lysine residues and still works in 5 M urea or 0.1% SDS. Glu-C cleaves at the carboxyl terminal side of glutamate or aspartate residues, depending on the buffer. Arg-C is a cysteine protease that cleaves peptide bindings at the carboxyl terminal side of arginine residues.

The proteases papain and chymotrypsin are rarely used. Chymotrypsin has broader specificity than trypsin. Under mild conditions (short incubation time), chymotrypsin preferentially cleaves peptide bindings near phenylalanine and tyrosine.

Proteases are proteins. They could digest themselves and often do this. Then the experimenter sequences protease peptides. It pays off to compare the preserved sequences to those of the protease.

Sources
1. Cleveland, D., et al. (1977). "Peptide Mapping by Limited Proteolysis in Sodium Dodecyl Sulfate and Analysis by Gel Electrophoresis," *J. Biol. Chem.* 252: 1102–1106.
2. Fernandez, J., et al. (1994). "An Improved Procedure for Enzymatic Digestion of Polyvinylidene Difluoride-bound Proteins for Internal Sequence Analysis," *Anal. Biochem.* 218: 112–117.
3. Kennedy, T., et al. (1988). "Sequencing of Proteins from Two-Dimensional Gels by Using In Situ Digestion and Transfer of Peptides to Polyvinylidene Difluoride Membranes: Application to Proteins Associated with Sensitization in *Aplysia*," *Proc. Natl. Acad. Sci. USA* 85: 7008–7012.
4. Leube, R., et al. (1987). "Synaptophysin: Molecular Organization and mRNA Expression as Determined from Cloned cDNA," *EMBO J.* 6: 3261–3268.
5. Lui, M., et al. (1996). "Methodical Analysis of Protein-nitrocellulose Interactions to Design a Refined Digestion Protocol," *Anal. Biochem.* 241: 156–166.
6. Prussak, C., et al. (1989). "Peptide Production from Proteins Separated by Sodium Dodecyl Sulfate Polyacrylamide Gel Electrophoresis," *Anal. Biochem.* 178: 233–238.

7.6.3.2 Bromine Cyanide and Acid Cleaving

The digestion of a protein by cyanogen bromide transforms the methionine residues into homoserine residues and cleaves the amino acid chain at the carboxyterminal side of the methionine residue. The cleaving is generally completed to 90 to 95%, and the reagents are easily removed because they are volatile. Diluted hydrochloric acid (0.03 N at 105° C) cleaves the proteins at the carboxyterminal side of aspartate residues.

Source
1. Morrison, J. R., et al. (1990). "Studies on the Formation, Separation and Characterization of Cyanogen Bromide Fragments of Human AI Apolipoprotein," *Anal. Biochem.* 186: 152–154.

7.6.4 The Edman Degradation

Classical microsequencing was invented in 1946/47 by the post-doctoral candidate Pehr Edman (1916–1977) and was worked in detail by 1950. Thus, the method is over half a century old. Still, the Edman degradation is not on its way out, but celebrates a revival. Indeed, its strongest competitor (the mass spectrometer) can do a lot. It can identify peptides and can even sequence peptides (see Section 7.6.5). However, for the latter (at least for the time being), it is not as well suited as the good old Edman degradation. From 0.1 to 10 µg of pure protein it yields sequences that are 20 to 30 amino acids long. If the sequencer has a good day, it shows you 40 amino acids from 50-ng protein.

As everybody knows, a sequencing machine digests the amino acid chain starting at the N-terminus and it identifies the amino acid derivatives via a connected HPLC. A requirement is a free N-terminus. The Pierce Catalog, for example, describes the chemistry of the Edman degradation of peptides with phenylisothiocyanate. Baumann (1990) compares the effectiveness of different sequencing techniques.

The most important task of the Edman degradation is still delivering sequence information for the synthesis of oligonucleotides. This preparatory work for protein cloning used to be done as follows. The protein purifier prepared a clean product and handed the preparation to a group specializing in Edman degradation.

Sources
1. Baumann, M. (1990). "Comparative Gas Phase and Pulsed Liquid Phase Sequencing on a Modified Applied Biosystems 477 A Sequencer," *Anal. Biochem.* 190: 198–208.
2. Fischer, P. (1992). "25 Jahre Automatisierte Proteinsequenzierung," *Nachr. Chem. Techn. Lab.* 40: 963–971.

7.6.5 Carboxyterminal Sequencing

A protein has two ends, and instead of from the N-terminus it could be also sequenced from the C-terminus. This can even be easier because the C-termini are generally not blocked. C-terminal sequencing was developed in the 1920s (Schlack and Kumpf 1926). For this, the

C-terminal amino acid is transformed with thiocyanate to thiohydantoine, which is then split off and identified. This method has its problems, however. It detects only a few amino acids, and fails with Asp and Pro residues. The initial yield lies at 10 to 15% (Edman: 20 to 80%), and the per-step yields are lower than with the Edman degradation. They are generally below 70% (Edman: 80 to 96%). With 1 nM peptide you can often identify only three residues. However, the success of C-terminal sequencing depends less on the amount of protein as on its primary structure.

Three Swedish scientists from the Karolinska Institut in Stockholm believe they have ended the sad state of C-terminal sequencing. Bergman et al. (2001) claim to be able to routinely determine five residues with their C-terminal sequencer, sometimes even up to 11 residues. Furthermore, they claim to be able to skip proline residues and to have pushed the sensitivity down to the 10-pM level. The beauty of the method is, they say, that it can easily be combined with N-terminal sequencing. You first sequence N-terminally and then place the washed sample onto the C-terminal sequencer and start nibbling on the other end.

In the tradition of their N-terminal compatriot Pehr Edman, these researchers have already been working on C-terminal sequencing for years and they have C-terminally sequenced hundreds of proteins and peptides. You should take advantage of the enthusiasm of these aficionados. If you don't get anywhere with your sample, why don't you call them in Stockholm and ask them for a run in the optimized C-terminal sequencer? You have nothing to lose.

Sources
1. Bergman, T., et al. (2001). "Chemical C-terminal Protein Sequence Analysis: Improved Sensitivity, Length of Degradation, Proline Passage, and Combination with Edman Degradation," *Anal. Biochem.* 290: 74–82.
2. Inglis, A. (1991). "Chemical Procedures for C-terminal Sequencing of Peptides and Proteins," *Anal. Biochem.* 195: 183–196.
3. Schlack, P., and Kumpf, W. (1926). "Über Eine Neue Methode zur Ermittelung der Konstitution von Peptiden," *Hoppe-Seyler's Z. Physiol. Chemie.* 154: 125–170.

7.6.6 Ladder Sequencing of Peptides

The MW determination by means of MALDI-TOF is so exact that the mass difference between a peptide with n amino acids and the same peptide minus one amino acid allows the identification of the missing amino acid. Even Asp (MW 115.1) and Asn (MW 114.1) can be determined this way. High-end devices even detect whether the peptide is lacking Lys (MW 146.190) or Gln (MW 146.146). However, the peptides should not be too large (maximum 30 amino acid residues). Following Chait et al. (1993), ladder sequencing is based on the accuracy of MALDI-TOF and a modified Edman reaction (Figure 7.10).

The peptide is transformed with phenylisothiocyanate (PICT), which contains 5% phenylisocyanate (PIC). PICT forms a phenylthiocarbamyl peptide with the N-terminal amino acid, PIC the corresponding phenylcarbamyl peptide. After this coupling reaction, superfluous reagent is extracted. The remaining (derivatized) peptides are washed repeatedly and dried in the vacuum centrifuge. Then TFA is added. Under the influence of TFA, the PICT peptide cyclizes, and the N-terminal amino acid splits off (Edman reaction). TFA does nothing to the PIC peptide. Thus, the solution contains the PIC peptide and the peptide shortened by one amino acid. The latter has a free N-terminal amino group. The products are dried in the vacuum centrifuge. All reactions, and washing and drying processes, are performed in one vial.

The cycle can be repeated at will. The result is a ladder of PIC peptides with n, n-1, n-2, n-3 amino acids and a residue peptide with a free N-terminus. This one also becomes blocked with PIC after the last cycle. Now, the MALDI-TOF separates the PIC peptide ladder and measures the MW of the individual PIC peptides. By their MW differences, you can identify the cut-off amino acids. Then, the peptide sequence can be read directly from the spectrum (Figure 7.10).

The laborious parts of the method of Chait et al. (1993) are the extraction and the washing and drying processes. In addition, you lose peptide. Bartlet-Jones et al. (1994) work more elegantly. They also create the peptide ladder via successive cleaving of the N-terminal amino

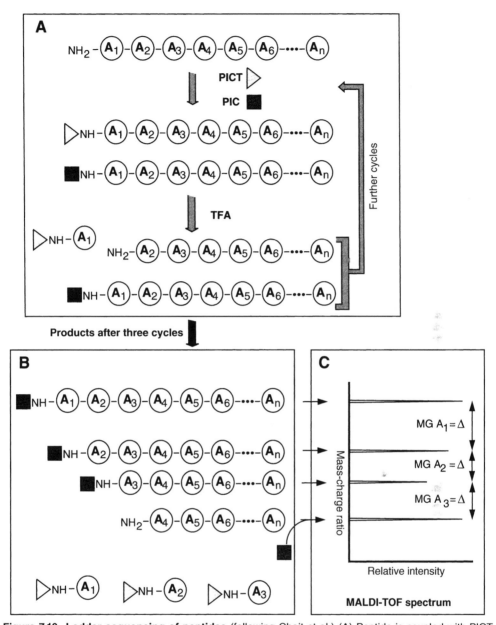

Figure 7.10. Ladder sequencing of peptides (following Chait et al.) (A) Peptide is coupled with PICT and PIC. TFA cleaves the PICT-coupled amino acid. The PIC-derivatized peptide remains unchanged. (B) Several cycles (here three) of PICT/PIC coupling and subsequent acid cleavage generate a peptide ladder with PIC N-termini, a residue peptide with free N-terminus, and the PICT-derivatized amino acids. For the analysis of the peptide ladder, you also block the residue peptide with PIC.

acids, but they use volatile reagents: the volatile coupling reagent trifluoroethylisothiocyanate (TFEITC) and volatile buffers (Figure 7.11). The extraction steps do not apply. Buffer and superfluous reagents are volatilized via vacuum pumps (Figure 7.12). Another advantage: you always pipette into the reaction vessel and you don't take anything out until the end. This diminishes contamination and loss. Blockers for the N-terminus such as PIC are not needed by Bartlet-Jones et al. They cleave the resulting derivatized amino acid—for pipetting reasons—with heptafluorobutyric acid instead of with TFA. A peptide ladder with free N-termini forms (not a ladder of PIC-blocked peptides, as with Chait).

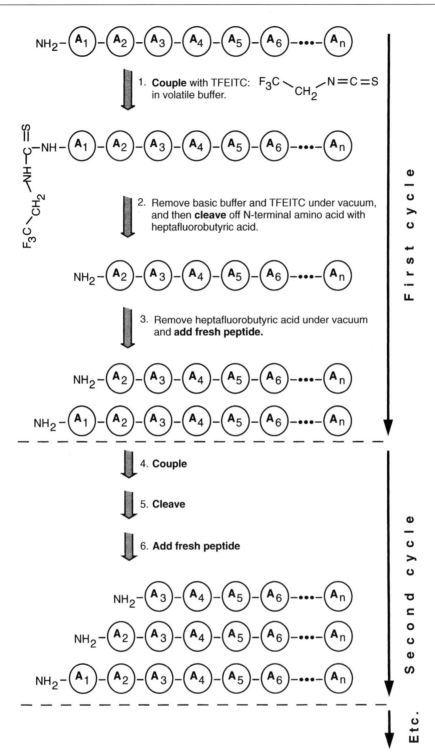

Figure 7.11. Peptide ladder following Bartlet-Jones et al.

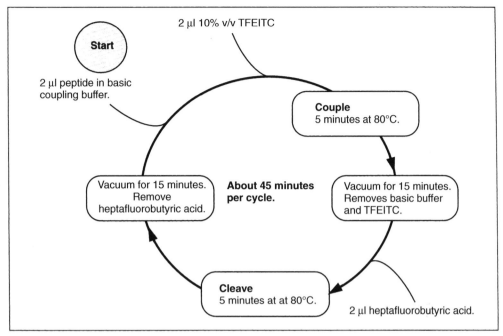

Figure 7.12. We make a peptide ladder! You need a 100-μl glass tube, a heating block, two vacuum pumps, and two solvent traps: one for trimethylamine/trifluoroethanol (coupling buffer) and TFEITC and one for heptafluorobutyric acid (cleaving reagent). Aerial oxygen does not interfere, and nitrogen gassing is unnecessary.

Repeat steps 1 through 10 in Table 7.3 n-1 times for n cycles. After the last cycle, add 2 μl peptide and 5 μl water. Then dry for at least 1 h over NaOH pills.

Dissolve the peptide ladder in 3 to 5 ml 50% aq. acetonitrile/0.1% (v/v) TFA and ultrasonicate for 5 minutes. Pipette several aliquots (0.3 to 0.5 μl) one after the other onto the carrier and let them air dry for 5 minutes each. Finally, pipette 0.3 μl matrix solution (1% (w/v) α-cyano-4-hydroxy cinnamic acid in 50% aq. acetonitrile/0.1% (v/v)/TFA) onto the dried sample and air dry again. Insert this preparation into the MALDI mass spectrometer and analyze it. Important:

- After each coupling step, the basic buffer must be completely removed. Otherwise, the salts of the heptafluorobutyric acid accumulate over the cycles. These inhibit the cleaving of the amino acids (buffer effect) and suppress the signal in the MALDI-TOF. Bartlet-Jones et al. recommend two independent vacuum systems with acidic or basic traps for drying (Figure 7.12).
- You ideally carry out the reactions in glass minivials, because you lose peptide in polystyrene or polypropylene tubes.
- You increase the sensitivity as follows. Transform the peptides of the ladder (which have free amino termini) with quaternary N-alkyl compounds before the analysis in the MALDI-TOF. This way, you give all peptides a positive charge, which improves the signal. The proton transmission with MALDI is, after all, not quantitative. Bartlet-Jones et al. (1994) describe the synthesis of a suitable N-alkyl compound.

Once you have the peptide ladder, you get the sequence of your peptide within minutes from a MALDI-TOF run. Because you can produce ladders from different peptides (in different tubes) at the same time, sequencing with peptide ladders and MALDI-TOF is faster than with the traditional Edman method Table 7.4. In addition, you need less peptide. This is true only, however, if it works, because as any other method ladder sequencing also has its problems.

- With peptides with more than 30 amino acids, the measured MW differences become inexact (you are dealing with small differences between large MWs). The cleaved amino acids cannot be determined with certainty anymore. Fortunately, most protein cleavings result in peptides with 20 to 30 amino acids.

Table 7.3. Protocol for sequencing according to Bartlet-Jones.

1. Dissolve peptide in coupling buffer (2,2,2-trifluoroethanol; water: 12.5% trimethyl ammonium carbonate pH 8.5 (5:4:1, (v/v)), in 2 µl per cycle + 4 µl (thus, 14 µl with 5 cycles).

2. Pipette 2 µl peptide into a 100-µl HP glass minivial and immediately add 2 µl 10% (v/v) TFEITC.

3. Close vial.

4. Heat for 5 minutes to 80° C (heating block).

5. Dilute with 5 µl water.

6. Attach the first vacuum system: 15 minutes at 4×10^{-2} mbar with toluol sulfonic acid in the trap and P_4O_{10} as a drying agent.

7. Add 7 µl heptafluorobutyric acid.

8. Close vial.

9. Heat for 5 minutes to 80° C (heating block).

10. Attach the second vacuum system: 15 minutes at 4×10^{-2} mbar with NaOH as acid trap.

* Source: J. Zimmermann, University College, London.

Table 7.4. Comparison of the sequencing methods.

	Traditional Edman Degradation	Peptide Ladder Sequencing (Following Bartlet-Jones Et Al.)
Source material	Peptide and proteins of any size with free N-terminus	Peptides with 20 to 30 amino acids and free N-terminus
Amount of peptide	1 to 10 pM	0.5 to 1 pM
Number of sequenced amino acids	20–30	5–10
Can several samples be sequenced at the same time?	No	Yes
Time	1 day	1 day for 10 cycles. The MALDI-TOF run takes only minutes. If you produce ladders for several samples at the same time, you can sequence several samples in a day.
Are phosphorylized amino acids recognizable?	Not directly	Yes
Device costs	$200.000–250.000	$50.000–300.000

- Leu and Ile have the same MW (i.e., the MALDI-TOF mass spectrometer cannot distinguish between Leu and Ile). Only a very good MALDI-TOF recognizes whether Lys or Gln was removed from a peptide.
- Both Chait and Bartlet-Jones produce their peptide ladders by means of Edman chemistry. Thus, they presuppose a free N-terminus. You can avoid this problem. Peptide ladders also originate from acid cleaving or partial digestion with exoproteases. There are exoproteases that attack at the N-terminal end, and others that cleave off amino acids from the C-terminal end. One can avoid an N-terminal blockage with exoproteases attacking from the C-terminal end. With a free N-terminus, you can sequence the peptide from both sides (Woods et al. 1995; Thiede et al. 1995). The partial digestion is applied directly to the sample carrier of the mass spectrometer. Some pM of peptide are enough for the sequence.

Sources

1. Bartlet-Jones, M., et al. (1994). "Peptide Ladder Sequencing by Mass Spectrometry Using a Novel, Volatile Degradation Reagent," *Rapid Communications in Mass Spectrometry* 8: 737–742.
2. Chait, B., and Kent, S. (1992). "Weighing Naked Proteins: Practical, High-accuracy Mass Measurement of Peptides and Proteins," *Science* 257: 1885–1894.
3. Chait, B., et al. (1993). "Protein Ladder Sequencing," *Science* 262: 89–92.

4. Thiede, B., et al. (1995). "MALDI-MS for C-terminal Sequence Determination of Peptides and Proteins Degraded by Carboxypeptidase Y and P," *FEBS Lett.* 357: 65–69.
5. Woods, A., et al. (1995). "Simplified High-sensitivity Sequencing of a Major Histocompatibility Complex Class I-associated Immunoreactive Peptide Using Matrix-assisted Laser-desorbtion Ionization Mass Spectrometry," *Anal. Biochem.* 226: 15–25. The authors explain the details of their procedure badly, or not at all. The paper is useful only as a clue for your own trial experiments.

7.7 Strategy

It is nice that proteomics has become so popular lately, but should you delve into it as a hopeful doctoral candidate? I don't think so. At least not for long. Two-dimensional gels will be produced on the assembly line, one digestion will immediately be followed by the next, and mass spectra will fly from the printers like moths from old closets. The future of proteomics reeks of repetitive work, of standardization, of boredom—and of a vast amount of numbers.

In fact, some companies (e.g., Micromass and Bio-Rad) already offer proteome robots that fully automatically prep 2D gels: punch out protein spots, destain, dehydrogenate, digest, extract, and run mass spectra. The only thing the researcher still has to do himself is run the 2D gels and stain them. Otherwise, he sits in front of the monitor and picks out spots. With all sympathy, I cannot call this original work.

The typical proteome experiment generally does not strike you as original. It is always the same, consisting of comparing gels, comparing conditions, and searching for spots.

Chapter 8 Subunits

Many proteins consist of subunits: several polypeptides held together by disulfide bridges or noncovalently in a defined structure. In homooligomers the subunits are identical. In hetero-oligomers they differ. Examples of homooligomers are the enzymes catalase and aldolase (both with four subunits). Examples of heterooligomers are neurotransmitter receptors such as the acetylcholine receptor (five subunits $\alpha_2\beta\gamma\delta$) or voltage-dependent K^+ channels (four subunits).

With homooligomer proteins you need to determine the number of subunits per oligomer. With heterooligomer proteins, you also need to figure out the number of different subunits in the oligomer (the stoichiometry indices). This task requires instinct, judgment, and experimental skill.

> And so he jogged on, so occupied with his thoughts and easy in his mind that he forgot all about the hardship of travelling on foot.

8.1 Number and Stoichiometry of Subunits

8.1.1 About the Difficulties with Stoichiometry Determinations

You cannot find the number of subunits of (for example) a homooligomer by determining the MW of the oligomer and dividing by the MW of the subunit. The inaccuracy of most MW determinations is so high that (for example) with a result of 200 kd for the oligomer and of 50 kd for the monomer it is not possible to distinguish between a trimer, tetramer, or pentamer. Hydrodynamic measurements of oligomer MW are at best to 10% precise. With membrane proteins, the insecurity factor is even higher (between 20 and 40%) because of the additional bound phospholipid and detergent molecules (see Section 3.2.2).

The MW of the subunits ascertained by SDS gel electrophoresis easily deviates from the real one by about 10 to 40%, because glycosylation, unusual amino acid composition, phosphorylation, sulfation, and so on let proteins run atypically. After covalent cross-linking of all subunits, SDS gel electrophoresis also delivers an estimate of the oligomer's MW. This estimate is even more uncertain than that for the individual subunits, because additional cross-linking molecules increase the MW, and cross-linking of the polypeptide chains sometimes changes their run speed in SDS gels and with it the apparent MW.

Furthermore, even high concentrations of cross-linker often yield only small quantities of the cross-linked oligomer. Instead, you get lots of undefined high-molecular material that makes the stained SDS gels look like pieces of modern art and their interpretation similarly difficult. Thus, the band with the highest MW does not have to be the completely cross-linked oligomer, because higher-molecular bands could still exist, whose protein amounts, however, lie below the sensitivity threshold of the stain. Finally, a high-molecular band can also be an artifact from intermolecular cross-linking.

A precise determination of the MW of the subunits is only possible by means of the MALDI-TOF mass spectrometer (see Section 7.5). However, as long as the MW of the oligomer cannot be determined with similar precision, number and stoichiometry of subunits (N&S) remain uncertain and should be confirmed with other methods. There are three classical strategies for determining the subunit composition of an oligomer: X-ray structural analysis, hybridization, and cross-linking. Amino acid analyses and subunit-specific antibodies likewise only provide clues about N&S.

8.1.2 N&S with X-ray Structural Analysis

The most precise method of determining the N&S of a protein is X-ray structural analysis. A prerequisite for this technique is good crystals of the protein. Their production is difficult even for soluble proteins. With rare membrane proteins, it is almost impossible. Furthermore, the X-ray structural analysis is the domain of a few specialists and requires detailed knowledge of physics.

Source
1. Deisenhofer, J., et al. (1985). "Structure of the Protein Subunits in the Photosynthetic Reaction Center of *Rhodopseudomonas viridis* at 3 A Resolution," *Nature* 318: 618–624.

8.1.3 N&S with Hybridization Experiments

With homooligomers, the following considerations form the basis for the stoichiometry determination by hybridization. Assume that for a homooligomer with n subunits α it is possible to produce subunits *a* that resemble α and can also combine with α to form the oligomer. However, the subunit *a* differs in one property (e.g., charge, enzyme activity, ligand binding) from α. Because *a* hybridizes with α into an oligomer according to the laws of combinatorics, a mixture of *a* and α forms the oligomers $\alpha_i a_{n-i}$, where i goes from 0 to n. n + 1 oligomer species are formed (e.g., for a tetramer the 5 species α_4, $\alpha_3 a$, $\alpha_2 a_2$, αa_3, and a_4, and for a dimer the three species $\alpha\alpha$, αa, and aa, where n − 1 oligomers are hybrids of α and *a*). The number of subunits can be inferred from the number of oligomers resulting from hybridization.

The quantitative portion of oligomer species $\alpha_i a_{n-i}$ in the oligomer mixture also follows from combinatorial considerations. Let fα and fa be the relative molar amount of α and *a* (fα + fa = 1). Then the relative molar amount f$\alpha_i a_{n-i}$ of the oligomer species $\alpha_i a_{n-i}$ in the oligomer mixture is calculated according to the formula shown in Figure 8.1. The sum of the relative molar amounts f$\alpha_i a_{n-i}$ of all oligomer species $\alpha_i a_{n-i}$ is equal to 1 (see formula shown in Figure 8.2).

Assume the oligomers α_n and a_n and their hybrids have an activity A (enzyme activity, translocator activity) that is inhibited by the inhibitor I, which binds only to subunit *a*. Then the activity of the oligomers $\alpha_i a_{n-i}$ (i from 0 to n − 1) is equal to 0 in the presence of saturating concentrations of I. In this case, the slope of the linear function lnA against ln(fα) according to the formula shown in Figure 8.3 yields the number of subunits of the oligomer (MacKinnon 1991).

Similar considerations are valid if the inhibitor I inhibits only a_n (see formula in Figure 8.4).

$$f\alpha_i a_{n-i} = \binom{n}{i} \cdot f\alpha^i \cdot fa^{n-i}$$

Figure 8.1. The relative molar amount $f\alpha_i a_{n-i}$ of oligomer species $\alpha_i a_{n-i}$ at relative amounts $f\alpha$ and fa of subunits α and *a*.

$$1 = \sum_{i=0}^{i=n} f\alpha_i a_{n-i} = \sum_{i=0}^{i=n} \binom{n}{i} \cdot f\alpha^i \cdot fa^{n-i}$$

Figure 8.2. Sum of the relative molar amounts $f\alpha_i a_{n-i}$ of the oligomer species $\alpha_i a_{n-i}$.

Let A be a functional parameter of the oligomers (e.g., an enzyme activity)

Let the inhibitor I bind selectively to subunit a. Then:

In absence of the inhibitor:
$$A = \sum_{i=0}^{i=n} k_i \bullet \binom{n}{i} \bullet f\alpha^i \bullet fa^{n-i}$$

In the presence of a saturation concentration of inhibitor:
$$A = k_n \bullet f\alpha^n$$

From which follows: $\ln A = \ln(k_n) + n \bullet \ln f\alpha$

Figure 8.3. Dependence of the activity of an oligomer mixture $\alpha_i a_{n-i}$ on the relative amount of subunit α ($f\alpha$) in presence of an inhibitor that binds to subunit a.

$$\overset{+}{A} = \sum_{i=1}^{i=n} k_i \bullet \binom{n}{i} \bullet f\alpha^i \bullet fa^{n-i} \quad \text{and} \quad \overset{-}{A} = \sum_{i=0}^{i=n} k_i \bullet \binom{n}{i} \bullet f\alpha^i \bullet fa^{n-i}$$

from this follows

$$\overset{-}{A} - \overset{+}{A} = k_0 \bullet fa^n \quad \text{and} \quad \ln(\overset{-}{A} - \overset{+}{A}) = \ln(k_0) + n \bullet \ln fa$$

Figure 8.4. Dependence of the activity of an oligomer mixture $\alpha_i a_{n-i}$ on the relative amount of subunit a (fa) in the presence of an inhibitor that selectively inhibits a_n. A^+ and A^- are the activities of the hybridized oligomer mixture in the presence or absence of a saturating concentration of inhibitor.

These combinatorics can also be applied to herterooligomers and yield similar equations. In the 1960s, hybridization experiments with enzymes were fashionable. You either produced a from α or you isolated an isoenzyme, whose monomers a differed from α in their charge and nevertheless hybridized with α into an oligomer. For the hybridization, the oligomers were mixed in a certain proportion, split into subunits, and afterward associated into oligomers again. With native gel electrophoresis or IEC, the experimenter determined the number of the hybrids and with it the number of subunits.

Oligomers are split into their subunits via SDS, alkaline, or acidic pH; by succinylation (treatment with succinic acid anhydride); high or low ion strength, 6 M urea, 4 to 5 M guanidine hydrochloride; and sometimes also ligands. The effect of SDS, pH extremes, and succinylation is based at least partially on a similar increase of the net charge of the subunits. The electrostatic repulsion causes the oligomers to disintegrate. In hybridization experiments, the oligomers are reversibly cleaved. Reversible cleaving often succeeds with pH changes, ion strength extremes, urea, or ligands.

The classical hybridization technique fails with families of heterooligomers whose rarity and variety in vivo makes it impossible to isolate a defined oligomer. The experimenter also has the cards stacked against him if the oligomer does not allow itself to be reversibly cleaved into subunits. For example, many oligomer membrane proteins disintegrate only under denaturing conditions (e.g., SDS, irreversibly into their subunits).

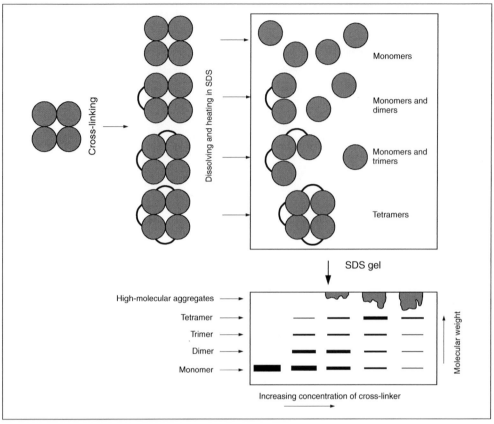

Figure 8.5. Cross-linking of homooligomers.

Molecular-biological technologies offer an elegant way out. If there is cDNA for α and *a*, you let a living cell do the hybridization (e.g., *Xenopus*-oocytes). cRNAs from each subunit are injected into the cell. As a result, the cell synthesizes the α und *a* subunits and combines them to form the oligomer species. If α-cRNA and a-cRNA are translated with the same efficiency, the amounts of α-cRNA and a-cRNA control the amounts fα and fa from α and *a*.

Sources
1. Cooper, E., et al. (1991). "Pentameric Structure and Subunit Stoichiometry of a Neuronal Nicotinic Acetylcholine Receptor," *Nature* 350: 235–238.
2. Lebherz, H., and Rutter, W. (1969). "Distribution of Fructose Diphosphate Aldolase Variants in Biological Systems," *Biochemistry* 8: 109–121.
3. MacKinnon, R. (1991). "Determination of the Subunit Stoichiometry of a Voltage-activated K⁺ Channel," *Nature* 350: 232–235.
4. Penhoet, E., et al. (1967). "The Subunit Structure of Mammalian Fructose Diphosphate Aldolase," *Biochemistry* 6: 2940–2949.

8.1.4 N&S with Cross-linking Experiments

Section 2.4 describes cross-linkers and their handling. Figures 8.5 and 8.6 show how simple the determination of the subunit composition of an oligomer theoretically is with cross-linkers. A detection assay is required for the oligomer, for the purified oligomer, and for specific antibodies against the subunits.

First, the experimenter estimates the MW of the oligomer (hydrodynamic measurements, radiation inactivation) and the MW of the subunits (SDS gel electrophoresis of the purified oligomer). She then treats the oligomer with cross-linkers (e.g., from the bismuthate family)

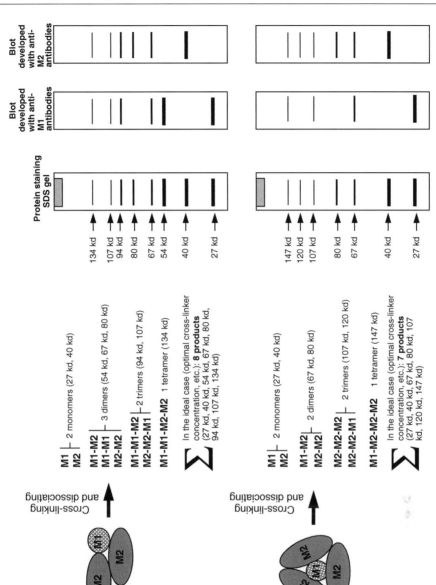

Figure 8.6. Cross-linking of heterooligomers. Here, heterotetramers with subunit M1 (MW = 27 kd) and subunit M2 (MW = 40 kd).

of increasing concentration and chain length. In the process, covalently linked intermediate stages develop from the oligomer (covalently linked dimers, covalently linked trimers, and so on). The intermediate stages can be separated on an SDS gel and stained on blots with antibodies against the respective subunits. From the number of different subunits and the band pattern on the SDS gels or blots, your can determine the number of subunits of the oligomer. With luck, the experimenter can even infer the stoichiometry indices of the subunits (Figure 8.6). Instead of adding cross-linker, you can try to link the polypeptides by sulfhydryl oxidation via disulfide bridges (Kobashi 1968).

The most important tool for cross-linking of oligomers is SDS gel electrophoresis. Because the products of cross-linking distribute over a wide MW range, the effort pays off for gradient gels (e.g., 5 to 12%). Adding 6 M urea to the Lämmli sample buffer or substituting mercaptoethanol with DTT improves the run of the sample during electrophoresis. Photoactivateable cross-linker, photo flash, and working in diluted solutions reduces intermolecular cross-linking to a minimum.

Source

1. Kobashi, K. (1968). "Catalytic Oxidation of Sulfhydryl Groups by O-phenanthroline Copper Complex," *BBA* 158: 239–245.

8.1.4.1 Cross-linking of Homooligomers

In spite of the theoretical simplicity, the determination of the number of subunits is susceptible to artifacts and is confusing even with homooligomers.

- High concentrations of cross-linkers or protein lead to intermolecular cross-linking, or the oligomer in solution partially aggregates to high-molecular complexes. Both lead to an overestimated number of subunits.
- Cross-linking denatures or changes the oligomer, which may also occur during purification.
- The oligomer is not a homooligomer but a heterooligomer if its different subunits exhibit a similar MW on SDS gels.
- Cross-linking of the oligomer's monomers is not continuous but stepwise with increasing concentration of cross-linker. The intermediate steps dimer, trimer, and so on do not occur then. On the SDS gel, you only see the monomers and/or high-molecular adducts.

The following sources give examples of how difficult the determination of the subunit number of a homooligomer can be.

Sources

1. Johnston, P., and Südhof, T. (1990). "The Multisubunit Structure of Synaptophysin: Relationship Between Disulfide Bonding and Homooligomerization," *J. Biol. Chem.* 265: 8869–8873.
2. Rehm, H., et al. (1986). "Molecular Characterization of Synaptophysin, a Major Calcium Binding Protein of the Synaptic Vesicle Membrane," *EMBO J.* 5: 535–541.
3. Thomas, L., et al. (1988). "Identification of Synaptophysin as a Hexameric Channel Protein of the Synaptic Vesicle Membrane," *Science* 242: 1050–1053.

> I say this because we all know one another, and it will not do to throw false dice with me; and as to the enchantment of my master, God knows the truth; leave it as it is; it only makes it worse to stir it.

Controls protect against disaster.

- Check the composition of the cross-linked bands in the SDS gel via experiments with cleavable cross-linker (e.g., SASD in Figure 2.21). Cut the band of the gel, cleave the cross-linking (e.g., with DTT), and examine the cleaving products in the second SDS gel electrophoresis and a blot.
- Test for intermolecular cross-linking via dilution experiments. The quantitative ratios of the cross-linked products should not change with constant cross-linker concentration and

decreasing oligomer concentration. In particular, no bands should disappear with decreasing oligomer concentration.
- Different cross-linkers should yield qualitatively identical results.
- The MW of the oligomer from cross-linking studies and from hydrodynamic measurements should be about the same.
- Artifacts often appear during the purification of the oligomer. Hence, you should cross-link under conditions that are as native as possible: with membrane proteins in membrane vesicles or in the living cell, with soluble proteins in the raw cell extract. The results should not contradict those with the purified oligomer.

> He who is prepared has his battle half fought; nothing is lost by my preparing myself, for I know by experience that I have enemies, visible and invisible, and I know not when, or where, or at what moment, or in what shapes they will attack me.

Sources
1. Darawshe, S., et al. (1987). "Quaternary Structure of Erythrocruorin from the Nematode *Ascaris suum*," *Biochem. J.* 242: 689–694.
2. Waheed, A., et al. (1990). "Quaternary Structure of the Mr 46000 Mannose 6-phosphate Specific Receptor: Effect of Ligand, pH, and Receptor Concentration on the Equilibrium Between Dimeric and Tetrameric Receptor," *Biochemistry* 29: 2449–2455.

8.1.4.2 Structure of Homooligomers

Once the experimenter has determined the number of subunits of a homooligomer, its structure is not defined by any means yet because the subunits of even a tetramer can be put together in different ways. The structure of a homooligomer must satisfy two conditions (Klotz et al. 1970).
- The subunits must hold equivalent positions. Thus, linear orderings are not possible, except for dimers, because the position of the subunits at the ends is different from the position of the subunits inside the chain.
- The arrangement must be a closed structure (i.e., there cannot be any free subunit binding sites). Otherwise, the oligomer is unstable and forms large adducts.

Theoretically, there are three classes of structures: cyclic, dihedral, and cubic. Cubic structures are only possible from 24 subunits upward. Only oligomers with an even number of subunits exhibit dihedral structures. Oligomers with an odd number of subunits have cyclic build.

Let n be the number of subunits. Then there is only one possible arrangement for n = 2 and n = 3, two for n = 4 (cyclic and tetrahedral), one for n = 5 (cyclic), and three for n = 6 (one cyclic, two dihedral ones) (Figure 8.7).

The quantitative analysis of cross-linking data yields evidence about the structure of the homooligomer. The homooligomer is incubated with different concentrations of cross-linker, the reactions are stopped after a certain time, and the solutions are separated on an SDS gel. Monomers, cross-linked dimers, and so on appear in the gel. The relative amount of monomers and cross-linked oligomers is determined for each cross-linker concentration (e.g., by scanning the gel). From the amounts, the cross-linking probabilities between the single subunits are computed (and from this the structure).

In this experiment, the number of bands and their intensity are telling. In the SDS gel, the color intensity of different bands is similar between homooligomers (which provides clues about the state of the solution). This is not true for the blot. High-molecular cross-linking products blot worse than low-molecular ones and are hence quantitatively underrepresented on the blot. Longer blot times compensate for this, but also harbor the danger that the small proteins sneak through by the blot membrane and thereby are partially lost. The protein stain of the blotted gel shows whether any proteins remained in the gel. The second blot membrane, which is laid under the first, shows whether small proteins came through.

Number of Subunits		Symmetry	Number of Bindings	Type of Bindings
2	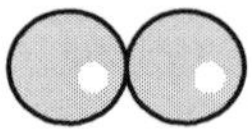	C2 (linear)	1	Equal
3	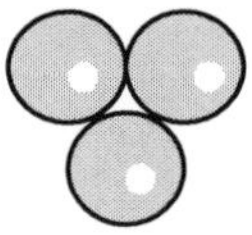	C3 (triangle)	3	Equal
4	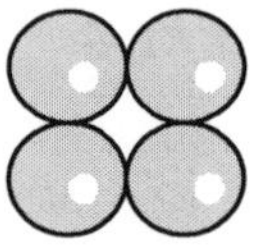	C4 (square)	4	Equal
4		D2 (tetrahedon)	6	3 different
5	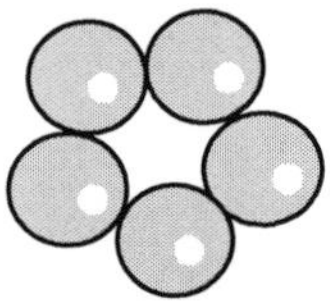	C5 (pentagon)	5	Equal
6	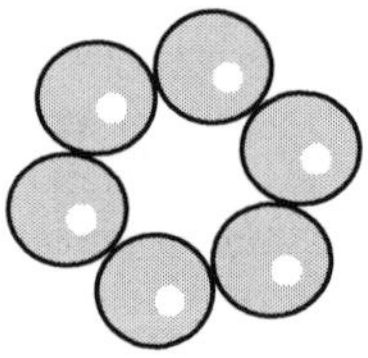	C6 (hexagon)	6	Equal
6		D3 (prism)	9	2 different
6	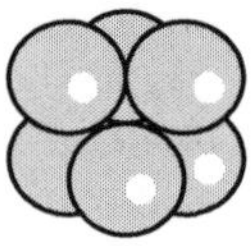	D3 (octahedron)	12	3 different

Figure 8.7. The structure of homooligomers (after Klotz et al. 1970).

The data analysis of such cross-linking experiments is difficult. A look into the appendix of Hucho et al. (1975) shows how complicated the mathematical treatment of just the tetramers is.

Sources
1. Darawshe, S., and Daniel, E. (1991). "Molecular Symmetry and Arrangement of Subunits in Extracellular Hemo-globin from the Nematode *Ascaris suum*," *Eur. J. Biochem.* 201: 169–173.
2. Hucho, F., et al. (1975). "Investigation of the Symmetry of Oligomeric Enzymes with Bifunctional Reagents," *Eur. J. Biochem.* 59: 79–87.
3. Klotz, I., et al. (1970). "Quaternary Structure of Proteins," *Ann. Rev. Biochem.* 39: 25–62.

8.1.4.3 Cross-linking of Heterooligomers

First it needs to be determined which subunits the heterooligomer has. For this, the experimenter purifies the heterooligomer. No protein can be presented 100% pure. Some of the bands shown by the purified heterooligomer in the SDS gel can be due to contaminations. On the other hand, some polypeptides that belong to the oligomer are lost during the purification procedure (e.g., due to protease effects or because of missing ligands or a wrong buffer). Section 5.4 describes purification assays. Specific antibodies against the subunits provide an additional check of the composition of the oligomers. The antibodies precipitate the oligomer from radioactively marked raw extract (IP; see Section 6.2), and the experimenter can compare the precipitates of different antibodies (against different subunits) in the autoradiogram of an SDS gel.

Figure 8.6 shows that cross-linking studies also determine the stoichiometry of heterooligomers. However, the state of affairs is more complicated. More bands appear, and subunit-specific antibodies are required for their identification. Also, the color intensity of the bands cannot be compared in the gel because in the cross-linking of heterooligomers different bands consist of different proteins. However, the color intensity of Coomassie or silver varies by a factor of 2 or 5, respectively, between identical amounts of different proteins.

The experimenter secures the won data with controls. Nevertheless, he will often wish for more clarity in the results of cross-linking experiments with heterooligomers. Only lucky people, utilitarian optimists, or people who already know what the result should be infer a heterooligomer's stoichiometry or number of subunits from cross-linking experiments only. If you want to be sure, you calm your conscience with additional data that is independent from cross-linking.

Sources
1. Gaffney, B. (1985). "Chemical and Biochemical Crosslinking of Membrane Components," *BBA* 822: 289–317.
2. Hucho, F., et al. (1978). "The Acetylcholine Receptor as Part of a Protein Complex in Receptor-enriched Membrane Fragments from *Torpedo californica* Electric Tissue," *Eur. J. Biochem.* 83: 335–340.
3. Langosch, D., et al. (1988). "Conserved Quaternary Structure of Ligand-gated Ion Channels: The Postsynaptic Glycine Receptor Is a Pentamer," *Proc. Natl. Acad. Sci. USA* 85: 7394–7398.

8.1.5 N&S with Amino Acid Analyses or Antibodies

The attempt to determine the stoichiometric ratio of two subunits of a protein complex from the measurement of their color intensity in SDS gel is like comparing apples to oranges. However, the ratio of the molarities of the subunits' terminal amino acids can be determined exactly via calibration curves from Edman degradation (see Chapter 7). This ratio is equal to the stoichiometric ratio of the subunits (see Table 8.1). However, if the terminal amino acids are completely or partly blocked, which occurs often, the method fails (see Section 7.2).

Kapp et al. (1990) determine the subunit structure of heterooligomer proteins via amino acid analyses. The idea is that the amino acid composition of the oligomer (known) depends on the amino acid composition of the subunits (known) and on their stoichiometry (unknown). The experimenter needs the MW of the oligomer, the number and MW of the subunits, and amino acid analyses of oligomer and subunits. The amino acid composition of the subunits can also be inferred from their sequence (if available). The mols of the amino acid A_i per mol heterooligomer are the sum (over all subunits) of the mols A_i per mol subunit multiplied by the number of the respective subunits in the oligomer (stoichiometry indices). Only the latter quantities, the stoichiometry indices, are unknown. A similar equation can be established for each amino acid (i.e., theoretically, about 20 equations). The number of subunit types and stoichiometry indices (the unknown quantities) is usually less than 20. The stoichiometry indices can also be computed. The method seems to be experimentally easy. It is also robust in the face of small uncertainties in the MW determination of heterooligomer and subunits. However, having to deal with 20 equations takes an accomplished number cruncher.

An immunological method likewise provides conclusions about the stoichiometry of the subunits of an oligomer, independently of cross-linking data (Figure 8.8) (Pestka et al. 1983;

Table 8.1. The subunit structure of a heterooligomer determines the ratio of the molarities of its subunits' terminal amino acids.

Subunit Structure	Molar Ratio Terminal Amino Acid (α-subunit) to Terminal Amino Acid (β-subunit)
$\alpha\beta$; $\alpha_2\beta_2$; $\alpha_3\beta_3$	1.0
$\alpha_2\beta_3$	1.5
$\alpha\beta_2$; $\alpha_2\beta_4$	2.0
$\alpha\beta_3$	3.0
$\alpha\beta_4$	4.0

Whiting et al. 1987). Let an oligomer have an unknown number of α subunits. An anti-α antibody that binds to a narrowly defined epitope of the α subunit (monoclonal antibodies or polyclonal antibodies against a short amino acid sequence) is covalently coupled to a matrix. The idea is that the oligomer binds to the immunomatrix only with an α subunit. If there is more than one α subunit per oligomer, some epitopes remain free on the oligomer. These epitopes bind to a free anti-α antibody. The experimenter loads the oligomer onto the immunomatrix and washes off the unbound oligomer. Then she examines how much free anti-α antibody is bound by the matrix loaded with the oligomer. It is advisable to label free anti-α antibody with [125]iodine. Immunomatrix without oligomer serves as control (or, better, immunomatrix loaded with oligomer in the presence of a binding inhibitor). A suitable binding inhibitor would be, for example, the peptide against which the antibody was produced. If the amount of the oligomer on the matrix cannot be determined exactly, only a qualitative conclusion is possible. Furthermore, the method is sensitive to disruptions. Nobody can guarantee that the oligomer binds only via one antibody to the immunomatrix. If the antibody density on the matrix is high, an oligomer with more than one α subunit may bind to the matrix via several α subunits and several antibodies. The binding of free antibody is prevented and the stoichiometry index of the α subunit is underestimated. Finally, with more than two subunits per oligomer and suitable steric ratios an antibody could saturate two free α epitopes. Again, the stoichiometry index would end up being too low.

Sources
1. Kapp, O., et al. (1990). "Calculation of Subunit Stoichiometry of Large Multisubunit Proteins from Amino Acid Compositions," *Anal. Biochem.* 184: 74–82.
2. Pestka, et al. (1983). "Specific Immunoassay for Protein Dimers, Trimers, and Higher Oligomers," *Anal. Biochem.* 132: 328–333.
3. Whiting, P., et al. (1987). "Neuronal Nicotinic Acetylcholine Receptor β-subunit Is Coded for by the cDNA Clone α4," *FEBS Lett.* 219: 459–463.

8.2 What Holds Our World Together?

Once the structure of an oligomer is determined, the question remains why the oligomer has precisely this and no other structure. How does this specificity come about? Which areas of the subunits are responsible for the oligomer formation? How are these contact sites arranged geometrically? The following help provide answers.

- If the contact sites consist of partial sequences and are not scattered in a few locations over broad areas, the experimenter can synthesize the peptides that correspond to half of the contact site (i.e., of a binding site). The peptides should bind to the contact site and inhibit the formation of oligomers. The same holds for antibodies against the sequence of the contact site. How do you find the right peptide from the hundreds of possible ones? Either guess or

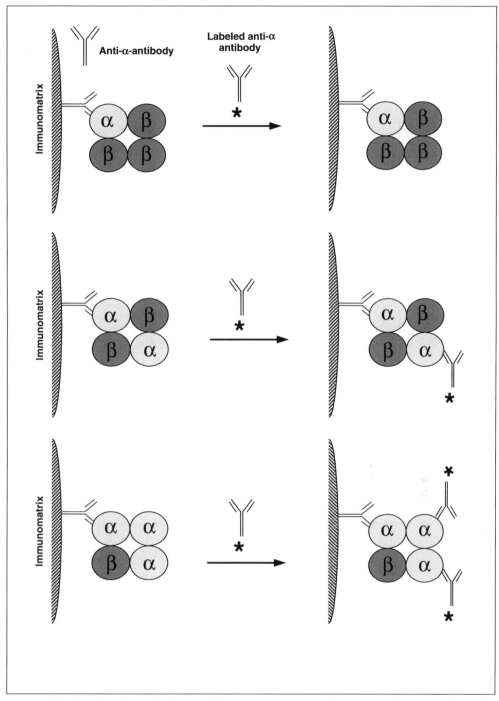

Figure 8.8. Investigation of the stoichiometry of heterooligomers with subunit-specific antibodies.

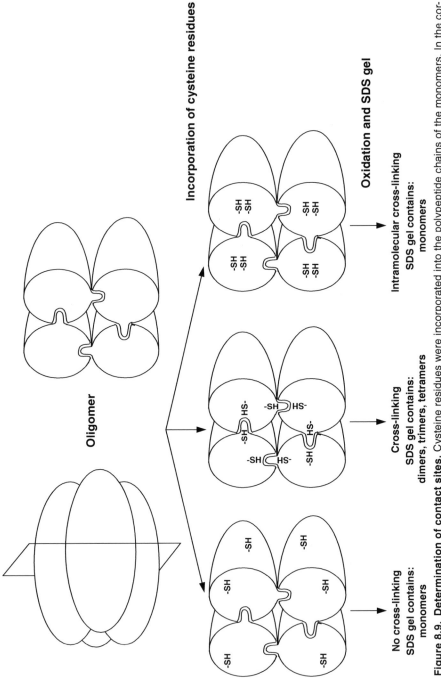

Figure 8.9. Determination of contact sites. Cysteine residues were incorporated into the polypeptide chains of the monomers. In the corresponding oligomer, the cysteine residues are oxidized. If the monomer's cysteine residues are situated near the contact sites of the monomers, the oxidation causes covalent cross-linking of the oligomer (disulfide bridges). Dimers, trimers, and so on appear in the SDS gel (nonreducing conditions).

divide. Halve the protein (or express two halves), see which half binds or forms an oligomer, halve this half again, and so on.

- The contact sites are the touch points of the oligomer's subunits (i.e., they are near each other). Neighboring partial sequences can be identified via cross-linking studies with artificially introduced cysteine residues. This requires that the cysteine residues do not change the function and conformation of the protein (Figure 8.9).
- Parts of the subunit are replaced by corresponding parts of a related molecule, which do not form oligomers with the remaining subunits. Then you examine (for example) via coimmunoprecipitation which chimaera form oligomers and which do not.

Sources
1. Li, M., et al. (1992). "Specification of Subunit Assembly by the Hydrophilic Amino-terminal Domain of the Shaker Potassium Channel," *Science* 257: 1225–1230.
2. Maniolos, M. (1990). "Transmembrane Helical Interactions and the Assembly of the T Cell Receptor Complex," *Science* 249: 274–277.
3. Pakula, A., and Simon, M. (1992). "Determination of Transmembrane Protein Structure by Disulfide Cross-linking: The *E. coli* Tar Receptor," *Proc. Natl. Acad. Sci. USA* 89: 4144–4148.
4. Sumikawa, K. (1992). "Sequences on the N-terminus of Ach Receptor Subunits Regulate Their Assembly," *Mol. Brain Res.* 13: 349–353.

Chapter 9 Glycoproteins

"Well," said Don Quixote, "if that be thy determination, . . . let us leave these phantoms alone and turn to the pursuit of better and worthier adventures; for, from what I see of this country, we cannot fail to find plenty of marvellous ones in it."

9.1 How, Where, and for What Purpose Are Proteins Glycosylated?

After the synthesis of a protein in the endoplasmic reticulum, the cell does not sit still but continues working on its product. It cuts the polypeptide chains; phosphorylizes or sulfatizes some serines, threonines, or tyrosines; and provides certain proteins with oligosaccharide chains. The latter is referred to as glycosylation. Many of the extracellular proteins in serum, urine, saliva, lymph, and cerebrospinal fluid are glycosylated, but also integral membrane proteins (on the extracellular side). Length, charge, and sequence of the sugar chain are not constant for a certain protein. Instead, they depend on species, tissue, age, and the state of the organism (e.g., illness, pregnancy, and so on). A glycosylated protein generally consists of a row of glycoforms (i.e., proteins with identical polypeptide sequence but different sugar residues).

The sugars of N-glycosylated proteins (e.g., synaptophysin, ovalbumin, transferrin) are attached to the asparagine of an Asn-X-Thr/Ser sequence motive. However, not every glycosylation motive is glycosylated. The sugar of O-glycosylated proteins (e.g., human IgA, plasminogen, fetuin) are attached to the polypeptide chain via serines or threonines. The glycosyl transfer to serine or threonine does not require a sequence motive. A specific glycoprotein can be N-glycosylated and O-glycosylated (e.g., fetuin, human IgA$_1$, and IgD).

For a long time it was not known what role protein glycosylation plays in the cell metabolism. Only the discovery of glycosylation inhibitors and the availability of selective high-purified glycosidases made it possible to answer this question.

- Glycosylation protects proteins against proteolytic digestion.
- The glycosylation of some receptors (e.g., insulin receptor) changes their affinity to the ligands, and the glycosylation of some ligands (e.g., thyrotropin, human chorionic gonadotropin) in turn influences their affinity to the receptor. The activity of certain hormones and enzymes (e.g., extrinsic tissue plasminogen activator) depends on their glycosylation.
- The glycosylation of a protein serves as a signal for intracellular transport. Glycosylated proteins are transported into the extracellular medium or to the cell membrane.
- The interaction of the sugar chains of membrane proteins and extracellular matrix material with lectins is regulated by the migration and distribution of certain cells in the organism (e.g., lymphocytes).

The research possibilities in the area of glycoproteins are thus not exhausted by determining the sequence and structure of a protein's sugar chains. The ambitious doctoral candidate better leave this laborious job to others, especially when the work predictably culminates in pure description and does not allow for functional conclusions. On the other hand, you open a possible gold mine with investigating the participation of the sugar chains of cell membrane proteins in the control of cell migrations and the formation of cell-cell contacts (and with it the development of the organism).

Sources
1. Kobata, A. (1992). "Structures and Functions of the Sugar Chains of Glycoproteins," *Eur. J. Biochem.* 209: 483–501.
2. Rademacher, T., et al. (1988). "Glycobiology," *Ann. Rev. Biochem.* 57: 785–838.

9.2 Detecting Glycoproteins in Gels

The traditionally minded biochemist stains glycoproteins in SDS gels with PAS. For this, the gel with the periodate-oxidized glycoproteins is successively treated with arsenite and Schiff's reagent (Fairbanks et al. 1971). Gerard (1990) describes a modification of the PAS stain. The PAS stain requires 3 µg protein/band and can be characterized by two words: *inconvenient* and *insensitive*.

A somewhat more sensitive method is the thymol-sulfuric acid stain (Gerard 1990). This stain also takes about one day and requires at least 1 to 2 µg glycoprotein. Furthermore, at RT the stain is only stable for a few hours (at −20° C it keeps for a few days).

Sources
1. Fairbanks, G., et al. (1971). "Electrophoretic Analysis of Major Polypeptides of the Human Erythrocyte Membrane," *Biochemistry* 10: 2606–2617.
2. Gerard, C. (1990). "Purification of Glycoproteins," *Methods Enzymol.* 182: 529–539.

9.3 Detection of Glycoproteins on Blots

9.3.1 Nonselective Glycoprotein Stain

Sensitive glycoprotein stains oxidize the vicinal hydroxyl groups of the oligosaccharides and transform the resulting aldehyde groups with hydrazides (Figure 9.1). The oxidizer is usually a periodate. The reactions can be done before electrophoresis or on the blot. For the periodate oxidation of glycoproteins on the blot, it is best to blot on PVDF membranes, because oxidized nitrocellulose creates a background stain.

Heimgartner et al. (1989) oxidize blotted glycoproteins with periodate and transform the resulting aldehyde groups with polyhydrazides and then with periodate-oxidized peroxidase. The method registers submicrogram amounts of glycoproteins. The polyhydrazides they use are not available in retail and the stain is not proportional to the sugar content of the proteins. Furthermore, the stain depends on the structure of the sugar chain.

O'Shanessy et al. (1987) transform periodate-oxidized glycoproteins with biotin-aminocaproyl hydrazide. The biotinilated glycoproteins are subjected to SDS gel electrophoresis, blotted on nitrocellulose, and detected with peroxidase streptavidin. The reagents are available in retail. The detection threshold is at 1 ng glycoprotein/band. It was not shown how proportional the stain was to the sugar content of the glycoprotein or how independent it was of the sugar chain structure.

Analogously to the O'Shanessy method, the glycoproteins are detected with digoxigenin-3-O-succinyl-ε-aminocaproic acid hydrazide (Böhringer, now Roche Diagnostics). The glycoproteins are oxidized first with periodate and then transformed with the hydrazide. After SDS gel electrophoresis and blot, they are detected with anti-digoxigenin antibodies coupled to alkaline phosphatase.

Sources
1. Heimgartner, U., et al. (1989). "Polyacrylic Polyhydrazides as Reagents for Detection of Glycoproteins," *Anal. Biochem.* 181: 182–189.
2. O'Shanessy, D., et al. (1987). "Quantitation of Glycoproteins on Electroblots Using the Biotin-streptavidin Complex," *Anal. Biochem.* 163: 204–209.

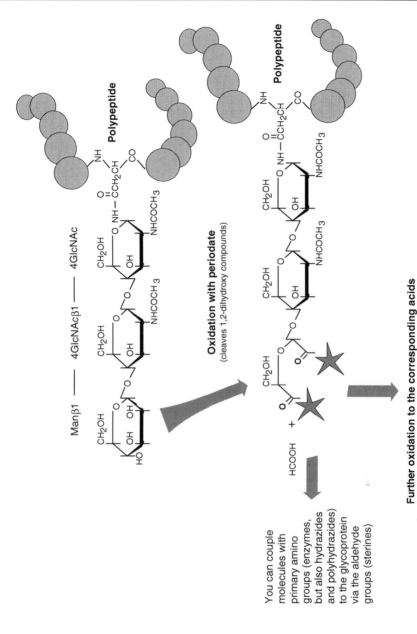

Figure 9.1. Oxidation of the sugar chains of glycoproteins. The vicinal hydroxyl groups of the sugar residues are oxidized with periodate to aldehyde groups. Hydroxyl groups in cis configuration and the acyclic diol groups in sialic acid oxidize especially easily.

9.3.2 Selective (Lectin) Stain

With labeled lectins, less than 10 ng glycoprotein/band is detectable on blots. Because certain lectins bind only to certain sugars, a number of different lectins exhibit a typical binding pattern with a specific glycoprotein. For example, the binding of RCA is evidence of terminal Gal, that of SBA for terminal GalNAc, and the binding of Con A for N-glycosylated proteins (for specificity and origin of the lectins, see Table 9.1). O-glycosylated proteins selectively bind to jacalin (Hortin and Timpe 1990), and after removal of terminal sialic acids through neuraminidase digestion to PNA (Kijimoto et al. 1985). If the glycoprotein has several sugar chains, the lectin-binding pattern does not allow any inferences regarding the structure of individual oligosaccharides.

For the characterization of glycoproteins with lectins, the sample is run on an SDS gel with an approximate 10-cm-wide pocket and then blotted. The blot (on PVDF membrane) is stained on protein and the approximately 10-cm-long protein band is cut crosswise into approximately 20 narrow strips. After blocking (e.g., with 2% polyvinylpyrrolidon 360 in 50 mM Tris-Cl pH 7.5 and 0.5 M NaCl), the strips are incubated in twos with a labeled lectin (Table 9.1) with or without inhibiting sugar, and are then developed.

Lectin blots are blocked with substances free from lectin-binding glycoproteins (Tween, polyvinylpyrrolidon 360, purified BSA, periodate-oxidized BSA, hemoglobin). Milk powder and the cheaper BSA preparations contain glycoproteins. A control strip of the blot should be incubated with lectin and competing sugar and show no stain. Don't forget: many lectins bind only in the presence of cofactors (e.g., Ca^{2+}, Mg^{2+}, or Mn^{2+}).

If you stain the blots with peroxidase-labeled lectins, you should know that the peroxidase substrate diaminobenzidine is carcinogenic and binds Con A peroxidases. A covalent labeling of Con A is thus unnecessary. Finally, you will have little luck if you transform the lectins with nitrocellulose blots, in that they bind to cellulose (alternative: PVDF membranes). Table 9.1 shows a selection of commercially available labeled lectins.

Sources
1. Hortin, G., and Timpe, B. (1990). "Lectin Affinity Chromatography of Proteins Bearing O-linked Oligossaccharides: Application of Jacalin-agarose," *Anal Biochem.* 188: 271–277.
2. Kijimoto, S., et al. (1985). "Analysis of N-linked Oligosaccharide Chains of Glycoproteins on Nitrocellulose Sheets Using Lectin-peroxidase Reagents," *Anal. Biochem.* 147: 222–229.
3. Ogawa, H., et al. (1986). "Characterization of the Carbohydrate Moiety of *Clerodendron trichotomum* Lectins," *Eur. J. Biochem.* 161: 779–785.
4. Rohringer, R., and Holden, D. (1985). "Protein Blotting: Detection of Proteins with Colloidal Gold, and of Glycoproteins and Lectins with Biotin-conjugated and Enzyme Probes," *Anal. Biochem.* 144: 118–127.

9.4 Deglycosylation

You deglycosylate proteins in order to determine the MW of the polypeptide chain or the activity of the deglycosylated protein. Sugar chains often also inhibit the protease digestion of a protein to defined peptides for the sequence analysis. Three ways lead to deglycosylated proteins: glycosylation inhibitor, endoglycosidases, and chemistry. The methods discussed here focus on the intact protein. The integrity of the replaced sugar chains is neglected. You get intact sugar chains with the technologies described in Section 9.5.2.1.

9.4.1 Glycosylation Inhibitor

The addition of a glycosylation inhibitor to cell cultures stops the glycosylation of the proteins at certain points in the glycosylation chain. The result is proteins with shorter or missing sugar chains. Table 9.2 enumerates the most important N-glycosylation inhibitors. Tunicamycin and amphomycin prevent N-glycosylation by inhibition of dolichol metabolism (Figure 9.2).

Table 9-1. Labeled Lectins

Label	Lectin	Specificity	Cofactors	Vendor	Specific Activity
^{3}H	WGA (Triticum vulgaris)	(D-GlcNAc)n	Ca²⁺, Mn²⁺	NEN	100µCi/mg
	Con A (Canavalia ensiformis)	α-D-Man, α-D-Glc	Ca²⁺, Mn²⁺	Amersham	30–90Ci/mmol
^{125}I	Con A (Canavalia ensiformis)	α-D-Man, α-D-Glc	Ca²⁺, Mn²⁺	NEN	30000–40000µCi/mg
^{14}C	Con A (Canavalia ensiformis)	α-D-Man, α-D-Glc		Sigma	50–50µg
	LCA (Lens culinaris)	terminal α-D-Man & α-D-Glc		Sigma	5–50µCi/mg
Peroxidase (P)[a], Biotin (B)[b] oder FITC (F)	AAA (Aleuria aurantia)	-α(1-6)Fuc		Böhringer (B)	3–5Mol biotin/Mol lectin
	APA (Abrus precatorius)	D-Gal		Sigma (B,F)	3Mol biotin/Mol lectin
	BPA (Bauhinia purpurea)	β-D-Gal(1-3)-D-GalNAc		Sigma (B,F)	4Mol biotin/Mol lectin
	BS-I (Bandeiraea simplicifolia)	α-D-Gal, α-D-GalNAc		Sigma (B,F,P)	5Mol biotin/Mol lectin
	CAA (Caragana arborescens)	D-GalAc	Ca²⁺, Mg²⁺, Mn²⁺	Sigma (B,F)	6Mol biotin/Mol lectin
	Con A (Canavalia ensiformis)	α-D-Man, α-D-Glc	Ca²⁺, Mn²⁺	Sigma (B,F,P), Böhringer (B), Vector Lab. (B)	4–8Mol biotin/Mol lectin (Sig); 1–3Mol biotin/Mol lectin (Böh)
	DBA (Dolichos biflorus)	α-D-GalNAc		Sigma (B,F,P); Vector Lab. (B)	4–8Mol biotin/Mol lectin (Sig)
	DSA (Datura stramonium)	β(1-4) crosslinked oligomers of GlcNAc		Böhringer (B)	2–4Mol biotin/Mol lectin
	ECA (Erythrina cristagalli)	β-D-Gal(1-4)-D-GlcNAc		Sigma (B,F,P)	2Mol biotin/Mol lectin
	ECorA (Erythrina corallodendron)	β-D-Gal(1-4)-D-GlcNAc		Sigma (B,F)	5Mol biotin/Mol lectin
	GNA (Galanthus nivalis)	terminal Man		Böhringer (B)	0.5–3Mol biotin/Mol lectin
	HAA (Helix aspersa)	terminal α-D-GalNac		Sigma (B,F,P)	2Mol biotin/Mol lectin
	HPA (Helix pomatia)	terminal α-D-GalNac		Sigma (B,F,P)	2–4Mol biotin/Mol lectin
	LCA (Lens culinaris)	terminal α-D-GalNac and α-D-Glc		Sigma (F,B)	2–3Mol biotin/Mol lectin
	MAA (Maackia amurensis)	α(2-3) crosslinked sialic acid		Böhringer (B)	3–5Mol biotin/Mol lectin
	MPA (Madura pomifera)	terminal α-D-Gal, α-D-GalNac		Sigma (F,G)	2Mol biotin/Mol lectin
	PHA-L (Phaseolus vulgaris)			Sigma (F,G)	6–12Mol biotin/Mol lectin
	PNA (Arachis hypogaea)	β-D-Gal(1-3)-D-GalNAc		Sigma (B,F,P), Böhringer (B), Vector Lab. (B)	3Mol biotin/Mol lectin (Sig); 10–15Mol biotin/Mol lectin (Böh)
	RCA 120 (Ricinus communis)	terminal β-D-Gal		Böhringer (B), Vector Lab. (B)	2–4Mol biotin/Mol lectin
	SBA (Glycine max)	D-GalNAc		Sigma (B,F,P)	3Mol biotin/Mol lectin
	SNA (Sambucus nigra)	β-D-Gal(1-4)-D-Glc; terminal Gal α(2-6)NeuAc; GalNAcα(2-)NeuAc		Böhringer (B)	6–10Mol biotin/Mol lectin
	UEA 1 (Ulex europaeus)	α-L-Fuc		Sigma (B,F,P), Vector Lab. (B)	6Mol biotin/Mol lectin (Sig)
	WGA (Triticum vulgaris)	(D-GlcNAc)n		Sigma (B,F,P), Böhringer (B), Vector Lab. (B)	2–4Mol biotin/Mol lectin (Böh)
Digoxigenin[c]	AAA (Aleuria aurantia)	-α(1-6)Fuc		Böhringer	1–3Mol Dig/Mol Lektin
	Con A (Canavalia ensiformis)	α-D-Man, α-D-Glc	Ca²⁺, Mn²⁺	Böhringer	1–3Mol Dig/Mol Lektin
	DSA (Datura stramonium)	(D-GlcNAc)2		Böhringer	1–3Mol Dig/Mol Lektin
	GNA (Galanthus nivalis)	terminal Man		Böhringer	1–3Mol Dig/Mol Lektin
	MAA (Maackia amurensis)	α(2-3) crosslinked sialic acid		Böhringer	1–2Mol Dig/Mol Lektin
	PNA (Arachis Hypogaea)	β-D-Gal(1-3)-D-GalNAc		Böhringer	1–6Mol Dig/Mol Lektin
	RCA 120 (Ricinus communis)	terminal β-D-Gal		Böhringer	1–3Mol Dig/Mol Lektin
	SNA (Sambucus nigra)	β-D-Gal(1-4)-D-Glc; terminal Galα(2-6)NeuAc, GalNAcα(2-)NeuAc		Böhringer	1–3Mol Dig/Mol Lektin
	WGA (Triticum vulgaris)	(D-GlcNAc)n		Böhringer	1–3Mol Dig/Mol Lektin

FITC: Fluorescein Isothiocyanate

a: the specific activity, measured in purpurogallin units per mg protein, varies with peroxidase-labeled lectins from 10–60

b: detected with avidin and biotinylated peroxidase

c: detected with anti-dogoxigenin coupled to fluroscein, peroxidase or alkaline phosphatase

Dig: digoxigenin

Table 9.2. Inhibitors of N-glycosylation.

Name	Source	Mechanism	Effective Concentration	Special Properties
Tunicamycin	*Streptomyces lysosuperificus*	Inhibits GlcNAc-1-P-transferase	0.2 to 1 μg/ml	Soluble in DMSO and dimethylformamide. Insoluble in watery solutions with pH < 6. Exists in several isomer forms. Tunicamycin is toxic!
Amphomycin	*Streptomyces canus*	Forms complexes with dolicholphosphate and thereby inhibits its glycosylation	25 to 100 μg/ml	Effectiveness depends on Ca^2 in the media.
Swainsonin	*Swainsona canescens, Astragalus*	Inhibits α-mannosidase II	0.2 to 1 μg/ml	Swainsonin only dissolves in water and chloroform.
Castanospermine	*Castanospermum australe*	Inhibits glucosidase I	10 to 50 μg/ml	
Nojirimycin	*Streptomyces*	Inhibits glucosidase I		
1-Deoxynojirimycin	*Bacillus*	Inhibits glucosidase I and in high doses also glucosidase II	150 to 800 μg/ml	
Bromoconduritol	Synthetic	Inhibits glucosidase II	500 to 1,000 μg/ml	Decomposes in water (T1/2 = 15 min.).
Deoxymannojirimycin	Synthetic	Inhibits mannosidase II	10 to 50 μg/ml	

Impure tunicamycin preparations also slightly inhibit the protein synthesis. Swainsonin, castanospermine, and so on prevent trimming of the N-glycosidically bound sugar and thereby change the glycosylation pattern (Figure 9.2). I do not know specific inhibitors for O-glycosylation.

As nice as working with glycosylation inhibitors is, you must first find a cell line that expresses the sought-after protein. However, this is difficult, certainly for rare proteins or proteins that characterize highly differentiated cells, such as the function proteins of nerve cells. Also, in my experience collecting and screening of cell lines comes with many phone calls, political difficulties, and traveling long distances.

Sources
1. Elbein, A. (1987). "Glycosilation Inhibitors for N-linked Glycoproteins," *Methods Enzymol.* 138: 661–709.
2. George, S., et al. (1986). "N-glycosilation in Expression and Function of β-adrenergic Receptors," *J. Biol. Chem.* 261: 16559–16564.
3. Rehm, H., et al. (1986). "Molecular Characterization of Synaptophysin, a Major Calcium Binding Protein of the Synaptic Vesicle Membrane," *EMBO J.* 5: 535–541.
4. Schwarz, R., and Datema, R. (1984). "Inhibitors of Trimming: New Tools in Glycoprotein Research," *TIBS* 9: 32–34.

9.4.2 Endoglycosidases

Endoglycosidases are the restriction enzymes of the sugar researcher (Table 9.3) because they cleave oligosaccharides only at specific sites (Figure 9.3). The endoglycosidases F and H cleave the terminal chitobiose unit in certain N-glycosidically bound sugar chains, so that only one monosaccharide residue remains in the protein. PNGases, on the other hand, cleave between

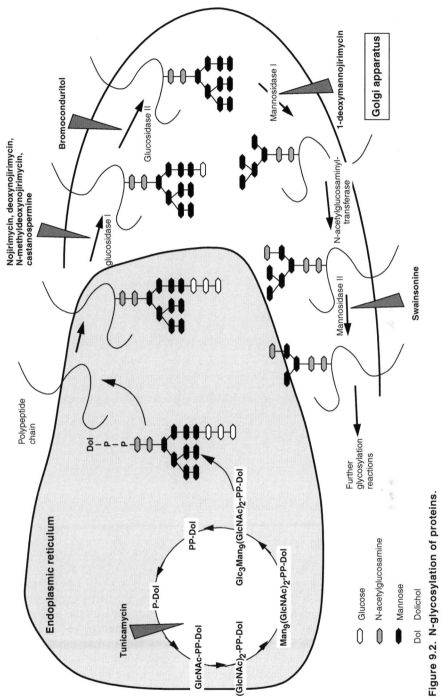

Figure 9.2. N-glycosylation of proteins.

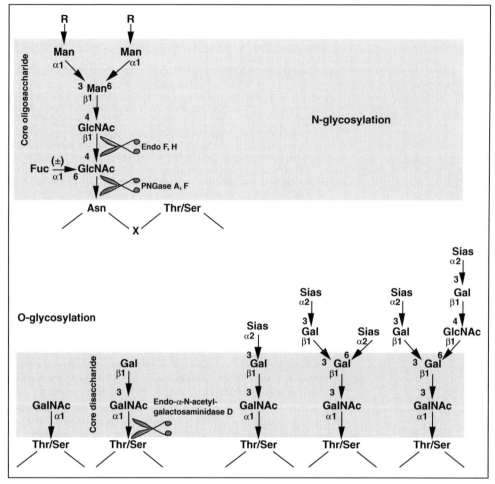

Figure 9.3. The structure of N- and O-glycosidically bound sugar chains and cleaving sites of different enzymes.

the Asn of the protein and the terminal GlcNAc of the sugar chain (Figure 9.3). In addition, they transform Asn into Asp and thus change the charge and the behavior of the protein.

For endoglycosidase to be effective, its cleaving site must be present, but the enzyme activity is also dependent on the conformation and size of the polypeptide as well as the number, arrangement, and type of the remaining sugars of the oligosaccharide. Maley et al. (1989) show a table of different sugar chains and their sensitivity for the most important endoglycosidases. Most endoglycosidases only cleave N-glycosidically bound sugars (Table 9.3). The enzymes neuraminidase, PNGase F, and Endo D,F,H are available in retail. Alexander and Elder (1989) describe isolating and detection methods for endo F and PNGase F.

Enzymes that split off O-glycosidically bound sugars are rare, but they exist (a biological truth is: there is nothing that does not exist). The GLYKO company offers (for example) recombinant O-glycanase of *Streptococcus pneumonia*. The enzyme cleaves the O-glycosidically bound disaccharide Galβ1-3GalNAc of the Galβ1-3GalNAcα1-Ser/Thr group of proteins. Substitution of the disaccharide core with sialic acids, fucose, N-acetylglucosamine, or N-acetylgalactosamine residues prevents the cleaving. It is thus advisable to digest the protein with neuraminidase before the O-glycanase treatment. You do not have a guarantee that the O-glycanase cleaves all of your O-glycosidically bound sugar residues, even after neuraminidase pretreatment. The company recommends a digestion time of 3 h at 37° C. In the literature you commonly find incubation times up to 72 h. And another hint: the enzyme is inhibited by EDTA, Mn^{2+}, and Zn^{2+}.

Table 9.3. Endoglycosidases.

Name	Source	MW (kd)	pH Optimum	Specificity	Special Properties
Endo-N-acetyl-β-glucosaminidase H (Endo H) EC 3.2.1.96	*Streptomyces plicatus*	29	5–6	Cleaves the β(1–4) binding between the first two GlcNAc of N-glycosidically bound sugar chains. The oligosaccharide remains intact with one single GlcNAc at the reducing end.	Resistant to protease. 0.5 M NaSCN increases the activity of the enzyme.
Endo-N-acetyl-β-glucosaminidase F (Endo F)	*Flavobacterium meningosepticum*	32	4–6	The other GlcNAc stays with the protein (like Endo H).	
Endo-N-acetyl-β-glucosaminidase D (Endo D)	*Diplococcus pneumoniae*	280	6.5	Cleaves like Endo H, but only at Man5GlcNAc2Asn and Man3GlcNAc Asn residues. Cleaves the N-glycosidic binding between sugar chain and asparagine of the polypeptide.	
Peptide N-(N-acetyl-β-glucosaminyl)-Asparaginamidase (PNGase F) EC 3.5.1.52	Endo-N-acetyl-β-glucosaminidase F (Endo F)	35.5	8.6	Cleaves the N-glycosidic binding between sugar chain and asparagine of the polypeptide to NH₃, polypeptide, and complete oligosaccharide (2 GlcNAc at the reducing end)—like PNGase F—but with lower specific activity.	Stable in 2.5 M urea. Citrate inhibits. Some preparations small amounts of a metal protease. Also available in cloned form.
Peptide N-(N-acetyl-β-glucosaminyl)-asparaginamidase (PNGase A)	Tonsils	67–80	4–6		Stable in TRITON-X-100, EDTA, 0.75 M NaSCN. Removes the GlcNAc leftovers after Endo H treatment.
Endo-α-N-acetylgalactosaminidase D	*Diplococcus pneumoniae*	160	7.6	Removes the unsubstituted disaccharide Galβ1-3GalNAc from o-glycosylated proteins removes sialic acid chains with at least 5 sections.	
Endo-N-acetylneuraminidase	*Bacillus sp.*				

If a glycoprotein has several sugar chains, different glycoforms with smaller MW result from incomplete digestion with an endoglycosidase. The glycoforms form a ladder in the SDS gel. The usual endoglycosidases cleave the sugar chains in or shortly before the contact site with the protein (Figure 9.3) and they otherwise leave the chains intact. Thus, if you know from other experiments that the cleaved sugar chains are of equal size you can infer from the number of bands in the SDS gel the number of the (endoglycosidase-sensitive) sugar chains of the glycoprotein. For example, if only three bands appear in the gel after the digestion of a glycoprotein, the protein has two sugar chains with identical MW. With four bands, it has three chains (i.e., in general: with n bands in the gel the protein has n-1 sugar chains). If, on the other hand, the sugar chains are of different sizes, it gets complicated: two chains yield 4 bands, three chains 6 or 8, and so on.

Although endoglycosidases also deglycosylate native glycoproteins, the enzymes work best with denatured and reduced glycoproteins. The sample is heated in 0.1% SDS in the presence of 1% mercaptoethanol, and before the enzyme is added the SDS is weakened via addition of 1% TRITON-X-100 or Nonidet P 40. This procedure also inactivates possible proteases. If the endoglycosidase shows no effect, a pretreatment with neuraminidase often helps, which removes terminal sialic acid residues. An alternative is endoglycosidase (effect patterns of endoglycosidases in Maley et al. 1989). Also, more enzyme for a longer time (12 to 36 h) at a higher temperature (37° C) often leads to success. A lot helps a lot.

If the deglycosylation of the protein is checked via the decrease of its MW in SDS gels, neither the used glycosidases nor the glycoprotein preparation may contain proteases. A recommendable but not sufficient way to check the glycosidase preparation for protease contamination is incubating the glycosidase at 37° C with a protease-sensitive nonglycosylated protein (e.g., purified BSA). The preparation is then analyzed in the SDS gel for proteolysis products. The glycoprotein preparation is checked for proteases by incubating for several hours under deglycosylation conditions, but without enzyme. The band pattern of the preparation in the SDS gel must stay the same. Efficient protease inhibitors are PMSF and 0.1 mM EDTA (see Table 5.1).

Sources

1. Alexander, S., and Elder, J. (1989). "Endoglycosidases from *Flavobacterium meningosepticum* Application to Biological Problems," *Methods Enzymol.* 179: 505–518.
2. Elder, J., and Alexander, S. (1982). "Endo-β-N-acetylglucosaminidase F: Endoglycosidase from *Flavobacterium meningosepticum* That Cleaves Both High-mannose and Complex Glycoproteins," *Proc. Natl. Acad. Sci. USA* 79: 4540–4544.
3. Maley, et al. (1989). "Characterization of Glycoproteins and Their Associated Oligosaccharides Through the Use of Endoglycosidases," *Anal. Biochem.* 180: 195–204.
4. Rehm, H. (1989). "Enzymatic Deglycosilation of the Dendrotoxin Binding Protein," *FEBS Lett.* 247: 28–30.
5. Thotokura, N., and Bahl, O. (1987). "Enzymatic Deglycosilation of Glycoproteins," *Methods Enzymol.* 138: 350–359.

9.4.3 Chemical Deglycosylation

This type of deglycosylation has been revitalized by mass spectrometry. At the moment, there are five methods (and none of the five is ideal). You can never be sure that you got rid of all oligosaccharides, and you are never sure whether you have not possibly cleaved the peptide. Treating glycoproteins with trifluoromethanesulfonic acid cleaves N-glycosidically bound sugar chains as well as O-glycosidically ones, and O-glycosidic bindings are a little bit more resistant. However, the core GalNAc (Figure 9.3) sticks to the serine or threonine. The method requires large amounts (> 100 μg) of protein, is inconvenient, denatures the protein, and destroys the oligosaccharide. The deglycosylation is accomplished to about 90%. However, if you perform the treatment at higher temperatures (25° C) you have to assume that peptide bindings are cleaved. Membrane proteins are often lost with the procedure at low temperatures. This loss can reach extreme quantities. During my experiments with the synaptic vesicle protein synaptophysine, of 100 μg protein I did not even have enough left for a silver stain after treatment with trifluoromethanesulfonic acid. The protein must have aggregated quantitatively (Rehm et al. 1986).

The once popular deglycosylation with hydrofluoric acid requires longer response times in comparison to trifluoromethanesulfonic acid and has smaller yields. In addition, special equipment is necessary for handling the extremely aggressive hydrofluoric acid (Mort and Lampart 1977).

Gerken et al. (1992) recommend a combination of oxidation and β-elimination, which is supposed to remove the O-glycosidically bound sugar carefully and completely (including the GalNAc bound to the amino acids). β-elimination runs at pH 10.5, which is a low value compared to other methods for eliminating β, which use pH 12 to 13. At pH 10.5, between 90 and 95% of the sugar chains are supposedly removed (at pH 8.4, still 60%). Unfortunately, Gerken et al. have tested the method only for a certain type of glycoprotein (mucines from saliva and respiratory tract), of which they apparently had gram amounts and which have a simple sugar structure. Whether the method also works as wonderfully with your protein, dear experimenter, is doubtful. Even Gerken et al. warn that sugar chains are not removed when the GalNAc that hangs on the amino acid is substituted at C3. However, Gerken et al. assure that this problem can be avoided by alternate treatment with trifluoromethanesulfonic acid and their method. Is the protein also of this opinion? Or won't the alternating acidic and alkaline baths destroy the protein? The method is also a little bit inconvenient. First, the glycoprotein is oxidized with periodate. Then the periodate is destroyed, and then β-eliminated with NaOH. The entire thing takes about two days.

The fourth method comes from Rademaker et al. (1998). These people remove O-glycosidically bound sugar chains via β-elimination with ammonium hydroxide. This sounds complicated, but is simple. You incubate the protein or glycopeptide in 25% ammonium hydroxide for 18 h at 45° C. The protein remains intact except for the fact that the sugar chain is replaced with NH_3.

The Rademaker method has been refined by Hanisch et al. (2001). Franz-Georg Hanisch β-eliminates with ethylamine or methylamine. This has the advantage that the sugar chain bearing threonines and serines is labeled with the alkylamines, because an alkylamine is incorporated in place of the cleaved oligosaccharide. The alkylaminylated amino acid differs in MW strongly from the source amino acid (serine or threonine) and thus you can determine the location of the protein glycosylation with the mass spectrometer.

Incubating the glycoprotein with 40% methylamine at 50° C for six h completely splits off O-glycosidically bound sugar chains. Incubating with 70% ethylamine at 50° C for 18 h splits off about 70% of the sugar chains. In both cases, the degradation of the peptide chain is "weak to moderate" or "weak"—whatever that is supposed to mean. The alkylaminylated protein can be digested with proteases such as trypsin and clostripain.

Thus, you can choose among five methods: trifluoromethanesulfonic acid, hydrofluoric acid, oxidation and β-elimination, and two β-elimination methods (overview: Table 9.4). I advise against other methods. Hydrazinolysis (see Section 9.5.2.1) may split the sugar from glycoproteins, but it also cleaves peptide bindings. The same is true for treatment with NaOH/boron hydride (Rehm et al. 1986).

Table 9.4. Chemical deglycosylation.

Method	Removes	Advantages	Disadvantages
Trifluoromethanesulfonic acid	N-glycosidic and O-glycosidic sugar chains	Removes up to 90% of the sugar chains.	Inconvenient. Peptide bindings are cleaved at temperatures over 25° C. Membrane proteins aggregate.
Hydrofluoric acid	N-glycosidic and O-glycosidic sugar chains	—	Unhealthy. Low yield. Long reaction times.
Oxidation plus β-elimination	O-glycosidic sugar chains	Gentle. At least sometimes almost complete cleaving.	Lengthy and inconvenient. Sugar chains substituted to GalNac are resistant.
β-elimination after Rademaker	O-glycosidic sugar chains	Gentle and almost complete.	Peptide bindings may be cleaved. Takes 18 h.
β-elimination after Hanisch	O-glycosidic sugar chains	Almost complete. The attachment sites of the sugars are labeled.	Peptide bindings may be cleaved. Takes 18 h.

How to proceed? This depends on what you want to get out of it. If you want to determine the glycosylation site, I recommend Hanisch et al. If you only want to determine the MW of the polypeptide chain, Rademaker et al. would also do. However, I would not put too much faith into their conditions, but would run (for example) several preparations at different temperatures. You can skip the latter if you are only interested in a partial sequence (i.e., if you remove the sugar only to expose the protein to, for example, trypsin).

Sources

1. Edge, et al. (1981). "Deglycosilation of Glycoproteins by Trifluoromethanesulfonic Acid," *Anal. Biochem.* 118: 131–137.
2. Gerken, et al. (1992). "A Novel Approach for Chemically Deglycosilating o-linked Proteins: The Deglycosilation of Submaxillary and Respiratory Mucins," *Biochemistry* 31: 639–648.
3. Hanisch, F., et al. (2001). "Glycoprotein Identification and Localization of o-glycosilation Sites by Mass Spectrometric Analysis of Deglycosilated/alkylaminylated Peptide Fragments," *Anal. Biochem.* 290: 47–59.
4. Mort, A., and Lamport, D. T. A., Init. (1977). "Anhydrous Hydrogen Fluoride Deglycosilates Glycoproteins," *Anal. Biochem.* 82: 289–309.
5. Rademaker, et al. (1998). "Mass Spectrometric Determination of the Sites of a-glycan Attachment with Low Picomolar Sensitivity," *Anal. Biochem.* 257: 149–160.
6. Rehm, H., et al. (1986). "Molecular Characterization of Synaptophysin, a Major Calcium Binding Protein of the Synaptic Vesicle Membrane," *EMBO J.* 5: 535–541.
7. Sogar, H., and Bahl, O. (1987). "Chemical Deglycosilation of Glycoproteins," *Methods Enzymol.* 138: 341–350.

9.5 The Sugar Chains

9.5.1 Monosaccharide Composition

The analysis of the monosaccharide composition of the sugars corresponds to the amino acid analysis of polypeptides. N-glycosidically bound sugars in mammals can contain GlcNAc, Man, Glc, NeuNAc, Gal, and Fuc. O-glycosidically bound sugars in mammals can contain GalNAc, Gal, NeuNAc, and NeuNgly. Acid hydrolyzes oligosaccharides to monosacchariden. For example, 2.5 M of trifluoroacetic acid at 100° C for 6 h releases neutral sugars and hexosamines. For sialic acids, a treatment with 50 mM sulfuric acids at 80° C is enough. The resulting monosaccharides are determined and quantified via HPLC (Ogawa et al. 1990). It has become unfashionable to separate charged sugars such as sialic acids or the borate complexes of uncharged sugars via chromatography on ion exchanger columns.

Sources

1. Elwood, P., et al. (1988). "Determination of the Carbohydrate Composition of Mammalian Glycoproteins by Capillary Gas Chromatography/Mass Spectrometry," *Anal. Biochem.* 175: 202–221.
2. Ogawa, H., et al. (1990). "Direct Carbohydrate Analysis of Glycoproteins Electroblotted onto Polyvinylidene Difluoride Membrane from Sodium Dodecyl Sulfate Polyacrylamide Gel," *Anal. Biochem.* 190: 165–169.

9.5.2 Structure and Sequence

N-glycosidically bound sugars differ in their degree of branching (unforked, bifurcated, trifurcated, tetrafurcated) and in their monosaccharide composition. In complicated sugars, the 3 Man in the core oligosaccharide are substituted with GlcNAc groups, and sometimes a Fuc hangs on the GlcNAc neighboring the Asp (Figure 9.3). Mannose sugars contain 6 or more Man and hybrid sugars 4 or 5 Man, which are partly substituted with GlcNAc. Similarly complicated is the situation with O-glycosidically bound sugars (Figure 9.3). If you are interested in sequence and structure of the sugar chains of glycoproteins, you run into two problems.

- A given glycoprotein generally contains several different sugar chains.
- Interesting glycoproteins, and thus the oligosaccharides attached to them, occur only in small amounts.

We, the true knights-errant, measure the whole earth with our own feet, exposed to the sun, to the cold, to the air, to the inclemencies of heaven, by day and night, on foot and on horseback; nor do we only know enemies in pictures, but in their own real shapes; and at all risks and on all occasions we attack them.

9.5.2.1 Releasing Intact Oligosaccharides from Glycoproteins

For structural determinations, an oligosaccharide must be available in pure form. To be able to purify the oligosaccharide chain of a glycoprotein, you first have to release the oligosaccharide from the polypeptide, in such a way that the oligosaccharide chain remains intact. The method of choice is endoglycosidase digestion (e.g., with PNGase F). In particular, the endoglycosidase digestion also spares the polypeptide chain (see Section 9.4.2).

If the oligosaccharide chain is O-glycosidically bound, you can try your luck with O-glycanase or with neuraminidase and O-glycanase. Some companies (e.g., GLYKO and Oxford Glycosystems) also offer enzyme mixtures that remove both O-glycosidic and N-glycosidic sugars. You cannot count on this, though. Now you can assume that the chains will all be removed, and completely removed. Also, the digestions take up to 72 h and during this time all types of things can happen—among other things reduction or modification of the oligosaccharide chains by enzymatic contaminations in your sample.

A thorough proteolysis of the protein content (e.g., with pronase, a mixture from endoproteases and exoproteases) also releases the sugar chains. At the reducing end, the chains still carry an Asn or short peptides. Because, furthermore, the digestion with proteases is always incomplete, especially with glycoproteins, this method is ill suited for the purification of oligosaccharides (for optimists: Finn and Krusins 1982).

Chemical methods have the advantage that their effect does not depend as much as that of enzymes on the conformation of the glycoprotein and the structure of the sugar chains. In spite of the busy biotech sales strategists, there are still chemistry aficionados around. They like to release the oligosaccharide chains via hydrazinolysis. This method is old (Takasaki et al. 1982), but it used to be applicable only to N-glycosidically bound sugars. Furthermore, some GlcNAc lost their N-acetyl groups, and some other unwanted reactions took place.

Patel et al. (1993) have refined the hydrazinolysis such that it can hold its own against the enzymatic methods. With their method, N-glycosydically as well as O-glycosidically bound sugar chains can be released. It is even possible to release (relatively) selectively O-glycosidically bound chains. Patel et al. dialyze the glycoprotein against 0.1% trifluoroacetic acid, freeze-dry it, and then incubate in anhydrous hydrazine at 60° C for 5 h (for O-glycosidically bound sugars) or at 95° C for 4 h (for N- and O-glycosidically bound sugars). Then they cool to RT and let the hydrazine evaporate. Patel et al. remove the protein residues via paper chromatography, but you can also use an HPLC. The oligosaccharides remain largely intact. You just need to watch out with certain N- and O-substituted sialic acid residues.

Treatment of glycoproteins with 50 mM NaOH, 1 M NaBH$_4$ releases the O-glycosidically bound oligosaccharides. These stay intact, except for the terminal sugar unit. The latter is reduced to the corresponding alcohol (Muir and Lee 1969; Rehm et al. 1989).

Sources
1. Finne, J., and Krusins, T. (1982). "Preparation and Fractionation of Glycopeptides," *Methods Enzymol.* 83: 269–277.
2. Muir, L., and Lee, Y. (1969). "Structures of the D-galactose Oligo Saccharides From Earthworm Cuticle Collagen," *J. Biol. Chem.* 244: 2343–2349.
3. Patel, T., et al. (1993). "Use of Hydrazine to Release in Intact and Unreduced Form Both N- and O-linked Oligosaccharides from Glycoproteins," *Biochemistry* 32: 679–693.
4. Rehm, H., et al. (1986). "Molecular Characterization of Synaptophysin, a Major Calcium Binding Protein of the Synaptic Vesicle Membrane," *EMBO J.* 5: 535–541.
5. Takasaki, et al. (1982). "Hydrazinolysis of Asparagine-linked Sugar Chains to Produce Free Oligosaccharides," *Methods Enzymol.* 83: 263–268.

Table 9.5. Detection methods for sugar.

Method	Sensitivity	Specificity	Disadvantages
Refraction index	Up to about 1 nM	None	Inconvenient.
UV absorption	Up to about 0.1 μM	None	—
Pulsed amperometry	Up to about 10 pM	Registers hydroxyl groups	Strong alkaline conditions. Has to be recalibrated for each sugar.
Color reaction with phenol, anthron, or orcinol	0.1 to 1 μM	Not very high	Only in a narrow range proportional to the sugar amount.
Color reaction with resorcinol	1 to 100 nM	Not very high	—

9.5.2.2 Detection of Oligosaccharides

From small amounts of glycoproteins you get even tinier amounts of oligosaccharides and thus have a big measuring problem. Conventional detection procedures for sugar are not very sensitive (see Table 9.5). For example, the detection methods with sulfuric acid and phenol, anthron, or orcinol require large amounts of sugar (0.1 to 1 μM), and only in a narrow range are they proportional to the amount of sugar. The chemical detection of neutral sugars with resorcinol following Monsigny et al. (1989) is apparently a little more sensitive (1 to 100 nM). Lectin blots are not suited for oligosaccharide detection, because oligosaccharides do not stick to blot membranes.

There are still the physical methods. For example, there is the refraction index and UV adsorption. Both are somewhat sensitive, even 1 nM can be detected, but they are completely unspecific. The method of choice, especially for alkaline ion exchanger column runs, is the pulsed amperometry (PAD). This method measures presence and amount of hydroxyl groups. Because sugars have many hydroxyl groups, the PAD exhibits a certain specificity in addition to its high sensitivity (10 pM).

PAD is based on the following. At high pH, some of the hydroxyl groups dissociate to hydroxy anions and protons. If you apply a potential, these ions deliver a measurable current. The current is proportional to the concentration of hydroxyl groups or sugars. This, however, lasts only for a short time, because the sugars are oxidized in the anode and the oxidation products contaminate the electrode surface. For this reason, the traditional amperometry— where a steady potential is applied and the corresponding current is measured—fails with sugar and with polyalcohols. For the PAD, on the other hand, several potentials are applied via gold or platinum electrodes. Within one second, an equilibrium potential, a measuring potential, a purification potential, and a regeneration potential occur in succession. Then the cycle starts over. The purification potential cleans the oxidation products from the electrode, and the regeneration potential reduces the resulting metal oxide back to metal. The electrode thus remains shiny, as the layman would say.

The PAD is of moderate specificity for sugar because hydroxyl groups also appear in other molecules (e.g., in some amino acids, buffers, and so on). Furthermore, it has the disadvantage that the signal strength (PAD signal/M of sample) strongly depends on the type of sugar. Thus, you have to specially calibrate the device for every sugar. Finally, the strongly alkaline conditions in some sugars trigger epimerization and "peeling" reactions.

Because of the shortcomings of the detection methods for oligosaccharides, it is still common to label the oligosaccharides in some way and then to measure the label. This is discussed in Section 9.5.2.3.

Source

1. Monsigny, et al. (1988). "Colorimetric Determination of Neutral Sugars by a Resorcinol Sulfuric Acid Micromethod," *Anal. Biochem.* 175: 525–530.

Table 9.6. Labeling oligosaccharides.

Method	Sensitivity	Specificity	Disadvantages
Reduction with NaB(^{3}H)$_4$	In the nM range	Labels aldehydes	Unhealthy, time consuming, and inconvenient.
2-aminopyridine	In the pM range	Labels aldehydes	Sialic acids are lost. Reaction not quantitative.
UV chromophore low	Low	Labels aldehydes	
Reductive amination following Towbin et al. (1988)	In the nM range	Labels aldehydes	Reaction not quantitative. Some oligosaccharides do not survive the procedure.

9.5.2.3 Labeling Oligosaccharides

Labeling is a solution to your detection problem. It allows you to comfortably measure oligosaccharides. The possibilities for labeling oligosaccharides are colorful (in the true sense of the word) and numerous (Table 9.6).

Small amounts of glycoprotein release even smaller amounts of oligosaccharides and thus create a big measuring problem. Conventional detection procedures for sugar are not very sensitive. For example, the detection methods with sulfuric acid and phenol, anthron, or orcinol require large amounts of sugar (0.1 to 1 µM), and only in a narrow range are they proportional to the amount of sugar. The chemical detection of neutral sugars with resorcinol following Monsigny et al. (1989) is apparently a little more sensitive (1 to 100 nM). Lectin blots are not suited for oligosaccharide detection either, because oligosaccharides do not stick to blot membranes.

You solve your detection problem by labeling your oligosaccharides. Often, you allow the protein producing cells to grow in a medium that contains radioactive monosaccharide substrates (e.g., ^{3}H-Man, ^{3}H-Gal, ^{3}H-Fuc, or ^{3}H-GlcN). The desired ^{3}H labeled glycoprotein is isolated from the cell lysate (e.g., by immunoprecipitation). Apart from the fact that this method needs substantial amounts of radioactivity (10 µCi to 1 mCi/ml culture medium), the glycosylation of a protein in a cell line differs significantly from glycosylation in situ and presumably depends also on the conditions in the culture. Furthermore, it is not the case for every protein that there is a cell line that expresses it. In the data analysis, you need to take into consideration that some sugars (e.g., ^{3}H-Man) are partially converted into other tritiated monosaccharides after entering the cell.

The tritiation of oligosaccharides with borohydride (NaB(^{3}H)$_4$) is reliable, sensitive, and unhealthy. NaB(^{3}H)$_4$ reduces the terminal sugar unit to alcohol and thus introduces ^{3}H (Mellis and Bänziger 1983a). With the tritiation of O-glycosidic sugars, the NaB(^{3}H)$_4$ must already be present when the sugar is cleaved off the polypeptide (Mellis and Bänziger 1983b). The labeling requires several days, a column run, a paper chromatography, and for oligosaccharides from 1 to 5 µg glycoprotein up to 20 mCi of NaB(^{3}H)$_4$.

Free from radioactivity and very sensitive is a method that transforms the reducing end of the oligosaccharides with 2-aminopyridine into fluorescent pyridylamino oligosaccharides (Tomiya et al. 1987; Hase et al. 1978). The detection of fluorescent compounds requires equipment that does not exist in every laboratory. Furthermore, the sialic acids get lost during this labeling. Finally, in order to make the reaction quantitative 2-aminopyridine has to be added in large excess to be removed afterward on a column.

For labeling larger amounts of reducing oligosaccharides (from more than 20 µg of glycoprotein), UV chromophore (Kakehi et al. 1991) can also be used. The mild reaction conditions do not cleave the sialic acids off the oligosaccharides, and the excess of UV chromophor is removed after the reaction by simple extraction. UV chromophor is not available in retail and its production takes a few days (Kakehi et al. 1991).

Towbin et al. (1988) transform the reducing end of oligosaccharides with chromophor 4′-N,N-dimethylamino-4-aminoazobenzene. Chromophor is colorful (dark green to yellow-orange, depending on pH) and makes the sugar more hydrophobic, which may be advantageous with HPLC separations in reversed-phase columns. However, the reaction is described only for larger amounts of sugar and is not optimized for oligosaccharides cleaved from glycoproteins.

Sources

1. Hase, S., et al. (1978). "Structure Analysis of Oligosaccharides by Tagging of the Reducing End Sugars with a Fluorescent Compound," *BBRC* 85: 257–263.
2. Kakehi, K., et al. (1991). "Precolumn Labeling of Reducing Carbohydrates with 1-(p-methoxy)phenyl-3-methyl-5-pyrazolone: Analysis of Neutral and Sialic Acid-containing Oligo Saccharides Found in Glycoproteins," *Anal. Biochem.* 199: 256–268.
3. Mellis, S., and Banziger, J. (1983a). "Structures of Oligo Saccharides Present at the Three Asparagin-linked Glycosilation Sites of Human IgD," *J. Biol. Chem.* 258: 11546–11556.
4. Mellis, S., and Bänziger, J. (1983b). "Structures of the O-glycosidically Linked Oligosaccharides of Human IgD," *J. Biol. Chem.* 258: 11557–11563.
5. Tomiya, N., et al. (1987). "Structural Analysis of N-linked Oligosaccharides by a Combination of Glycopeptidase, Exoglycosidase and High-performance Liquid Chromatography," *Anal. Biochem.* 163: 489–499.
6. Towbin, H., et al. (1988). "Chromogenic Labeling of Milk Oligosaccharides: Purification by Affinity Chromatography and Structure Determination," *Anal. Biochem.* 173: 1–9.

9.5.2.4 Separation of Oligosaccharides

You have cleaved intact oligosaccharides from the glycoprotein. Now you want to separate the mixture by species. Oligosaccharide mixtures are separated either via normal phase HPLC columns (Mellis and Bänziger 1981), reversed-phase HPLC columns (Tomiya et al. 1987), ion-exchange HPLC columns (Wang et al. 1990; Townsend et al. 1989), or chromatography on different lectin columns (Cummings and Kornfeld 1982; Gesundheit et al. 1987). For separation according to size, you use gel filtration on BioGel P4. Paper and thin-layer chromatography are only suited for oligosaccharides with less than 6 sugar units.

To HPLC: Mellis and Bänziger (1981) recommend normal-phase HPLC on amino derivatized silica gel for neutral oligosaccharides. Tomiya et al. (1988) claim to be able to separate most oligosaccharides with a 2D separation technique (two different HPLC columns). By comparing with marker oligosaccharides, the authors determine the structure of the oligosaccharides at the same time. However, because of the variety of oligosaccharides the number of required labels lies between 20 and 120 (Tomiya et al. 1987; Tomiya et al. 1988), and most of those labels are not available in retail. It may take a diligent doctoral candidate half a year to produce 50 labels. Thus, the method will only be introduced by someone who is certain that it will be of use over the entire doctoral thesis. And because we are talking about labels: in order to be able to detect your sugar in the column eluate, it is advisable to label them before (see Section 9.5.2.3).

You can skip the labeling if you separate your sugar via alkaline anion exchange HPLC with connected PAD (Townsend et al. 1989). Maybe this is why this method has gained such popularity lately. It also separates well. Especially with negatively charged oligosaccharides (with sialic acid, phosphate, and sulfate residues), the alkaline anion exchange HPLC will give you much to celebrate. But careful: the high pH can have some nasty side effects (see Section 9.5.2.2).

To lectin columns: Because of the specificity of the lectins, the separation of oligosaccharide mixtures with lectin columns also provides clues regarding the oligosaccharide structure. Every oligosaccharide mixture requires a different combination of lectins for isolating its components. Nevertheless, a well-established first step is a Con-A-Sepharose column eluded first with 10 mM α-methylglucoside and then with 100 to 500 mM α-methylmannoside. The unbound oligosaccharides (pass-through) are trifurcated and tetrafurcated complex sugars, whereas the α-methylglucoside eluates contain bifurcated complex sugars and the α-methylmannoside eluate contains mannose and hybrid sugars. Afterward, each of three fractions is separated in WGA-sepharose, in E-PHA- and L-PHA-agarose, in PNA-sepharose, or in LCA-sepharose into homogeneous sugars (Gesundheit et al. 1987; Cowan et al. 1982; Cummings and Kornfeld 1982).

To BioGel P4: The gel filtration on this matrix delivers—for a gel filtration—astoundingly good results, especially with neutral oligosaccharides (Kobata et al. 1987). The method does not distinguish isomers, and it takes time to perform it with best possible results. You use BioGel P4 (for shorter oligosaccharides also P2) because BioGel consists of polyacrylamide and not of sugars such as various agarose or Sepharose gels. BioGel will thus not contaminate your sample with sugars. BioGel P4 separates in the range of 800 to 4,000 Dalton. You elude the column with water, and sometimes with water plus 0.02% Na-Azid.

As is well known, the separating effect of a gel filtration column depends on its length (see Section 5.2.2.1). According to my (nonrepresentative) review of the literature, the typical dimensions of a BioGel P4 column for the separation of oligosaccharides are a length between 50 and 140 cm and a diameter of 1 cm. The recommended flow rates are 3 to 4 ml per h. A short calculation shows that a column run would take between 20 and 36 h. You can gain some speed by running the column at higher temperatures (e.g., 55° C). This apparently also improves the resolution. High run temperatures can be achieved with a column heater. Columns with heating sleeve are in short supply in the laboratory, and even normal columns are difficult to find. Also, the heater requires a complicated and flood-prone construction, an additional pump, a thermostat, and (do not forget) preheated elution buffer. Furthermore, the tubes should be thermoinsulated. Otherwise, the elution buffer is at RT again when it reaches the column. Maybe it is better to set up column and elution buffer in a warm air incubator (pump and fraction collector stay outside). Then you can use normal tubes and a normal column and there are no temperature differences in the column that lead to convection currents. If you do not only separate the oligosaccharides, but also want to determine their size, you have to run an internal standard. A suitable series of glucose oligosaccharides is typically for this purpose. Attention: an internal standard contaminates your sample.

Sources

1. Cowan, E., et al. (1982). "Analysis of Murine Ia Antigen Glycosilation by Lectin Affinity Chromatography," *J. Biol. Chem.* 257: 11241–11248.
2. Cummings, R., and Kornfeld, S. (1982). "Fractionation of Asparagin-linked Oligosaccharides by Serial Lectin-agarose Affinity Chromatography," *J. Biol. Chem.* 257: 11235–11240.
3. Gesundheit, N., et al. (1987). "Effect of Thyrotropin-releasing Hormone on the Carbohydrate Structure of Secreted Mouse Thyrotropin," *J. Biol. Chem.* 262: 5197–5203.
4. Kobata, A., et al. (1987). "BioGel P-4 Column Chromatography of Oligosaccharides: Effective Size of Oligosaccharides Expressed in Glucose Units," *Methods Enzymol.* 138: 84–94.
5. Mellis, S., and Bänziger, J. (1981). "Separation of Neutral Oligosaccharides by High-performance Liquid Chromatography," *Anal. Biochem.* 114: 276–280.
6. Tomiya, N., et al. (1987). "Structural Analysis of N-linked Oligosaccharides by a Combination of Glycopeptidase, Exoglycosidase, and High-performance Liquid Chromatography," *Anal. Biochem.* 163: 489–499.
7. Tomiya, N., et al. (1988). "Analysis of N-linked Oligosaccharides Using a Two-dimensional Mapping Technique," *Anal. Biochem.* 171: 73–90.
8. Townsend, R., et al. (1989). "Separation of Oligosaccharides Using High-performance Anion Exchange Chromatography with Pulsed Amperometric Detection," *Methods Enzymol.* 179: 65–76.
9. Wang, W., et al. (1990). High-performance Liquid Chromatography of Sialic Acid-containing Oligosaccharides and Acidic Monosaccharides," *Anal. Biochem.* 190: 182–187.

9.5.2.5 Sequencing of Oligosaccharides

The structural information won from HPLC or lectin columns can and should be confirmed by independent methods. These include NMR, mass spectrometry, and sequencing of purified oligosaccharide chains with exoglycosidases.

Exoglycosidases remove monosaccharides from the nonreducing end of a sugar chain. There are exoglycosidases with narrow specificity and some with broad specificity, but the enzymes generally cleave only α or β bindings. The activity of exoglycosidases is also influenced by the oligosaccharide residue. The following exoglycosidases are suitable for sequence determination: neuraminidase (removes sialic acids), β-galactosidase (removes terminal β(1–4) bound Gal), β-N-acetyl-hexosaminidase (removes terminal β(1–4) bound GlcNAc), and α-L-fucosidase (removes α(−1–6) bound Fuc) (Mellis and Bänziger 1983a; Tomiya et al. 1987). Before you use an exoglycosidase for sequence determinations, you have to be well acquainted with the properties of the enzyme.

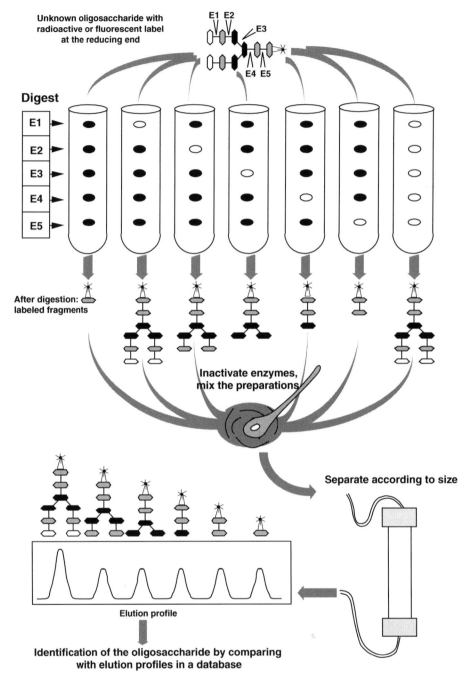

Figure. 9.4. Sequencing an oligosaccharide (following Edge et al. 1992). E1–E5: different exoglycosidases (note: E2 works only after E1 has cleaved its monosaccharide usw).

The sequencing works as follows. The purified oligosaccharide labeled at the reduced end is digested with an exoglycosidase. The experimenter releases the oligosaccharide residue and determines whether any monosaccharides were cleaved off and how many (usually via gel filtration). He adds new exoglycosidase to the oligosaccharide residue, and so on. The question is of course which exoglycosidase should be added in each case. Because the experimenter does not know, the urge to try out one exoglycosidase after the other makes the method a labor-intensive guessing game. Hence, any clues regarding the structure of the oligosaccharide are welcome (e.g., from lectin columns).

The sequencing method of Edge et al. (1992a) offers a brilliant way out of this guessing dilemma (Figure 9.4). The method consists of tagging the oligosaccharide at the reducing end and digesting aliquots with different batteries of exoglycosidases. The digested oligosaccharides are pooled and chromatographically separated (e.g., via a gel filtration in BioGel P4). The elution profile is compared, finally, to the theoretically possible elution profiles and the sequence of the oligosaccharide is inferred from this comparison (Edge et al. 1992a; Edge et al. 1992b). The Edge method should soon be available in automated form.

Sugar research often has a bitter taste. But sugar sequence investigations are a game of patience without risk. The sequence is there, as well as useful technologies. All it takes is diligence.

Sources

1. Edge, C., et al. (1992a). "Fast Sequencing of Oligosaccharides: The Reagent-array Analysis Method," *Proc. Natl. Acad. Sci. USA* 89: 6338–6342.
2. Edge, C., et al. (1992b). "Fast Sequencing of Oligosaccharides Using Arrays of Enzymes," *Nature* 358: 693–694.
3. Mellis, S., and Bänziger, J. (1983a). "Structures of Oligosaccharides Present at the Three Asparagin-linked Glycosilation Sites of Human IgD," *J. Biol. Chem.* 258: 11546–11556.
4. Tomiya, N., et al. (1987). "Structural Analysis of N-linked Oligosaccharides by a Combination of Glycopeptidase, Exoglycosidase and High-performance Liquid Chromatography," *Anal. Biochem.* 163: 489–499.

"And I say too," said the second, "that there are rare gifts going to loss in the world, and that they are ill bestowed upon those who don't know how to make use of them." "Ours," said the owner of the ass, "unless it is in cases like this we have now in hand, cannot be of any service to us, and even in this God grant they may be of some use."

Chapter 10 Treasure Island: Writing Papers

I, then, as it has fallen to my lot to be a member of knight-errantry, cannot avoid attempting all that to me seems to come within the sphere of my duties.

10.1 Of the Paper

What good is the best discovery if nobody knows about it? Even more important than the discovery is the announcement of discovery. In science it so happens that the discoverer becomes known with the discovery, or his professor does in any case. This is what is motivating. You make a discovery official by writing a paper about it (Krämer 1994).

The paper is the product of scientific work, and the papers of a researcher are the basis for the respect he enjoys among colleagues. A doctoral candidate, for example, must thus strive to write papers in graduate school. If this is not possible, she should at least appear as coauthor on the papers of others. However, it should be noted that number and quality of the papers do not necessarily correlate with the prospects for social promotion. At least the correlation has not been scientifically proven yet. The political and social circumstances of paper writing are described in Bär (1992).

It is not only important how many papers the doctoral thesis yields but which prestige these papers have. Prestige is a soft quantity, as hard to grasp and as nutritious as chicken soup. The prestige of a paper depends on the news value of the result, whether many laboratories work in the field, whether the results are commercially usable, or whether they contribute to the treatment of an illness. You also cannot underestimate what used to be called boasting and is nowadays referred to as marketing. Prestige is expressed in the magazine that publishes the paper. A *Nature* or *Cell* article weighs heavier than one dozen publications in third-rate journals. The respect for the journals pretty much parallels their "impact factor," a quantity whose definition and numerical values are provided by the *Science Citation Index*.

10.2 Of Writing a Paper

Diligent reading of papers sharpens your eye for these connections. Most journals organize their papers according to the following schema (Bär 1992): a paper is divided into title, authors' list, summary (abstract), introduction, methods, results, discussion, and references. The summary or the abstract contains the most important results of the work, briefly formulated in a few sentences. The introduction provides a short overview of the field of research and its development—with strong emphasis on contribution. In the results section, the results of the work are documented without evaluation with help of tables, figures, or photos. The discussion briefly summarizes the results, explaining them and comparing them to the results of other scientists. Finally, the discussion section offers the opportunity to present new hypotheses, to tear down old ones, or to smack your opponents in an elegant academic fashion. The argumentation is supported by references to papers of other scientists.

The formal details (such as how to cite, and so on) are different from journal to journal and are detailed in the "Instructions for Authors," which can generally be found at the beginning or end of every issue. These instructions often comprise only one to three pages, but some magazines (such as the *Journal of Biological Chemistry*) are exacting and detail their regulations in a small booklet the author must request.

Beginners (i.e., diploma students or doctoral candidates) typically write their papers together with an older post-doc or the professor (i.e., the doctoral candidate writes a raw draft, which is edited by the more experienced scientist). In some labs, the professor writes the paper by herself and the doctoral candidate serves only as a source of information. Nevertheless, you should insist on writing at least the crucial parts of the paper (methods and results) yourself. For the first paper, this is laborious and time consuming. But the earlier you start writing the faster you can do it. If you do not write, you are quickly known as an intellectual featherweight and a dumb measuring aide.

How do you start? As a model, you look for papers about a similar subject by American or English research groups (of course, from the journal to which you want to submit the manuscript). From this, you glean the arrangement of the material, the chapter layout, which controls are required, and so on. The papers also provide you with the standard formulations and jargon currently en vogue. I usually start with the graphs and tables. These summarize the results into the shortest form and do not require any formulating skill. I hang the text onto the scaffolding provided by the pictures and tables.

A paper is no overview article. The introduction does not have to be several pages long and list all references since the end of World War II. The same is true for the discussion. The results count, not their interpretation.

The references list the papers whose results were used in the work or papers that support assertions in the introduction and discussion. Also cite the competition. The reasons for this are that then they may also cite you and because this is the proper thing to do. Researchers love to be cited, and many see it as an affront if their work is ignored. Hence, the professors pay great attention to the references. By "proper" citing they try to appease possible reviewers, to keep their friends in good spirits, and so on.

Do not create a literary masterpiece! If your professor has this ambition, that is his thing. You should deliver a simple and clear piece of writing: short sentences, at most two statements per sentence, no swollen or unnecessary words (Gregory 1992). Your paper is read, if at all, only by the reviewers of the journal and by people who work on the same or a similar topic. They know the subject matter. Word bubbles do not impress them.

If you think your manuscript is perfect (i.e., not only in a readable state but ready for printing), let it sit for a few days. Then read it once again. Only if you have not touched the paper for a few days do you notice the countless overly complicated formulations and unnecessary phrases, and the incorrect and ambiguous statements. Improve the manuscript and let it sit again for a few days. After the second improvement at the earliest, the manuscript is ready for the eye of the master. By the way: for papers you do not get paid, in spite of the trouble. To the contrary, some journals take per-page fees from the author.

Sources
1. Bär, S. (1992). *Forschen auf Deutsch*. Frankfurt/Thun: Verlag H. Deutsch.
2. Gregory, M. (1992). "The Infectiousness of Pompous Prose," *Nature* 360: 11–12.
3. Krämer, W. (1994). *Wie Schreibe ich eine Seminar-, Examens- und Diplomarbeit?* Stuttgart: G. Fischer.

Chapter 11 Desert Planet: Researching the Literature

> In short, he became so absorbed in his books that he spent his nights from sunset to sunrise, and his days from dawn to dark, poring over them; and what with little sleep and much reading his brains got so dry that he lost his wits.

It can happen that you need literature not noted in this book. Maybe you want to know what is already known about a certain protein/process/technique, whether this or that protein has already been investigated with this or that method, and so on. There is, of course, MEDline and such: you enter keywords on your PC and it searches for suitable abstracts. This is good. This is useful. However, the snowball method (Krämer 1992) appears to be a necessary supplement to the electronic searching methods. What you get out of the computer depends on what you enter and what turns up. Only rarely do you stumble over unusual papers and new ideas. Also, going to the library cannot be avoided. Finally, depending on the system, the electronically available content often lags behind the journals by weeks or months.

What is the snowball method? The prerequisites are a good library, familiarity with the journals and their focus, and some curiosity. Example: you are looking for literature about the properties of the enzyme Ca-ATPase. First, you think about which journals publish biochemical or pharmacological papers about Ca-ATPases. You come up with the *Journal of Biological Chemistry*, the *European Journal of Biochemistry*, and *Annual Review of Biochemistry*. You prepare a tea for yourself and sit down with them in the library. There you page through the tables of contents of the last issues of the *Journal of Biological Chemistry*. A skilled paper reader, and that's what you have become by now, needs approximate five minutes to read the titles of an issue. After at most 10 issues you find an article about Ca-ATPase, which is detailed and in-depth like many articles in the *Journal of Biological Chemistry*. Maybe the paper does not contain the information you are looking for. However, the reference list of the paper lists other papers about Ca-ATPase, and the text provides clues about the thematic direction of these papers. Now you know several papers about the subject, and these papers in turn refer to other papers. The snowball becomes an avalanche that presumably contains the sought-after information.

This is not the only gain. Often you stumble over more interesting things than the information you were looking for. Also, the aimless browsing through the work of others engenders good ideas. Furthermore, during the brief hour you gain an overview of what the *Journal of Biological Chemistry* has published in the previous months. You know what is currently fashionable and get a feeling for what you could publish in the *Journal of Biological Chemistry*. Finally, you could right away look up the method section or the discussion and check whether it describes what you are looking for in the necessary detail and whether the paper is any good. The electronic searching methods often provide you only with title and abstract.

Of course, the snowball method also has disadvantages. For example, information can escape you because the library does not have the particular journal or because you do not like the journal. And, for some reason, authors sometimes place their article in a journal where it does not belong. Also, the method is time consuming for beginners who are not familiar with the thematic focus of the journals. Finally, every now and then you get caught in a citation circle. This is a ring of researchers who exclusively cite each other. The works of others are ignored and you thus do not find them with the snowball method. However, in the natural sciences exclusive citation circles are rare (they seem to be a specialty of the social and political sciences).

If you want to get an overview of a subject or you need to familiarize yourself with a subject, look for an overview article. *Science*, *Nature*, the *European Journal of Biochemistry*, *Scientific*

American, and *Biochemistry* publish good overview articles. However, it is difficult to search magazines. You are quicker with review books such as the *Annual Review of Biochemistry*, or the *Annual Review of Physiology*. These series cover practically every subject of biology. However, the review is often too old, and the cited papers even more so.

Overview articles have an additional disadvantage which is inseparably connected with their existence. Writing an overview article is time consuming. If the article is to be good, it takes the writer months. The magazines either do not pay anything for overview articles or just a ridiculously small amount. The question arises why a professor takes it upon himself to write the overview article. A partial explanation may be the boredom that comes with every mainly bureaucratic work. However, an important motive is also the following: the writer of overview articles writes the history of his field. He generally writes it in such a way that the contribution of the person who gave the impulses is finally properly appreciated. This person is usually he himself (in any case, that is the steadfast conviction of the review writer). Also, many review writers use the opportunity to give prominence to their friends and powerful supporters. It is thus advisable to read overview articles by different professors.

Source
1. Krämer, W. (1994). *Wie Schreibe ich eine Seminar-, Examens- und Diplomarbeit?* Stuttgart: G. Fischer.

> They say right that it takes a long time to come to know people, and that there is nothing sure in this life.

Last Things

The EXPERIMENTER is meant for students in the higher semesters, for diploma students, and doctoral candidates. Now, my heart contracts in my shirt every time I see the naiveté with which the latter choose their adviser. Romantic notions decide (I want to work on cancer because my aunt has died of it), or chance (recently I saw this flier in the XY institute), or fashionable topics that may impress their circle of friends (environmental care, malaria, and so on), or the eloquence and the likeable smile of the supervisor—and many simply say "I just want to get my doctorate. Where, I don't care. It doesn't matter anyway."

It is true, as a starting biologist/biochemist/chemist you do not have enough experience and knowledge of the literature to be able to assess the scientific value of an offered job. And the course of especially interesting projects is unpredictable. So what can you use for orientation? That depends on what you want!

Do you need the title for an industry career? Then choose a supervisor who guarantees a quick graduation. Or would you like to go into academic research? Look for a lab with a famous professor (holders of the Nobel Prize are very suitable), who is under 50, and whose research is going well. A well-run lab can likely offer an interesting doctoral thesis with reasonable risk. A medium-size lab is running well if it has published several *Cell*, *Nature*, and *Proc. Natl. Acad. Sci. USA* papers in the last two years. Search with patience and MEDline. After all, this is about three of your best years!

> "No one is born ready taught," said the duchess, "and the bishops are made out of men and not out of stones."

Appendix A Professional Resources

A.1 Suppliers

A.2 Suppliers by Product

Antibodies and products for immunotests: Becton Dickinson, BioProducts, BioSource, BIOLOGY TREND, Boehringer, Calbiochem, Cappel, Clonatec, Costar, DAKO, Slide novelties-Immunotech, Genzyme, Jackson Rennet., Janssen, neosystem, Nordic, Pierce, Promega, Takara, Vector Lab

Blot and blot development: Bio-Rad, Serva, USB, Pall Life Sciences

Cell culture material: Labotec, neo-Lab, Nunc, Serva

Centrifuges: Beckman, Eppendorf, Heraeus, Hettich Zentrifugen, Hitachi, Kendra, Kontran

Chemicals, enzymes, substrates: Aldrich, Bachem, BDH, Boehringer, Calbiochem, Cambridge Res. Biochemicals, Fluka, ICN Biomedicals, Merck, Qiagen, Research Biochemicals, Roth, Serva, Sigma, Strathmann, Wako

Chromatography column materials and matrices: Bio-Rad, IBF-biotechnics, Kem-en-Tec, PerSeptive Biosystems, Pharmacia, Pierce, Sigma, Spectrum, Supelco, TosoHaas

Chromatography devices: Bio-Rad, Gilson, Pharmacia, Supelco, TosoHaas

Cross-linkers: Pierce

Dialysis: Pierce, Spectrum

Electrophoresis, devices, and reagents: Bio-Rad, Hoefer, JKA-Biotech, Kem-en-Tec, Pharmacia, CBS Scientific Company, Serva

Filters: Millipore, Schleicher & Schull, Whatman, Pall Life Sciences

Fridges (and accessories): Bender & Hobein, Hüber, neo-Lab

Glassware: Bender & Hobein, Brand, Labotec

Homogenizators: Bender & Hobein, Braun, Labotec, neo-Lab, Zinsser

HPLC: TosoHaas, Zinsser

Lab equipment: Bender & Hobein, Fischer, Labotec, neoLab

Lab clothes: Roth

Lectins and glycosidases, etc.: BIOTREND, Boehringer, Genzyme, GLYKO, Oxford Glycosystems, Sigma

Lipids: Avanti Polar, Bachem, Biomol, Sigma

MALDI-TOF mass spectrometers: Bruker Analytische MeBtechnik, Ciphergen Biosystems, Micromass, Finnigan MAT, Hewlett-Packard, PerSeptive Biosystems

Microcalorimeters: Heath Scientifics

Microsequencing of peptides: Prosequenz, Toplab

Peptides: American Peptide, BioProducts, Calbiochem, Cambridge Res. Biochemicals, Neosystem, Peninsula Laboratory.

Peptide syntheses: Abimed, Biochrom, BIOTREND, Cambridge Res. Biochemicals, Multiple Peptide Systems, Neosystem, Orpegen, Toplab

pH meters: Aldrich, Ingold, Roth

Photography and image processing: Aldrich, Appligene, Cybertech, Labotec

Pipettes: Brand, Costar, Eppendorf, Gilson, Hamilton

Plastics (E-containers, test tubes, etc.): Brand, Labotec

Radioactively labeled substances: Amersham, Biotrend, NEN

Scales: Mettler, Sartorius
Stirrers: Bender & Hobein, Labotec, Roth
Sugars: Roche Diagnostics, Cambridge Res. Biochemicals, BioCarb
Toxins (animal): RBI Biochemicals, Sigma, Spider Pharm, Wako, Janssen
Toxins (bacterial): Sigma, Wako

Index

Printed and bound by CPI Group (UK) Ltd, Croydon, CR0 4YY

09/10/2024

01042621-0001